Physical Meteorology

Technology Press Books

PHYSICAL METEOROLOGY
By John C. Johnson

NATIONALISM AND SOCIAL COMMUNICATION
An Inquiry into the Foundations of Nationality
By Karl W. Deutsch

THE MATHEMATICS OF CIRCUIT ANALYSIS
Extensions to the Mathematical Training of Electrical Engineers
By E. A. Guillemin

MAGNETIC CIRCUITS AND TRANSFORMERS
A First Course for Power and Communication Engineers
By Members of the Electrical Engineering Staff, M.I.T.

APPLIED ELECTRONICS
A First Course in Electronics, Electron Tubes, and Associated Circuits
Second Edition by Truman S. Gray

ELECTRIC CIRCUITS
A First Course in Circuit Analysis for Electrical Engineers
By Members of the Electrical Engineering Staff, M.I.T.

FATIGUE AND FRACTURE OF METALS: A SYMPOSIUM
Edited by William M. Murray

METADYNE STATICS
By Joseph Maximus Pestarini

THE PREFABRICATION OF HOUSES
By Burnham Kelly

METHODS OF OPERATIONS RESEARCH
By Philip M. Morse and George E. Kimball

PRESSURES ON WAGE DECISIONS
A Case Study in the Shoe Industry
By George P. Shultz

THE DOLLAR SHORTAGE
By Charles P. Kindleberger

THE TRANSMISSION OF NERVE IMPULSES AT NEUROEFFECTOR JUNCTIONS AND PERIPHERAL SYNAPSES
By Arturo Rosenblueth

MID-CENTURY
The Social Implications of Scientific Progress
Edited and annotated by John Burchard

AN INDEX OF NOMOGRAMS
Compiled and edited by Douglas P. Adams

THE EXTRAPOLATION, INTERPOLATION, AND SMOOTHING OF STATIONARY TIME SERIES
with Engineering Applications
By Norbert Wiener

CYBERNETICS
Or Control and Communication in the Animal and the Machine
By Norbert Wiener

Q.E.D., M.I.T. IN WORLD WAR II
By John Burchard

INDEX FOSSILS OF NORTH AMERICA
By H. W. Shimer and R. R. Shrock

THE MOVEMENT OF FACTORY WORKERS
By C. A. Myers and W. R. Maclaurin

WAVELENGTH TABLES
Measured and compiled under the direction of G. R. Harrison

Physical Meteorology

JOHN C. JOHNSON

Research Associate, Research Laboratory
of Physical Electronics, Tufts College

PUBLISHED JOINTLY BY
The Technology Press of
The Massachusetts Institute of Technology
AND
John Wiley & Sons, Inc., New York

Chapman & Hall, Ltd., London

Printed in the United States of America

Library of Congress Catalog
Number: 54-7836

Preface

Physical meteorology is a fringe study, investigating those meteorological phenomena not directly linked with the circulation of the atmosphere. Admittedly wide, it connects meteorology with other branches of science. In this respect, it occupies a position analogous to that which physical chemistry occupies between chemistry and physics when we think in terms of the nineteenth century distinctions between these two sciences.

Much of the original research in the topics forming the field of physical meteorology was and still is being carried on by astronomers, radio physicists, chemists, and aeronautical engineers. Of their researches, those in which the atmosphere plays a significant part are collected together in this book.

The material presented is at a level that presupposes basic courses in differential and integral calculus plus a thorough grounding in general physics. Although designed primarily as a first course for students training to be professional meteorologists upon graduation, the text is suitable for an elective course in the physics curriculum to be taken by undergraduate students in their third or fourth year. The teacher of physics has an opportunity in a course of this kind to demonstrate the complex nature of seemingly simple physical problems and to indicate the devices and approximations that have been tried toward their solution. For a demonstration of these problems, the atmosphere is the laboratory and it is one to which all people have access. However, it is well to recognize that atmospheric phenomena are seldom pure in form. Compared with the usual laboratory preciseness, many of the data receivable from measurements of natural atmospheric phenomena are reminiscent of a laboratory experiment that has been poorly conceived and sloppily executed. Nonetheless, the theoretical explanation of many atmospheric processes is well advanced, and the theory is substantiated by firm experimental evidence.

Because this is a textbook bridging many fields of applied science, occasional difficulty has been experienced in choosing symbols that are consistent throughout the text and yet have some association with those symbols commonly used in the literature of a particular topic. In a few places, especially in Chapter 9, symbols such as V for volume and V for electric potential appear in the same equation. This unfortunate juxtaposition is neither an oversight nor an error. Inspection of the equation will make it clear that roman type is used for electrical quantities and italic type for nonelectrical quantities. Other deviations are made from a seemingly logical choice of symbols to avoid possible misinterpretation of literal characters.

The references to the literature were chosen more to assist the research worker than the beginning student. In view of the common practice of many of us to seek information on unfamiliar subjects in the order of textbooks, source books, and as a last resort the original literature, a number of excellent source books have been listed at the end of each chapter. These books contain extensive references to the literature in the field of which they treat. Original sources selected for mention in this book either are papers of outstanding merit, are in the nature of survey papers, or are additions to the literature between the date of the source book and the spring of 1953.

I wish to express my appreciation for the interest and encouragement given by friends, family, and colleagues while the manuscript was being prepared. The impetus for writing this book came from teaching the undergraduate course in physical meteorology at the Massachusetts Institute of Technology for the last several years. Considerable direction to the topics treated in this book came from classroom notes written by Professor Henry G. Houghton and used as the framework of the course. The manuscript was typewritten by Mrs. M. Dorothea Murphy, and the illustrations were drafted by Mr. Secondo A. Mazzuchelli. It has been a distinct pleasure to work so closely with these people.

J.C.J.

Marblehead, Massachusetts
December 9, 1953

Contents

LIST OF SYMBOLS

a	diameter of a sphere
A	area, visual albedo, prism angle
$\mathcal{A}$	absorptivity
B	brightness, coefficient for Rayleigh scattering
$B^*(\lambda)$	intrinsic brightness (wavelength dependent)
$B(z)$	flux density arising from molecular conductivity of heat (height dependent)
B_s	adiabatic bulk modulus
c	speed of electromagnetic radiation in vacuo
c_p	specific heat of air at constant pressure
c_v	specific heat of air at constant volume
C	capacity of a condenser
C	a constant, brightness contrast, concentration
C_D	drag coefficient
D	mean solar distance, density of water, angle of deviation
D'	density of ice
e	water vapor pressure
e	charge on the electron
e_s	saturation water vapor pressure
E	flux density
E_0	solar constant, initial flux density
$E(z)$	flux density of solar energy (height dependent—short wave)
E	magnitude of the electric-field vector, collection efficiency
f	force
F	flux
g	acceleration of gravity
G	gain of a radar receiver
h	Planck's constant, thickness coordinate
H	height of the homogeneous atmosphere, radiation function, rate of heating, scale height
i	electric current
i	angle of incidence, subscript indicating an integer, $\sqrt{-1}$
i'	angle of refraction in a medium whose index of refraction is changing slowly
I	intensity, number of ion pairs formed per second per cubic centimeter

$I_\perp$	intensity of radiation polarized perpendicular to the plane of polarization
$I_\parallel$	intensity of radiation polarized parallel to the plane of polarization
$\mathbf{j}$	electric current density
$J_i(\ )$	Bessel function of order i
k_{ij}	recombination factor for ozone, oxygen, and molecular oxygen
k	Boltzmann constant
k_a	absorption coefficient per unit length
k_s	scattering coefficient per unit length
$\hbar$	electromagnetic absorption coefficient (per wavelength)
K	diffusivity of water vapor in air, curvature
K_s	scattering area coefficient
K_s'	back-scattering coefficient
l	pulse length
$\bar{l}$	mean free path
L	latent heat
$L(z)$	flux density arising from eddy diffusion
m	index of refraction, mass per unit volume
M	mass of various particles, molar concentration
n	number of various particles per unit volume, an integer
N	total number of particles, number of particles crossing unit area in unit time
p	pressure, air pressure
P	dipole moment
q	magnitude of electric charge per unit volume
q	specific humidity, number of quanta incident on unit volume in unit time
Q	electric charge
Q	radiation function, number of quanta *absorbed* per unit volume per unit time, time rate of production of electrons per unit volume, quantity of heat
r	radial co-ordinate, radius, angle of refraction
R	radius of the earth
R_G	gas constant per unit mass of dry air
R_W	gas constant per unit mass of water vapor
Re	Reynolds number
R	electrical resistance
$\mathcal{R}$	reflectivity
$\mathcal{R}_\lambda$	reflectivity per unit wavelength
$\mathrm{R}(z)$	columnar resistance
s	curvilinear co-ordinate

S	area of a detector
$S(z)$	flux density of terrestrial radiation (height dependent—long wave)
t	time
T	absolute temperature
u	optical path length, particle velocity in a trajectory, group velocity
U	air velocity
v	relative velocity, phase velocity, specific volume
v_T	relative terminal velocity
$\dot{v}$	acceleration
V	volume
V	electric potential and potential difference
V_m	visual range
w	precipitable water, mass of liquid water in a unit volume of air
w_s	saturation mixing ratio
$W(z)$	flux density of moisture transport (height dependent)
x	co-ordinate distance
y	co-ordinate distance
z	height
z'	virtual height
Z	number of nuclear charges
α	$\pi a/\lambda$, elevation angle, scattering angle for polarized radiation, α-particle
α'	apparent elevation angle
α_i	fraction of light quanta absorbed
β	elevation angle, coefficient of Rayleigh scattering
β'	apparent elevation angle
γ	ambient lapse rate
γ_A	autoconvective lapse rate
Γ	polarization, recombination coefficient for electrons
δ	polarization defect, phase angle, non-selective scattering coefficient
ϵ	dielectric constant, threshold of brightness contrast, pyranometer excess, excentricity of an ellipse, threshold of radiant power
ζ	zenith angle
η	variable quantity, kinematic viscosity
θ	zenith angle
θ'	apparent zenith angle
Θ	potential temperature
κ	thermal diffusivity of air (conduction), mobility of an ion
κ_E	eddy diffusivity of air

κ_H	coefficient of heat transfer (conduction)
κ_s	thermal diffusivity of soil
λ	wavelength
$\bar{\lambda}$	specific conductivity
Λ	conductivity, nondimensional parameter for collection efficiencies
μ	micron, viscosity
ν	frequency, molecular collision frequency
ρ	air density
ρ_w	water-vapor density
σ	attenuation coefficient per unit length, collision cross section, surface tension, electric charge per unit area
$\sum$	summation sign
τ	transmissivity, nondimensional time co-ordinate
Ξ	nondimensional velocity coordinate
ϕ	forward scattering angle
χ	nondimensional co-ordinate distance
ψ	nondimensional co-ordinate distance, angular co-ordinate
ψ_λ	visibility (colorimetry definition)
Ψ	astronomical refraction
ω	solid angle

CHAPTER 1 Atmospheric Refraction

Introduction

Since the 1930's there has been an increasing tendency in the teaching of physics to stress the unity of the subject and its relation to other sciences, rather than to take the historical approach of dividing the subject into the topics of mechanics, heat, electricity and magnetism, optics, sound, and the catch-all modern physics. In the same way, physical meteorology now embraces such a wealth of topics in atmospheric physics and to some extent chemistry that a unified approach to the subject is possible. One such method of approach is to consider the general principles that govern the transmission, attenuation, polarization, and dispersion of electromagnetic energy through the atmosphere. After these general concepts have been developed, they may be easily particularized and simplified for the various domains of light, infrared radiation, microwaves, and ultra-high-frequency radio waves.

A particular phenomenon that can be so generalized is the study of atmospheric refraction. In general, the term refraction denotes the change in velocity of radiation (both in direction of path and in speed of propagation) on passing from one medium to another. A knowledge of this phenomenon is important in the domain of atmospheric optics, microwaves, and ultrahigh-frequency radio propagation through the atmosphere. Certain basic ideas are common to all these domains. Certain details can be neglected or are at least negligible when, for example, optics is being considered, but not when radio frequencies are under discussion. These differences will be brought out in subsequent sections.

First, let us generalize a system that generates, emits, collects, and detects electromagnetic energy. Between the emitter and collector, the electromagnetic energy traverses an atmospheric path that may often be of the order of kilometers or hundreds of kilometers. The physical appearance may vary as much as a system consisting

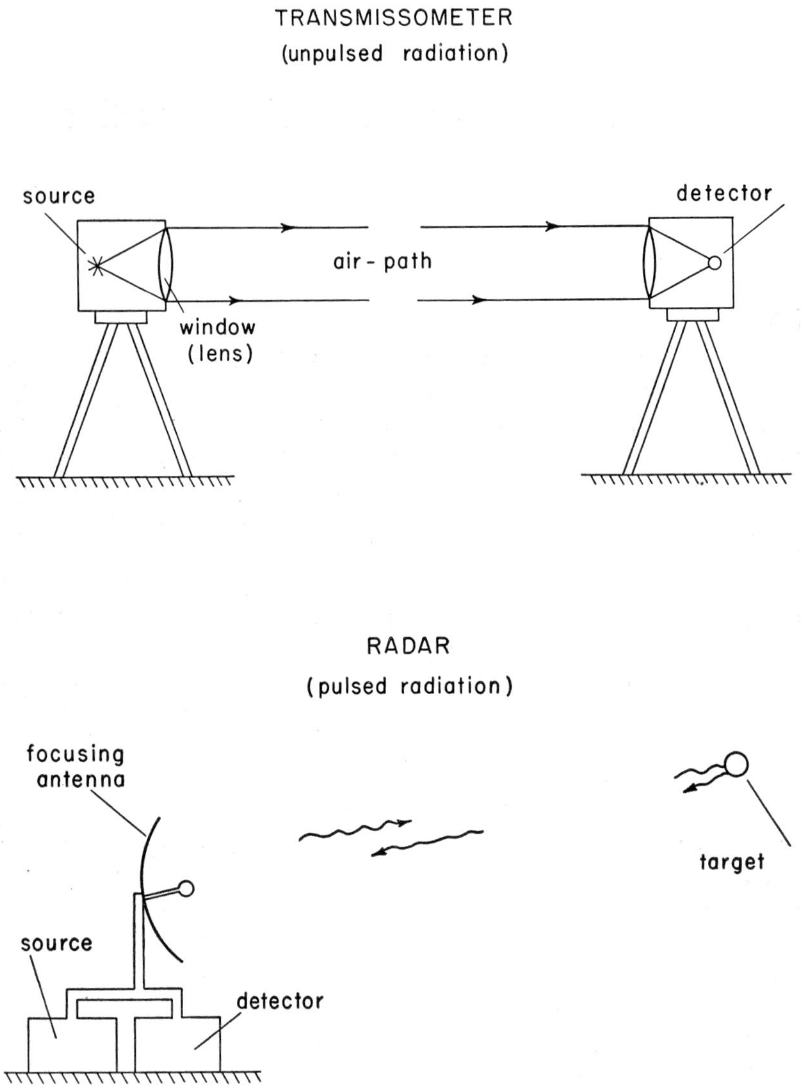

Fig. 1.1 Schematic diagram of two systems radiating electromagnetic waves. The transmission system uses a separate source and detector. The radar system uses the same antenna for both emitting and detecting radiation. A dual-purpose antenna is possible when pulsed radiation is used. The time interval between the pulses of emitted radiation is the time when the antenna is used to detect radiation reflected by the target.

of the sun as a source and a pyroheliometer as a detector differs from a system which includes a VHF beacon in an airplane and a VHF radio receiver on the ground. Figure 1.1 shows two other systems that are used. In all cases the radiation has a path, generally long, through the atmosphere. What happens to the radiation during this time is a concern of physical meteorology. It is illustrative to tabulate a few of these systems in the various regions of the electromagnetic spectrum. This is done in Table 1.1.

Table 1.1

A Few Typical Systems for Propagating Electromagnetic Energy through the Earth's Atmosphere.

(These systems are not used necessarily throughout the entire range.)

Region	Range of Wavelengths	Source	Disseminator	Collector	Detector
Ultra-violet	1200 to 4000 angstrom units	gas discharge	quartz windows	quartz windows	photo-graphic film
Visible	400 to 700 millimicrons	lamp	mirror or lens	mirror or lens	photo-cell
Infrared	1 to 400 microns	black body	NaCl window	NaCl window	thermo-couple
Micro-wave	1 millimeter to 1 meter	magnetron (cavity oscil-lator)	dipole antenna and reflector	reflector and dipole antenna	crystal recti-fier
High radio frequencies	(H.F.) (V.H.F.) (U.H.F.) 1 meter to 100 meters	vacuum tube oscil-lator	dipole or loaded wire antenna	dipole or loaded wire antenna	crystal diode or triode vacuum tube
Radio broadcast	200 meters to 600 meters	vacuum tube oscil-lator	wire antenna	wire antenna	tuned vacuum tube detector

Units: 1 meter = 10^2 centimeters = 10^3 millimeters = 10^6 microns = 10^9 millimicrons = 10^{10} angstrom units.

General considerations of refraction phenomena

Refraction effects arise because the atmosphere is neither a constant-density medium nor free space. Electromagnetic waves are propagated with speeds that are a function of the density and to a large extent the internal electrical structure of the molecules of gas in the atmosphere. The ratio of the speed of propagation of the radiation in free space to the speed of the radiation in a medium, in this case the atmosphere, is called the index of refraction, denoted by the symbol m. Letting the speed in free space be c, a universal constant equal to 2.998×10^8 meters per second, and the velocity in the atmosphere be v, we then have

$$m = \frac{c}{v}\cdot \tag{1.1}$$

The ratio is here stated in notation using the concept of optics in the definition, and this concept will be used throughout. For radio frequencies, the ratio is a function of the familiar dielectric constant of the air, ϵ.

The relationship in this notation at radio-broadcasting frequencies ($\sim$ 1 megacycle) takes the simple form

$$m^2 = \epsilon. \tag{1.2}$$

Equations will be developed to express the curvature of the path of the radiant energy. It will be shown that the curvature depends on the index of refraction of the medium through which the energy is propagated. The geometrical representation of this energy path is called a *ray,* or a *pencil of rays.* The method of analysis is called *ray tracing.* The application throughout the electromagnetic spectrum will be quite general, provided that certain scale restrictions as to what constitutes a ray are observed.

Two theoretical conditions restrict the regions in which ray tracing is valid and gives correct results. Both conditions depend on wavelength. It is the wavelength, then, that may be looked upon as being the scale factor.

The first condition that a ray pattern must satisfy for ray tracing to be valid is that the index of refraction must not change appreciably in a distance equal to one wavelength of the radiation. In the conventional symbols of the text, i.e., λ wavelength and z height, the condition may be stated for the atmosphere as (when $m^2 \to 1$)

$$\frac{\lambda}{2\pi}\left|\frac{\Delta m}{\Delta z}\right| \ll 1.$$

This condition implies that, if a ray pattern is drawn for a given vertical gradient of index of refraction, then as the wavelength of the radiation decreases this ray pattern becomes an increasingly better approximation to the actual state of affairs. Although by this condition ray tracing of optical wavelengths is more reliable than ray tracing of microwaves, in practice the limit on ray tracing does not occur until a limit of 3 meters wavelength has been reached.

Above about 9 meters wavelength, ray tracing becomes quite invalid because of the dominating influence of a conducting ionosphere and a semiconducting earth. An alternative way of stating this condition is to say that longer wavelengths do not react to the fine structure of density changes in the atmosphere, but do an averaging process over loosely defined units of height.

The second restriction on the use of *geometric optics,* a synonym for *ray tracing,* is that the fractional change in the spacing of rays shall be small in an interval of a wavelength. This restriction means that a ray pattern has questionable significance in regions where rapid convergence of the rays takes place. This is very true in the region where a ray pattern converges to a focal point. Another region where this restriction applies is at a caustic,† which is the envelope formed by all rays emanating from a single point, this envelope being coincident with the surface that appears to reflect these waves. Within a fixed geometric distance from these limiting points and/or surfaces, ray patterns become an increasingly poor approximation to the true situation that exists as the wavelength is allowed to increase. In the atmosphere, the focal restriction prevails near radars or other emitters sending out a beam with a measurable angular beam width. The caustic restriction applies at the walls of wave guides or their atmospheric analogue, the radio duct. These latter terms will be defined in a later section of this chapter, when the need for such terminology becomes apparent.

Theory of terrestrial refraction

Throughout a wide range of wavelengths, including both the visible and microwave regions of the electromagnetic spectrum, the index of refraction for dry air can be represented by[(5)]‡

$$(m - 1) \times 10^6 = \text{constant} \times \rho \tag{1.3}$$

where m is the index of refraction and ρ is the density of dry air. From the equation of state for a gas, ρ should be expressed as a ratio of the pressure p to the absolute temperature T. The form of Eq. 1.3 has theoretical justification.

When the air is not dry, but contains water vapor, the effect of the polar water molecules on the index of refraction is taken into account through the use of the vapor pressure e. By theory, Eq.

† See Figure 1.11.

‡ Reference numbers refer to the paper or source book of the same number listed at the end of each chapter.

1.3, when corrected for water vapor, assumes the following form(B2)

$$(m - 1) \times 10^6 = A \frac{p}{T}\left(1 + \frac{B}{T}\frac{e}{p}\right). \tag{1.4}$$

Equation 1.4 reduces to Eq. 1.3 when $e = 0$, the condition for dry air. A and B must be found by either theory or experiment.

For the visible region of the spectrum, $0.4\mu < \lambda < 0.7\mu$, the constants are

$$A = 77.5\left(1 + \frac{5.15 \times 10^{-3}}{\lambda^2} + \frac{1.07 \times 10^{-4}}{\lambda^4}\right)$$

$$\dagger\, \frac{B}{T} = -0.120$$

when p and e are in millibars, T in degrees absolute, and the wavelength λ in microns. When dispersive effects are not important, A may be taken as 79° abs per millibar. This value of A corresponds to the yellow line of sodium ($\lambda = 0.589$ micron) at standard conditions of temperature and pressure (1013.2 millibars, 273° abs).

For the microwave region greater than 2 centimeters wavelength, the constants become

$$A = 79° \text{ abs/mb}$$

$$B = 4800.$$

These latter constants apply throughout the microwave region to the point where refraction effects become negligible. This extreme is of the order of 3 meters wavelength (100 megacycles per second).

In the infrared and for microwaves $<$ 2 centimeters, absorption effects enter and anomalous dispersion occurs. As a result, the index of refraction is a rapidly changing function of frequency and a formula of the type of Eq. 1.3 no longer applies. Figure 1.2 shows these regions of absorption.

A simple method for handling the terrestrial refraction problem will now be developed. The solution of the problem is one that is illustrated most often in meteorological optics, and is the problem of explaining mirages. *Mirages* arise from *terrestrial refraction* and although this term is used most commonly in reference to the visual sense, presentation of an anomalous image on a radar scope may be a result of the same atmospheric effect.(B2)

Let us consider, therefore, a plane wave being propagated through an atmosphere of dry air in which the density of the air ρ changes with

† For visible light, B/T is so small that it may be considered a constant.

height z. As the densities of air at a distance dz apart are not the same, Eq. 1.3 predicts that their indices of refraction, and hence their velocities, are different. Thus, the change in nearly horizontal

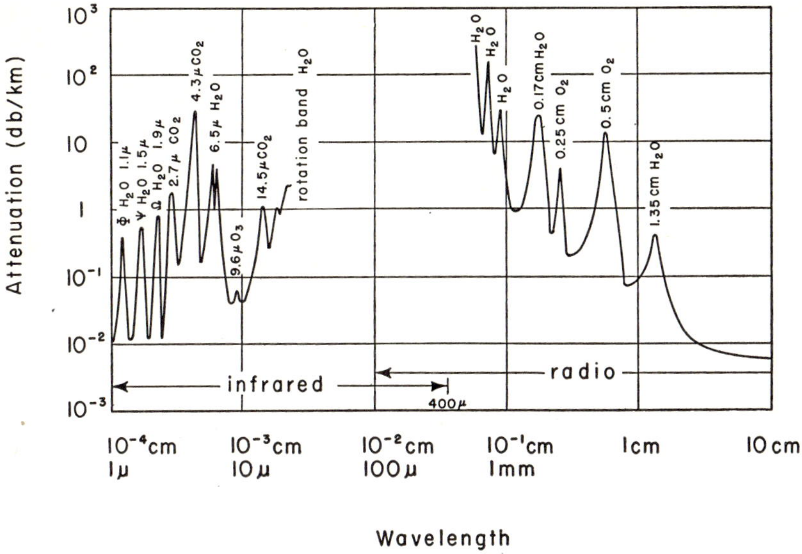

Fig. 1.2 The absorption spectra of the various atmospheric gases in the infrared and short-wave radio regions of the electromagnetic spectrum. The unit of attenuation by absorption is the decibel per kilometer in air at 20° C and 1000-millibar pressure. The carbon dioxide content is 300 parts per million by volume, and the water vapor density is 7.5 grams per cubic meter of air. The decibel is defined as

$$\text{Db} = 10 \log_{10} E/E_0$$

where E/E_0 is the fraction of power attenuated. E_0 is a constant power level, arbitrarily chosen.

velocity with height, dv/dz, results in the ray's taking a curved path whose instantaneous curvature is $1/r$, the reciprocal of the radius of curvature. The geometry is illustrated in Fig. 1.3. The angular velocity v/r is the same for all sections of the wave front so that

$$\frac{v}{r} = \text{constant} = \frac{dv}{dr}$$

on differentiation. The difference in angle between dr and dz is so small that we may take $dr = dz$, making

$$\frac{dv}{dz} = \frac{v}{r}. \tag{1.5}$$

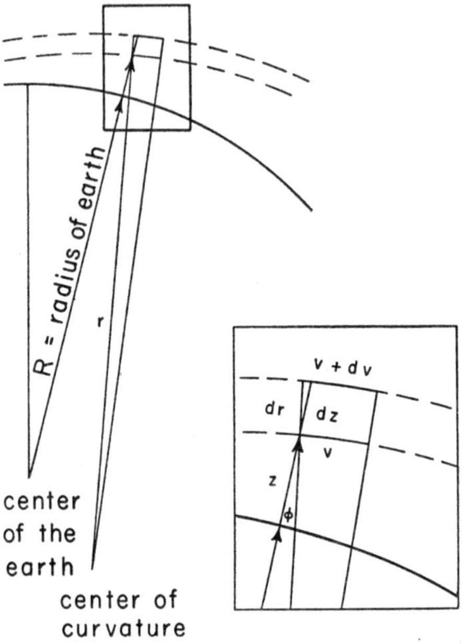

Fig. 1.3 The geometry and approximations used in defining terrestrial refraction. Instantaneous ray paths are the dashed lines. These paths are arcs of circles of radius r and $r + dr$ in which the energy travels at velocities v and $v + dv$. The instantaneous angular velocity is the same in each path, so that

$$\text{Angular velocity} = \frac{v}{r} = \frac{v + dv}{r + dr}$$

solving

$$vr + v\,dr = vr + r\,dv$$

and

$$\frac{dv}{dr} = \frac{v}{r}.$$

Also, from the figure,

$$dr = \cos\phi\,dz$$

when

$$\phi \to 0, \quad \cos\phi \to 1$$

and

$$dr \to dz.$$

These approximations are used when the ray paths are nearly concentric with the earth's surface, this special case being called terrestrial refraction.

From the definition of index of refraction,

$$v = \frac{c}{m}, \tag{1.6}$$

differentiation in respect to height gives

$$\frac{dv}{dz} = -\frac{c}{m^2}\frac{dm}{dz}. \tag{1.7}$$

c is the speed of light (2.998×10^8 meters per second) and is a constant in vacuo for all electromagnetic radiation. dv/dz is known from Eq. 1.5. dm/dz can be obtained from Eq. 1.4 by differentiation after setting $e = 0$, the condition for dry air. When these two equations that result from the indicated operations are substituted in Eq. 1.7, we obtain the following.

On differentiating Eq. 1.4

$$10^6 \frac{dm}{dz} = \frac{A}{T}\frac{dp}{dz} - A\frac{p}{T^2}\frac{dT}{dz}. \tag{1.8}$$

From the hydrostatic equation:

$$\frac{dp}{dz} = -\rho g = -\frac{pg}{R_G T}, \tag{1.9}$$

after substituting from the equation of state, $p = \rho R_G T$. R_G is the gas constant for dry air, and g is the acceleration of gravity.

Substitution of (1.9) in (1.8) and factoring give:

$$\frac{dm}{dz} = -A \times 10^{-6}\frac{p}{T^2}\left(\frac{g}{R_G} + \frac{dT}{dz}\right). \tag{1.10}$$

The parenthesis of Eq. 1.10 contains a term that admits of easy evaluation from standard meteorological data. From pressure and temperature measurements taken by weather balloons (radiosondes) at the site where refraction effects are to be investigated, a curve of temperature versus height can be drawn. Such a curve is called a sounding. The slope of this curve at any point is dT/dz,† the vertical temperature gradient. $-(dT/dz)$ is called the lapse rate of temperature and is signified by the symbol γ. g/R_G is a constant value of the vertical temperature gradient called the *autoconvective lapse rate*, γ_A. This lapse rate exists in an atmosphere where the density is constant with height.‡ It is to be noted that lapse rate is defined in such a way as to make the normal lapse of temperature positive.

† In general, lapse rate is equal to $-\partial T/\partial Z$. Under the implicit assumption of no horizontal temperature gradient of large magnitude and of steady-state conditions, the partial derivative may be replaced by the total derivative.

‡ Such an atmosphere is called a *homogeneous atmosphere*. The density is the surface density. The height is given by $H = R_G T_0/g$, where R_G is the gas constant for the air, T_0 is the surface temperature, and g is the acceleration of gravity. At standard conditions $H = 7990$ meters.

Substituting these definitions into Eq. 1.10 and indicating the constant A at standard conditions of temperature and pressure (null subscript) from Eq. 1.4 lead to

$$\frac{dm}{dz} = -(m_0 - 1)\frac{p}{p_0}\frac{T_0}{T^2}(\gamma_A - \gamma). \tag{1.11}$$

Substituting Eqs. 1.11 and 1.5 into Eq. 1.7 gives with the aid of (1.6)

$$K = \frac{1}{r} = \frac{(m_0 - 1)}{m}\frac{p}{p_0}\frac{T_0}{T^2}(\gamma_A - \gamma). \tag{1.12}$$

At standard conditions: $p = p_0 = 1013.2$ millibars, $T = T_0 = 273°$ abs; and Eq. 1.12 becomes

$$\frac{1}{r} = 1.07 \times 10^{-6}\,(34.1 - \gamma)\ \text{km}^{-1}.$$

At a normal lapse rate of the order of 6.5° C per kilometer, the curvature of the ray becomes

$$\frac{1}{r} = 2.96 \times 10^{-5}\ \text{km}^{-1}.$$

The mean radius of the earth R is equal to 6369.8 km, making the curvature of the earth

$$\frac{1}{R} = 15.7 \times 10^{-5}\ \text{km}^{-1}.$$

Inspection of these two equations shows that even in so-called line of sight transmission, there is a curvature of light rays in the atmosphere in the same sense as that of the earth's surface, equal to about one-fifth or one-sixth of the earth's curvature. This is because of the normal temperature gradient found in the atmosphere.

Equation 1.12 further predicts that when $\gamma = \gamma_A$, there will be no curvature, so that a ray tangent to the earth will be projected in a straight line. If $\gamma > \gamma_A$, the ray will curve upward. It is a negative curvature because of an increase in density with height. When $\gamma \approx -114°$ C per kilometer, which corresponds to a temperature increase of 6.3° F per 100 feet, the curvature of the ray is of the same order as the curvature of the earth. For larger temperature inversions, the rays can actually bend back to earth and be reflected from the earth into the atmosphere again. The effect is illustrated in Fig. 1.4.

So far in the treatment of terrestrial refraction, the water vapor content of the air has been neglected. The development of the theory of refraction in a moist atmosphere follows the line of attack just completed for dry air. However, in evaluating dm/dz from Eq. 1.4 to substitute in Eq. 1.7 to find the curvature of the ray, vapor pressure and its derivative with height de/dz must not be neglected. In fact, the term containing the vapor pressure in Eq. 1.4 is so large in the microwave region that lapse rates of mixing ratio of the order of $\frac{1}{2}$ gram per kilogram per 100 feet in an isothermal atmosphere

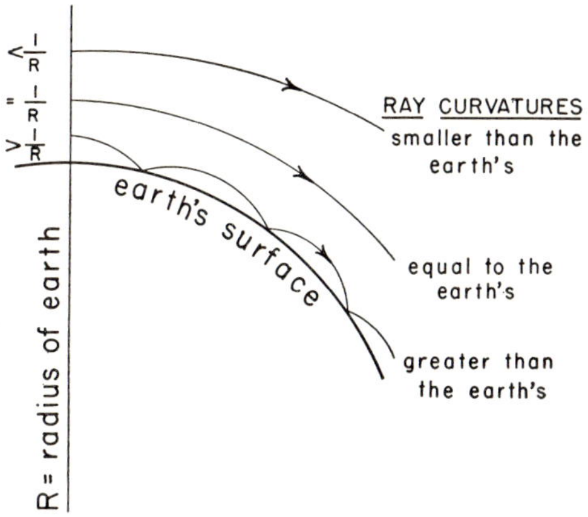

Fig. 1.4 Ray paths for several different curvatures taken in respect to the earth's curvature. The lowest ray path shows refraction in the atmosphere with reflection from the earth's surface. The three curvatures arise from three different lapse rate conditions and all can take place at the same geometric height, but not simultaneously.

can cause curvatures of the microwave of the order of the curvature of the earth. The details of the derivation will be left to the problems. With visible light, water vapor gives a relatively unimportant contribution to the curvature of a light ray, the effect being in the opposite sense and less than 1/100 the magnitude of the corresponding microwave case.

So far in the mathematical development of refraction, no distinction has been made between microwaves and visible light. The distinction is implicit in the measurement of the gradients of temperature and moisture, dT/dz and de/dz, respectively. Because of the wavelength restriction on ray tracing, these mathematical differential quantities become larger and larger geometrical intervals with increasing wavelength. Thus, a finite temperature difference over a given height interval may be a differential gradient for a microwave, several differential gradients for visible light, and a fraction of a differential gradient which must be averaged with several others above and below it for a longer radio wave. This limitation to ray tracing imposed by theory is borne out experimentally.

Before the theory of curvature of rays in the atmosphere is left, one fairly obvious fact should be stated. It is this. If a ray of radiation undergoes a curved path, the angle measured at the detector will be along a line tangent to the curved path at the detector. This statement is illustrated in Fig. 1.5.

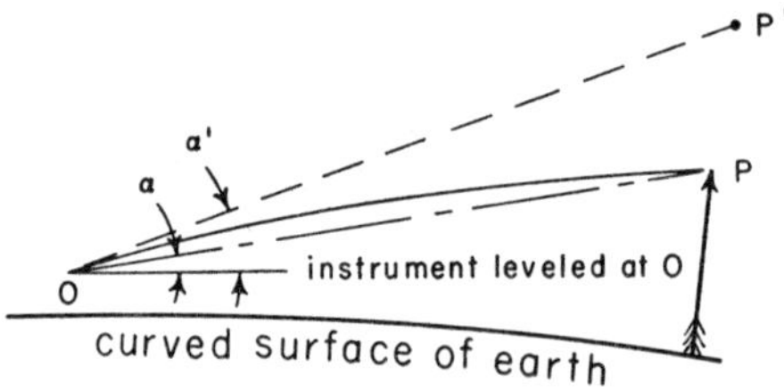

Fig. 1.5 An illustration of the angular position of the image point P' in a refracting atmosphere. α is the angular height of object point P in the absence of a refracting atmosphere. α' is a similar angle measured in a refracting atmosphere.

The solid line OP indicates the curved path of the ray. The center line notation (— – —) with angle α gives the true elevation of point P. Any detector, whether it is the eye, a telescope, a radar, or any other device, will measure P to be at P', by measuring the angle α'. Concave curvature of the ray would make $\alpha' < \alpha$ and put P' below P. Measurements of this type are the best we can do at point O if the detector responds to radiation traversing the path OP.

Retardation effects of index of refraction

The index of refraction enters into calculations involving the measurement of distances. These techniques involve the measurement of a time interval calibrated in terms of distance.

For example, assume that a target object (such as a balloon or airplane) at a known altitude emits a pulsed signal. The signal is of a single frequency in the frequency range lying between 220 and 350 megacycles per second (1.36 to 0.86 meters wavelength). The delay time from target to detector can be measured. From this measurement, the slant distance to the target can be determined. A system performing these functions is known as *shoran* (*sho*rt *ra*nge *a*ids to *n*avigation).

Retardation effects are also important for a system using frequencies between 3000 and 30,000 megacycles per second (10 centimeters to 1 centimeter wavelength). Radiation at these frequencies is the basis of a *radar* system (*ra*dio *d*etection *a*nd *r*anging) provided the geographical site of both the emitter and detector are the same. The radiation is reflected or backscattered from a distant target. In most cases, radars emit radiation in pulses, but continuous wave radar exists and is used on moving targets.

The retardation problem is always solved initially as though the radiation passed through a constant-density atmosphere. Let v_0 be the velocity of the radiation in this medium and v the velocity of the radiation in an atmosphere different from the constant-density atmosphere assumed. c is the speed of the energy in free space. Thus, in the constant-density atmosphere

$$s_0 = \int_0^t v_0 \, dt = \int_0^t \frac{c}{m_0} \, dt. \tag{1.13}$$

m_0 is the index of refraction of the constant-density atmosphere, s_0 the distance along the curved path in this atmosphere, and t the time required for the energy to travel a distance s_0. The design velocity v_0 is usually equal to 186,219 miles per second, the speed of an electromagnetic wave in an atmosphere where $(m_0 - 1) \times 10^6 = 293$. This is a standard sea-level refractivity. The true distance along the path is

$$s = \int_0^t v \, dt = \int_0^t \frac{c}{m} \, dt. \tag{1.14}$$

Here m is a variable index of refraction and v is the true velocity in the medium. For accurate work, a correction factor Δs equal to the true minus the assumed path is added algebraically to the assumed range s_0. This correction factor enters in such a way as to increase the range. Equation 1.13 subtracted from Eq. 1.14 gives the correction factor

$$\Delta s = s - s_0 = \int_0^t \left(\frac{c}{m} - \frac{c}{m_0} \right) dt$$

$$\Delta s = \int_0^{v_0 t} (m_0 - m) \, ds_0. \tag{1.15}$$

In the denominator, $m = 1$ to a high degree of approximation. In general, Eq. 1.15 is evaluated as a sum of short straight-line paths that are an approximation to the true curved path. Through taking a fixed interval Δs_0, and letting $\Delta m = (m_0 - m)$, Eq. 1.15 may be represented as

$$\Delta s = \Delta s_0 \sum_{j=1}^{j=n} \Delta m_{j-\frac{1}{2}}. \tag{1.16}$$

The process is illustrated in Fig. 1.6. The general technique for the ranging of targets here elaborated for electromagnetic waves is also

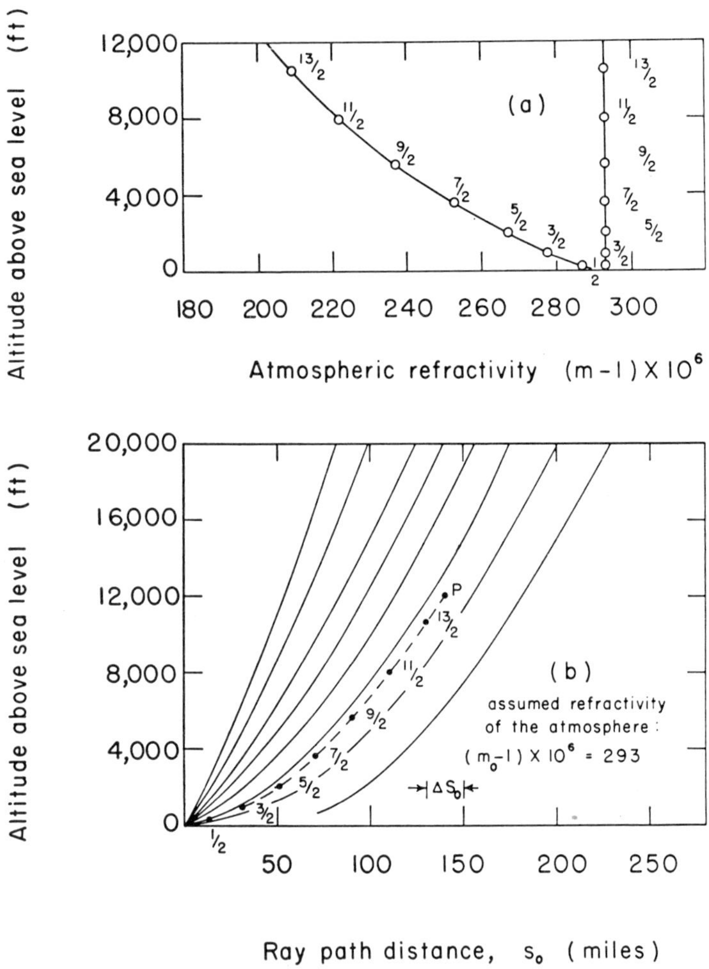

Fig. 1.6 (*After Aslakson and Fickeissen.*) The ray-path distance as a function of height for electromagnetic waves in an atmosphere where the refractivity of the atmosphere is a constant independent of height, $(m_0 - 1) \times 10^6 = 293$. The nomogram may be used for light waves as well as radio-frequency waves. s_0 is a curvilinear distance.

used in sound-ranging techniques. These techniques will be discussed under measurements of upper atmospheric temperatures.

The accuracy of measurement of this type is such that the average error in locating a target by shoran is of the order of 10 feet in 200 miles. Occasional abnormal atmospheric water vapor and temperature stratifications exist that increase this error over corrections computed for a United States standard atmosphere, but they can be corrected for by the above technique. Conversely, if the true geodetic range is known, the errors in distance over the United States standard atmosphere give some indication of the thermal and vapor structure of the atmosphere. Aside from purely military uses, techniques such as shoran have application in geodetic survey work in regions inaccessible except by aircraft in flight.

As an example of calculating the actual path length of a refracted ray in the atmosphere, let us use Eqs. 1.13, 1.14, and 1.16 as a guide to explaining Fig. 1.6. Before this calculation, the following information is available.

1. A knowledge of the pressure, temperature, and water-vapor structure of the atmosphere over the path the ray travels. This is obtained by radiosonde balloon or airplane soundings. This information gives both γ and Δm as a function of height. Figure 1.6*a* illustrates a typical curve of $(m - 1)$ versus elevation.

2. The elevation of the emitter (such as a VHF beacon in an airplane). This information may be telemetered from the beacon.

3. The altitude of the ground station.

4. The delay time from the beacon to the receiver. By means of Eq. 1.13 an uncorrected distance s_0 is computed.

Figure 1.6*b* is a nomogram(1) that expresses the distance measured along a ray path as a function of the height of an emitter. This family of curves is computed for an atmosphere having a constant index of refraction with height of $(m_0 - 1) \times 10^6 = 293$. Using this condition in Eq. 1.12 the curvature of the shoran ray can be computed, and by the proper geometry the height of the ray above the surface of the earth for any path length can be evaluated. Such a computation leads to Fig. 1.6*b*.

Figure 1.6*a* illustrates a typical variation of the index of refraction in the atmosphere with elevation. The straight line whose value is $(m_0 - 1) \times 10^6 = 293$ is the index of refraction for which Fig. 1.6*b* is computed. From this graph, it will be observed that a difference exists between the standard and ambient index of refraction. This difference amounts to $(293 - 222) \times 10^{-6}$ unit of $(m - 1)$ at 8000 feet. Such a variation indicates that Fig. 1.6*b* must be corrected for range because of the nonstandard index of refraction. The procedure follows.

Assuming a standard atmosphere, the delay time is multiplied by 186,219 miles/second. This step gives the shoran distance in an atmosphere with a uniform refractivity of $(m_0 - 1) \times 10^{-6} = 293$. This value is the abscissa while the height of the airborne emitter is the ordinate of Fig. 1.6*b*. A sea-level station will be assumed. This point will be called P. In the example chosen, the coordinates of P are (12,000 feet, 140.037 miles), the latter corresponding to a delay time of 752 microseconds. The dashed line corresponds to the path of the shoran ray in the uniform atmosphere. This shoran distance, which is s_0 of Eq. 1.13, has to be corrected for a nonuniform atmosphere. The correction will be of the form of Eq. 1.16.

Following the form of Eq. 1.16, the correction term may be written as

$$\Delta s = \Delta s_0 (\Delta m_{1/2} + \Delta m_{3/2} + \Delta m_{5/2} + \cdots + \Delta m_{n-1/2}).$$

For the example chosen, $\Delta s_0 = 20$ miles and the $\Delta m_{j-1/2}$ terms are the abscissa distances between the two curves of Fig. 1.6*a*. The heights were obtained by laying off equal intervals of Δs_0 on Fig. 1.6*b*, noting the midpoints, and reading the height at which the ordinate through these points intersects the dashed curve. The correction becomes

$$\Delta s = 20 \times 10^{-6} (6 + 16 + 26 + 40 + 56 + 71 + 84) = 0.006 \text{ mile}.$$

This correction must be added to the value of s_0 because the index of refraction in the ambient atmosphere is less than that in the uniform atmosphere. Therefore, the true distance along the ray from emitter to receiver is

$$s = s_0 + \Delta s = 140.037 + 0.006 = 140.043 \text{ miles}.$$

Once the true ray distance has been found, this distance may be projected onto the surface of the earth to give the true geodetic range of the source from the detector.

Applications of the theory of terrestrial refraction to atmospheric optics

Using the theory of refraction of electromagnetic radiation just developed, with reference particularly to Eqs. 1.3 or 1.4, 1.12, 1.13, and 1.15 as a guide, we can discuss in a qualitative manner a number of atmospheric phenomena. In principle, these effects are also capable of an exact quantitative treatment. However, the last sentence may imply so much work that the calculation is never done, or at best is done only for extremely idealized cases.

Looming, towering, and stooping

When a strong temperature inversion exists at the surface of the earth, the density decrease with height of the air is more rapid than usual. In accord with Eqs. 1.3 and 1.12, and in the absence of fog these conditions lead to the refraction of light. The amount of curvature is a function of the temperature gradient.

Figure 1.7*a* shows the paths of two representative rays through an atmosphere with a strong temperature inversion at the ground. The inversion is shown schematically in Fig. 1.7*b*.

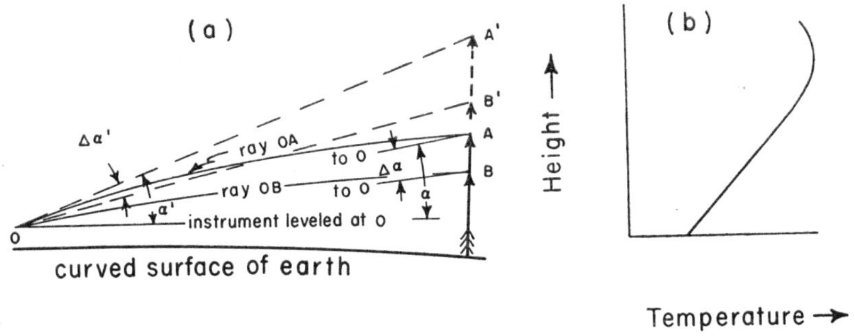

Fig. 1.7 Schematic path of two rays from object points A and B in an atmosphere having a strong temperature inversion at the ground. The image points are at A' and B'. As depicted, because $\Delta\alpha' > \Delta\alpha$ and $\alpha' > \alpha$, both looming and towering occur.

The actual paths of the two representative rays are AO and BO from the object points A and B. Geometrically, these two points have an angular separation AOB. The detector at O, the eye in the case of optics, perceives the rays as coming from the tangent directions OA' and OB'. The image $A'B'$ has an apparent angular subtense $A'OB'$ and the image to be appears elevated above the true position of the object AB. This phenomenon of elevation is called *looming.* Because of the actual temperature distribution, the angular size of the image $A'OB'$ may be less than, equal to, or greater than the true angular subtense of the object AOB. For $A'OB' > AOB$ the image is stretched vertically over the object size; the phenomenon is called *towering.* For $A'OB' < AOB$, the image is compressed vertically; this optical shortening phenomenon is called *stooping.* In the case of a linear decrease of density with height, $A'OB' = AOB$. When a nonlinear change of density with height occurs, so that ray AO has a greater curvature than ray BO, *towering* is observed. Conversely, when the curvature of AO is less than the curvature of BO, *stooping* occurs.

Sinking and shimmer

When the temperature distribution shows a density increase with height, the paths of two representative rays are as shown in Fig. 1.8*a* with the corresponding temperature distribution shown in Fig. 1.8*b*. These gradients may show temperature changes of 20 to 30° C in the first meter from the surface. This distribution is not uncommon over deserts in the summer, or over warm open water in winter. The geo-

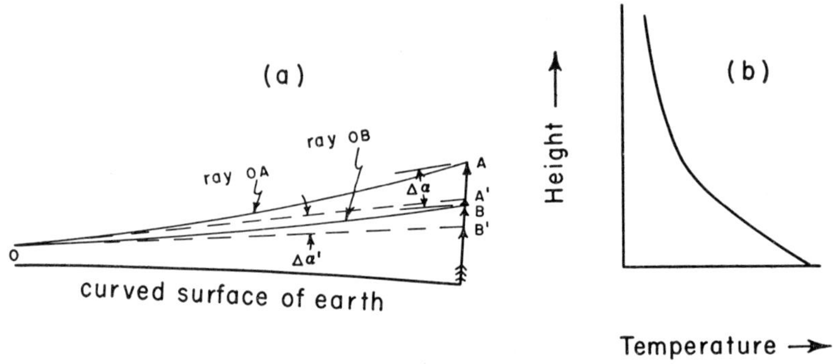

Fig. 1.8 Schematic path of two rays from object points A and B in an atmosphere having a highly unstable lapse rate at the ground. The image points are at A' and B'. As the figure shows, because $\Delta\alpha' < \Delta\alpha$ and the elevation angle of A' is less than A, both sinking and stooping occur.

metric angular subtense of the object is AOB and the image appears to have an angular subtense $A'OB'$. The image appears at an angular position below the object. This phenomenon is called *sinking*.

When such extremely high lapse rates are encountered, the atmosphere is in turbulent motion. The temporal variations in density coupled with the spatial density variations of the air cause the eye to perceive rapid fluctuations of image and intensity in any given direction, since the rays tangent to this direction at the eye originated from many different parts of an object situated at some distance. This effect is called optical *shimmer*. It is very noticeable when one is looking at distant objects through the hot exhaust gases behind a jet airplane.

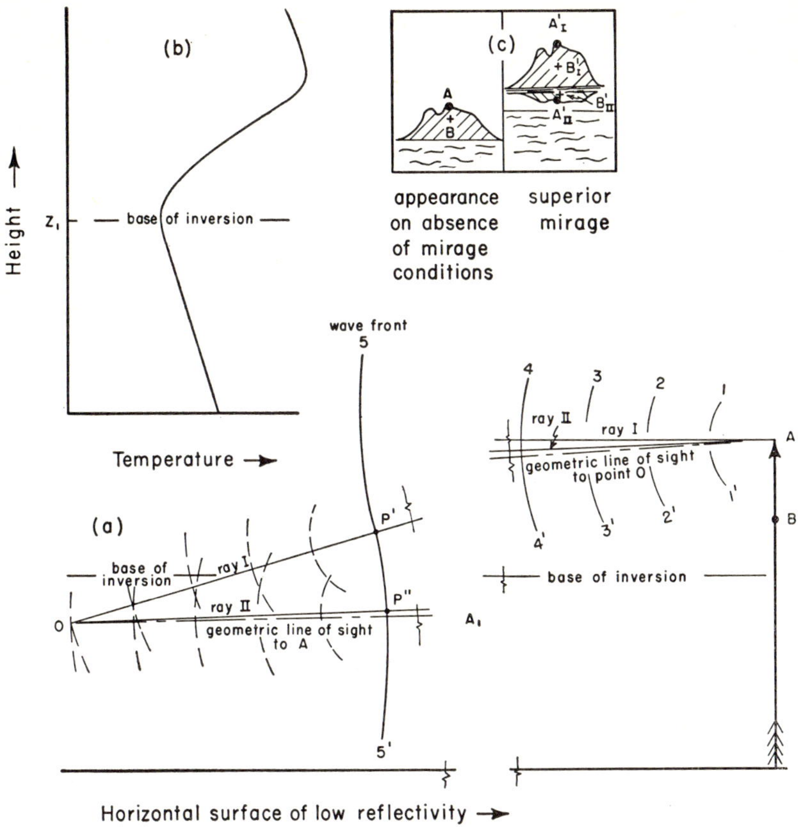

Fig. 1.9 Schematic ray paths, wave fronts, and temperature distribution encountered in a superior mirage. The image nearest the object always appears inverted.

Superior and inferior mirages

The stratification of the air may be such that radiation from the same point on an object a long distance away from an observer may travel two different paths through the atmosphere in arriving at the observer's eye. The observer perceives the image of this point at two widely separated points in space. If the image seems to be reflected from a layer in the sky, it is called a *superior mirage,* and if the sensation is that of apparent reflection from the earth's surface, it is called an *inferior mirage.* The location of object and observer in relation to the layer of the atmosphere containing the abnormal density distribution is important, and small changes in these relative positions may cause the effect to disappear.

Let us consider a stratification of the air such as is shown in Fig. 1.9*b*. The height of the base of the inversion z_1 may be of the order of 10 meters above the ground. Reflected light from point A on the object, denoted by the arrow in Fig. 1.9*a*, appears to emanate from this point. In accordance with the wave theory of light, this point acts as the center of a series of wave fronts, 11′, 22′, 33′, 44′, and after a very long distance 55′. Because of the density distribution and hence the speed of the light, these wave fronts (lines of equal phase) are not perfectly spherical. This fact means that at large distances, the wave front is not that of a plane wave but has the form taken by wave front 55′. The rays labeled I and II are the normals to the wave front that will ultimately reach the detector at O. Rays I and II therefore are also normal to wave front 55′, intersecting this wave front at P' and P'' respectively. Points P' and P'' are so close to the detector that there is no appreciable curvature of each ray for the remaining distance to the detector. Alternatively, these points may be considered as the source of two sets of spherical waves given by the dotted lines, concentric about P' and P''.

With either of these approaches, point A seems to lie both along the line OP' and the line OP''. In reality, point A lies along OA_1, a segment of the straight line OA, shown by the centerline marking.

Thus, the images of point A are elevated above the geometric elevation angle of point A. Consideration of the paths of other representative points, such as point B, will show there must be two images, one of which is upright and the other inverted. The entire mirage *looms* above the horizon. This phenomenon is called a *superior mirage* and conceivably might have the appearance shown in Fig. 1.9*c*.

The *inferior mirage* may arise from a temperature distribution as given in Fig. 1.10*b*. The paths of two representative rays from the same point are represented in the same manner as in Fig. 1.9. This temperature distribution causes point A to appear at points P' and P'' of Fig. 1.10*a*, giving both an erect and inverted image as illustrated in Fig. 1.10*c*. The images appear below the object as though reflected from the earth's surface. This sensation gives the name *inferior mirage* to the phenomenon. Because of the high degree of instability shown by this temperature distribution, the *inferior mirage* is apt to be more poorly defined and to exhibit more shimmer than the *superior mirage*.

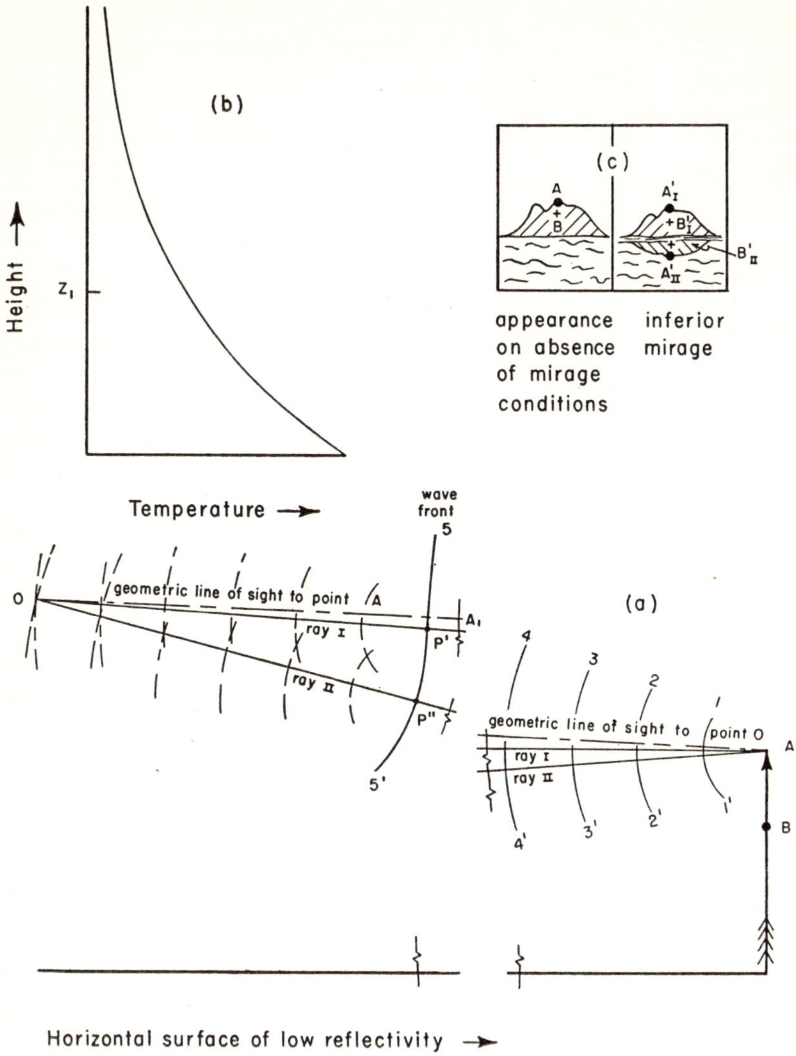

Fig. 1.10 Schematic ray paths, wave fronts, and temperature distribution encountered in an inferior mirage. The lower image appears to be reflected from the surface.

Radio ducts

Let us transfer our attention from optics to high radio frequencies. In this frequency range, 100 to 10,000 megacycles per second, moisture as well as temperature gradients becomes important in causing refractive effects during the propagation of energy through the atmosphere. Experience has shown that, for temperature inversions in excess of 9° C per 100 meters and/or a mixing ratio gradient of water vapor of 1.6 grams per kilogram per 100 meters decreasing upwards, the curvature of the ray may equal or exceed the earth's curvature. This fact leads to transmissions or reflections, as the case may be, far in excess of the normal "line of sight" conditions for which radars are calibrated and television channel frequencies are assigned. This condition is known as *anomalous propagation* or *super-refraction.*

In the atmosphere, there frequently occurs a layer of air that has a higher mean temperature than the air immediately above and below it. Such a layer, a temperature inversion, forms a part of a *radio duct,* so called because the strong refraction in this channel traps high radio frequencies. Anomalous propagation of these frequencies results. These radio ducts have many of the features of a wave guide, provided that the restrictions of ray tracing are strictly adhered to. A *wave guide* for our purposes is visualized as a duct with perfectly reflecting walls. In wave guides, there are phase† restrictions that must be met at the reflecting surfaces. These restrictions are very important when the cross-sectional dimensions of the wave guide are of the same order of magnitude as the wavelength of the wave that is being propagated. When the cross-sectional dimensions of the guide are fixed, there is an upper limit on the wavelength of the radiation transmitted. For wavelengths much smaller than these dimensions, there is essentially no restriction. In the intermediate region, propagation is seriously attenuated for those wavelengths that do not meet the required phase relations. The dotted lines of Fig. 1.11*a* show the ray patterns of a wave guide.

A radio duct in the atmosphere gives an analogous effect because of the refraction of the rays by an atmosphere stratified in a manner similar to Fig. 1.11*b*. The thickness of the duct extends between limits z_1 and z_2, and the ray pattern is similar to the solid lines in part *a* of the figure. This particular sounding might illustrate an upper level subsidence inversion, with the moisture content of the air decreasing sharply from the base to the top of the inversion. The caustics are indicated, being the surfaces $z = z_1$ and $z = z_2$. In the case of a duct at the surface of the earth, the bottom turning points

† The phase of a wave is the angular position of the amplitude of a wave. Two waves are said to be in phase when the angular positions of their maximum and minimum amplitudes coincide. When two waves have their maximum at angular positions θ_1 and θ_2, these waves are said to be out of phase by an amount equal to the phase angle $\delta = \theta_2 - \theta_1$.

of the ray would show true reflection from the earth's surface, especially over the water, a good reflector for microwaves as well as light. The reflection coefficient remains high over water (>50 per cent) up to grazing angles of 30°, approaching 100 per cent over smooth land as well as water for zero grazing angle.

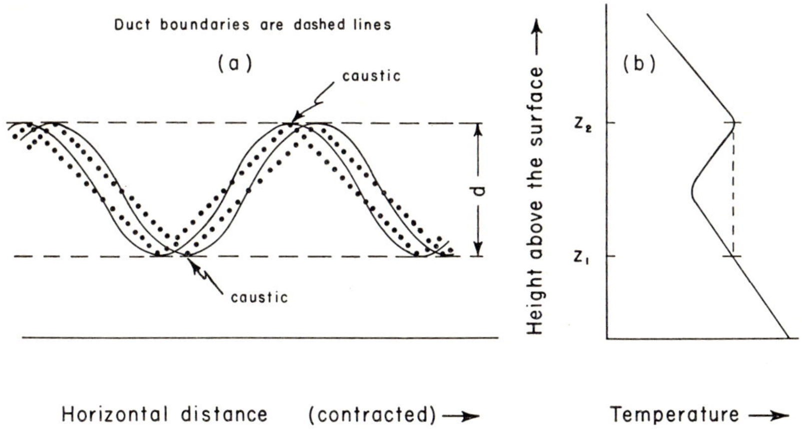

Fig. 1.11 Ray pattern and radiosonde balloon sounding associated with an elevated atmospheric radio duct.

From the theory, a convenient rule for the maximum wavelength that will be trapped by a duct extending from z_1 to z_2 has been established. Longer wavelengths will be affected but not trapped under these average conditions. The rule is that if $z_2 - z_1$ is set equal to d, measured in feet, the maximum wavelength measured in centimeters that will be trapped by a duct is

$$\lambda_{max} = 0.014d^{3/2}. \tag{1.17}$$

Table 1.2 gives some minimum duct widths for various regions of the spectrum.

Table 1.2

Maximum Trapped Wavelength as a Function of Duct Thickness

Region	Wavelength	Minimum Duct Thickness
Visible	0.5 micron	1 centimeter
Infrared	10 microns	2 inches
Microwave	10 centimeters	80 feet
Meter wave	10 meters	2000 feet

Since atmosphere ducts have depths somewhere between 50 and 500 feet in the free atmosphere, the duct phenomenon finds its greatest importance at microwave frequencies and in the visible range. At optical frequencies complicated ducts may exist near the surface of the ground, leading to extremely complicated optical effects.

In conclusion, atmospheric ducts close to the ground might cause dangerous conditions in blind-landing procedures for aircraft using such techniques as *ILS* (*I*nstrument *L*anding *S*ystem) on 3-meter wavelengths or *GCA* (*G*round *C*ontrolled *A*pproach) using microwave frequencies. Fortunately, in most cases where these radio aids have to be utilized, the moisture and temperature stratification is usually such as to preclude strong ducts.

Astronomical refraction

When radiant energy traverses the entire atmosphere, the refraction that results from this long path is called *astronomical refraction.* At the present time it is of quantitative importance to astronomers. Refraction causes errors in position that must be corrected for, especially when stars or planets are low on the horizon. The definition of astronomical refraction is precisely what the above usage suggests it should be; namely, it is the difference between the true and the measured zenith angles of a celestial object. The zenith angle itself is the angle between the object and a perpendicular to the earth's surface erected at the site of detection. These definitions show the zenith angles to be θ and θ' in Fig. 1.12, with a corresponding refraction of

$$\Psi = \theta - \theta'. \tag{1.18}$$

The general case of refraction is derived quite easily. Astronomical and terrestrial refraction are special cases of the general theory. The general case requires a knowledge of only two things, the first being a bit of elementary calculus, and the second a knowledge of the generalized form of Snell's law of refraction.

First, a consideration of the geometry; this is shown in Fig. 1.12. The right triangle relationship for a differential length ds of the ray path, located at a central angle ϕ and a distance r from the center of the earth is

$$ds^2 = (r\,d\phi)^2 + dr^2. \tag{1.19}$$

By factoring $(dr/r)^2$, and taking the square root of both sides after solving for $d\phi^2$, Eq. 1.19 can be written

$$d\phi = \sqrt{\left(\frac{ds}{dr}\right)^2 - 1}\,\frac{dr}{r}\,. \tag{1.20}$$

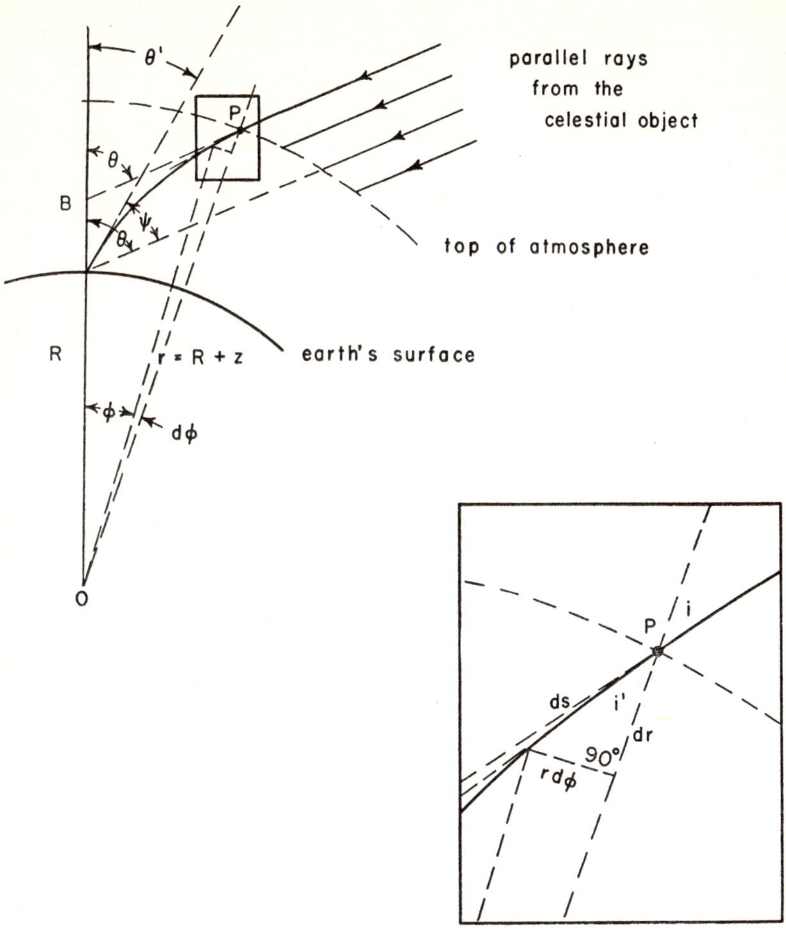

Fig. 1.12 The geometry of astronomical refraction.

From the differential triangle, $ds/dr = \sec i'$, making

$$d\phi = \sqrt{\sec^2 i' - 1}\,\frac{dr}{r} = \tan i'\,\frac{dr}{r}\cdot \tag{1.21}$$

A further inspection of the figure shows the angles of triangle OBP, to add to

$$\theta = \phi + i'. \tag{1.22}$$

Eliminating θ between Eqs. 1.18 and 1.22 and then differentiating shows

$$d\Psi = d\phi + di'. \tag{1.23}$$

$d\phi$ is known from Eq. 1.21. di' must be found by differentiating the form of Snell's law valid for density surfaces concentric with the earth's surface.(11) This form of the law states that

$$rm \sin i' = (R + z)m \sin i' = Rm' \sin \theta' = \text{constant} \qquad (1.24)$$

where R is the radius of the earth and z is the height of point P of Fig. 1.12 above the surface. $m \sin i'$ is evaluated at P and $m' \sin \theta'$ is evaluated at the point of observation on the earth's surface. The assumption of a vertical density gradient only is quite realistic.

Taking the natural logarithm of Eq. 1.24, differentiating, and solving for di' give

$$di' = -\tan i' \frac{dr}{r} - \tan i' \frac{dm}{m}. \qquad (1.25)$$

Substituting Eqs. 1.25 and 1.21 in 1.23 gives the astronomical refraction in differential form

$$d\Psi = -\tan i' \frac{dm}{m},$$

which on integration gives

$$\Psi = \int_m^{m'} \tan i' \frac{dm}{m}. \qquad (1.26)$$

To the degree of accuracy that depends only on the assumption that $(m'/m)^2 = 1$, this difference being of the order of less than 10^{-7}, $\tan i'$ may be written

$$\tan i' = \left[\frac{\csc^2 \theta'}{\left(1 + \frac{z}{R}\right)^2} - 1\right]^{-1/2} \qquad (1.27)$$

where only z is a function of the index of refraction, θ' being the measured zenith angle. The trigonometric manipulations to obtain Eq. 1.27 are left as a problem. When the celestial object is low on the horizon the path length through the atmosphere is large, θ' approaches 90°, and the form of the variation of the index of refraction with height is important in determining the astronomical refraction Ψ. For zenith angles $< 75°$, the form of the variation is relatively unimportant because of the shorter atmospheric path length.

As an illustration, consider a constant-density atmosphere. In this type of atmosphere, called a *homogeneous atmosphere*, the height z is of the order of 8 kilometers, making $\left(1 + \frac{z}{R}\right)^2 = 1$ for all practical purposes. This makes $\tan i' = \tan \theta'$. On integrating Eq. 1.26 with constant θ',

$$\Psi = \tan \theta' \ln m', \tag{1.28}$$

where $m = 1$ at $z = 8$ kilometers, the top of the homogeneous atmosphere. m' is so close to 1 that $\ln m' = m' - 1$, making

$$\Psi = (m' - 1) \tan \theta' \approx 293 \times 10^{-6} \tan \theta' \text{ (radians)}. \tag{1.29}$$

Evaluating Eq. 1.29 and comparing values derived from it with values of Ψ computed by H. Gyldén(16) from approximation formulas valid at normal temperature and pressure show good agreement up to a zenith angle of 75°, even with such a gross assumption as a constant-density atmosphere. Some values are shown in Table 1.3.

Table 1.3

Mean Values of Astronomical Refraction Ψ as a Function of Apparent Zenith Angle θ'

θ' (degrees)	Ψ (minutes) Gyldén	Eq. 1.29
0°	0′	0′
30°	0′ 33.56″	0′ 35″
50°	1′ 9.18″	1′ 12″
60°	1′ 40.37″	1′ 44″
70°	2′ 38.39″	2′ 46″
75°	3′ 33.67″	3′ 45″
80°	5′ 18.67″	5′ 42″
85°	9′ 51.23″	11′ 30″
88°	18′ 17.4″	28′ 48″
90°	34′ 40.9″	∞

This table indicates that celestial objects may be visible to the observer when they are actually below the geometric horizon. For example, the moon has an angular subtense of about 31 minutes. The table indicates that when the entire moon is just visible above the horizon, nearly the entire lunar disk is below the geometric horizon. Furthermore, the rapid change in astronomical refraction near the horizon aids in explaining the apparent rapid rise of the moon, sun, or stars from below to a few degrees above the horizon. Perspective effects also help.

Astronomical refraction also gives an explanation for the sun's rising earlier and setting later than would be expected through pure geometric considerations of a planet having no atmosphere. This extra duration of sunshine and twilight is important in polar regions where the sun is always low. The onset of sunrise amounts to as much as a day and a half at the poles after the months of winter darkness. The effect is obviously a minimum and relatively unimportant in equatorial regions where the sun rises at an angle nearly normal to the horizon.

Before the dispersive effects of the atmosphere are touched on, it might be well to indicate the relationship between terrestrial and astronomical refraction. As it is commonly used, Eq. 1.26 and its auxiliary, Eq. 1.27, are used to explain astronomical refraction. In this case, the lower limit of the index of refraction, which occurs at the top of the atmosphere, is exactly 1. For terrestrial refraction, the limits of the integral may be nearly or even exactly the same. This does not make the integral zero, because it is a path or line integral. A plot of tan i' versus ln m would define an area. It is convenient, therefore, to change the form of the integral into one that can be handled more easily. To do this, consider the geometry of Fig. 1.12. In this figure, the ray becomes nearly horizontal as

$$\tan i' = r\frac{d\phi}{dr} \rightarrow \frac{ds}{dz} \tag{1.30}$$

and

$$d\Psi \rightarrow d\phi \rightarrow \frac{ds}{r}$$

because $di' \rightarrow 0$. Substituting these conditions in the differential form of Eq. 1.26 shows

$$\begin{array}{ccc} d\Psi & = & -\tan i' \dfrac{dm}{m} \\ \downarrow & & \downarrow \\ \dfrac{ds}{r} & = & -\dfrac{ds}{dz}\dfrac{dm}{m}, \end{array}$$

giving

$$\frac{1}{r} = -\frac{1}{m}\frac{dm}{dz}. \tag{1.31}$$

Equation 1.31 is precisely Eq. 1.7, because $v = c/m$. Equation 1.7 as originally derived shows a more straightforward geometric analysis of the more important problem at the present time, that of terrestrial refraction.

Dispersion in the atmosphere

Refraction is important for explaining the colors of heavenly bodies close to the horizon. These considerations are not to be confused with explanations of the sun's or moon's color based on scattering phenomena, explained in a later chapter. The explanation given here depends on the dispersion effects (prismlike) of the atmosphere. These effects are so small, even for celestial objects close to the horizon, that the effects correspond to a prism whose faces are nearly parallel. The separation of the colors is so small angular wise, that a point source must be used. An extended source, such as the sun's disk, subtends an angle many times the angular dispersion of the atmosphere. Hence, an observation of any point on the solar disk from the earth's surface is in reality an observation of a small but finite area of the sun's surface, each part of which contributes the proper color to make this point on the sun appear white. The colors follow slightly different paths in the earth's refractive atmosphere so as to arrive at the eye along the same apparent line of sight. The mixture of these colors is white light, the observed color of the extended solar disk.

However, stars, planets, comets at low elevation, and the upper edge of the sun's disk before it disappears below the horizon offer essentially point sources. Chromatic separation can occur. Equation 1.4 allows computation of the index of refraction for various wavelengths in the visible. At a fixed pressure and temperature, this equation indicates the blue end to be refracted more than the red end of the spectrum. This is true even when the effect is integrated over the atmosphere. As a consequence, on the horizon one would expect these bodies to change in color continuously from red through orange, yellow, green to blue as they set, in reverse sequence as they rise. The refracted yellow and green appear occasionally at the sun's rim, but the blue is scattered out by the air molecules. The change is so rapid in a small angular distance, that only a startling change in color in a bright body is usually noticed. An excellent example of this is the *green flash* of the sun as it disappears below the horizon. It is seen only on very clear days, usually over the ocean, and the effect lasts only of the order of a few seconds. Likewise, scintillation of the stars both in intensity and chromaticity may be due to short-period density fluctuations in the atmosphere. This optical "noise" affects absolute measurements of intensity or color adversely.

In conclusion, it may be restated that the methods of this chapter are quite general. Although many of the examples have been drawn from atmospheric optics (in the visible) because the effects are readily observable to all, infrared and microwaves are subject to the same effects though usually in a less pronounced fashion.

References

1. Aslakson, C. I., and Fickeissen, O. O., "The Effect of Meteorological Conditions on the Measurement of Long Distances by Electronics," *Trans. Am. Geophys. Union, 31,* 816–826 (1950).
2. Aslakson, C. I., "New Determinations of the Velocity of Radio Waves," *Trans. Am. Geophys. Union, 32,* 813–821 (1951).
3. Bamford, A. J., "Atmospheric Distortion of Star Observations for Position and Time," *Roy. Astron. Soc.,* Monthly Notices *94,* 204–228 (1934).
4. Barrell, H., "The Dispersion of Air Between 2500 Å and 6500 Å," *J. Opt. Soc. Amer., 41,* 295–299 (1951).
5. Booker, H. G., "Meteorological Aspects of Propagation Problems," *Compendium of Meteorology,* 1290–1296 (1951), American Meteorological Society, Boston.
6. Brunt, D., "The Index of Refraction of Damp Air and Optical Determination of the Lapse-Rate," *Quart. J. Roy. Meteorol. Soc., 55,* 335–339 (1929).
7. Fleagle, R. G., "The Optical Measurement of Lapse Rate," *Bull. Am. Meteorol. Soc., 31,* 51–55 (1950).
8. Hulburt, E. O., "The Green Segment Seen from an Airplane," *J. Opt. Soc. Amer., 39,* 409 (1949).
9. Ives, R. L., "Meteorological Conditions Accompanying Mirages in the Salt Lake Desert," *J. Franklin Inst., 245,* 457–473 (1948).
10. Kroll, C. W., "A Rigorous Method for Computing Geodetic Distance from Shoran Observations," *Trans. Am. Geophys. Union, 30,* 1–4 (1949).
11. McLeod, A. R., "On Terrestrial Refraction," *Phil. Mag., 38,* 546–568 (1919).
12. Neuberger, H., "General Meteorological Optics," *Compendium of Meteorology,* 61–78 (1951), American Meteorological Society, Boston.
13. Rayleigh, Lord, "On the Theory of Stellar Scintillation," *Phil. Mag., 36,* 129–142 (1893).
14. St. Amand, P., and Cronin, H., "Atmospheric Refraction at College, Alaska, during the Winter 1947–48," *Trans. Am. Geophys. Union, 31,* 161–164 (1950).
15. Sweer, J., "Path of Ray of Light Tangent to Surface of Earth," *J. Opt. Soc. Amer., 28,* 327–329 (1938).
16. Wünschmann, F., "Über die Konstitution der Atmosphäre und die astronomische Inflexion in ihr," *Gerlands Beitr. Geophys., 31,* 83–118 (1931).

Source Books

B1. Humphreys, W. J., *Physics of the Air,* McGraw-Hill Book Co., New York, 3rd ed. (1940), Part IV, Chapters I and II.

B2. Kerr, D. E., "Propagation of Short Radio Waves," *M. I. T. Rad. Lab. Series, 13* (1951), McGraw-Hill Book Co., New York.

B3. Minnaert, M., *Light and Colour in the Open Air* (1940), G. Bell and Sons, London (translation by H. M. Kremer-Priest, revised by K. E. Brian Jay).

Problems

1.1 Derive an expression showing how the curvature of the earth and the refraction of the atmosphere determine the horizon distance when the horizon is viewed from height z. Show that in most cases of practical interest the derived expression can be simplified to

$$s = \sqrt{\frac{2Rz}{1 - (R/r)}}.$$

s is the horizon distance as measured along the earth's surface, z is the height of the observer above the horizon, and R and r are respectively the radii of curvature of the earth and atmosphere.

1.2 A sailor stationed as a lookout 100 feet above the surface of the ocean views the horizon on a day when the mean atmospheric lapse rate of temperature causes light to refract with a curvature one-sixth that of the earth. How far away is his horizon? What is the mean lapse rate of temperature? How far away would the horizon be to this observer if atmospheric refraction could be neglected?

1.3 Under the same conditions as Prob. 1.2, both atmospheric and geometric, the lookout observes a light on the horizon. The sailor knows that this is a signal light mounted 50 feet above water level on another ship. How far apart are the two ships when the light is exactly on the horizon? Would having both observer and signal light 75 feet above water level cause any significant change in maximum observable range?

1.4 On a certain cold night, the mean temperature and lapse rate in the lowest 100 feet of the atmosphere are 7° C and 8° C/km, respectively. An observer on the shore noticed rapid color fluctuations of a light from a ship that is on the horizon. The color variations were from blue-green to orange-red. Assuming the dominant wavelengths of these colors to be 0.500 and 0.620 micron, find the maximum short-period variation in lapse rate during this situation.

1.5 To a high degree of approximation, the density of the atmosphere can be represented by the empirical equation

$$\ln \rho = Cz + D.$$

$C = -1.37 \times 10^{-6}$ and $D = -6.72$ when z is height in centimeters and ρ is density in grams per cubic centimeters. The logarithm is to the base e. (*a*) Express the above equation as index of refraction as a function of height. Use visible light. (*b*) With the aid of the expression found in part *a*, express Eq. 1.23 in a form suitable for evaluation. Give numerical values for the constants. (*c*) Find the astronomical refraction for a star at an apparent zenith angle of 30°; another at 60°; and a third at 89°. Compare your results with Table 1.3. Use numerical integration if necessary. Let the index of refraction at the surface of the earth equal 1.000293.

1.6 In many microwave-propagation problems, the earth can be considered flat, provided that the index of refraction is replaced by the modified index N where

$$N = m + (z/R).$$

z is the height in the atmosphere and R is the radius of the earth. The refractive modulus is defined as

$$M = (N - 1) \times 10^6.$$

a. Express the variation with height of the refractive modulus as a function of pressure, temperature, and vapor pressure.

b. Express the curvature of a microwave ray as a function of the parameters in part *a*.

c. What is the curvature of a microwave ray in a layer of the atmosphere bounded by:

Pressure	Temperature	Vapor Pressure
1010 mb	4° C	7.8 mb
890 mb	4° C	4.2 mb

Assume a linear variation of these parameters through the layer.

1.7 An instrument landing system transmits a beam of radio-frequency energy at 335 Mc. The beam, called a glide path, is elevated at an angle of 3 degrees to the horizontal.

What will be the true height and angle of descent of the beam 6.4 km from the point where the beam intersects the ground when the atmosphere has a lapse rate of 6.5° C/km? Assume that the vapor pressure gradient decreases at a rate of 3 mb/km.

CHAPTER 2

Scattering in the Atmosphere

In Chapter 1, a number of useful explanations of atmospheric phenomena were developed through use of the theory that the atmosphere is a homogeneous medium in which the density varies in a smooth fashion from point to point. As an approximation, the atmosphere was considered to be thin concentric shells of homogeneous matter. This concept has limitations, however, and fails completely in explaining the blue of the sky, the change in color of distant objects, radar echoes from storms, the polarization of sky light, and a host of other phenomena. Such explanations have to be made on the basis of the fact that the atmosphere contains myriad particles which range from air molecules, dust, and condensation nuclei to cloud drops, rain, and snow. The study of the energy patterns produced by these discrete particles, both singly and in the aggregate, when illuminated by radiant electromagnetic waves is the study of the *scattering* phenomenon. This study may properly include much of physical optics, such as the study of diffraction, reflection, and some polarization phenomena. By custom, scattering is separated from diffraction, the division depending on the ratio of the size of the scatterer to the wavelength of the incident electromagnetic radiation. When the ratio is small, of the order of $< 1/10$, scattering is said to occur. For ratios of the order of unity, diffraction occurs, and for ratios > 10, ray reflection and geometric shadows begin to appear. These divisions are quite arbitrary, because there is a continuous transition from one region to another, shown beautifully in the theory of scattering developed by Gustave Mie[(5)] early in the twentieth century.

In order to fix our thinking, let us examine how scattering in the usual sense might arise. We are now confining our attention to particles whose largest dimension is small compared to a wavelength of the incident light. Under these conditions, it is the volume of the particle that is most important, the shape being of secondary importance. Within the limits of shapes usually encountered the shape factor can be neglected.

The concept of scattering to be presented is in agreement with experimental facts, many of which were known before the phenomenon was understood. Let us consider what happens when plane polarized radiation impinges on a scattering particle located at O, the origin of

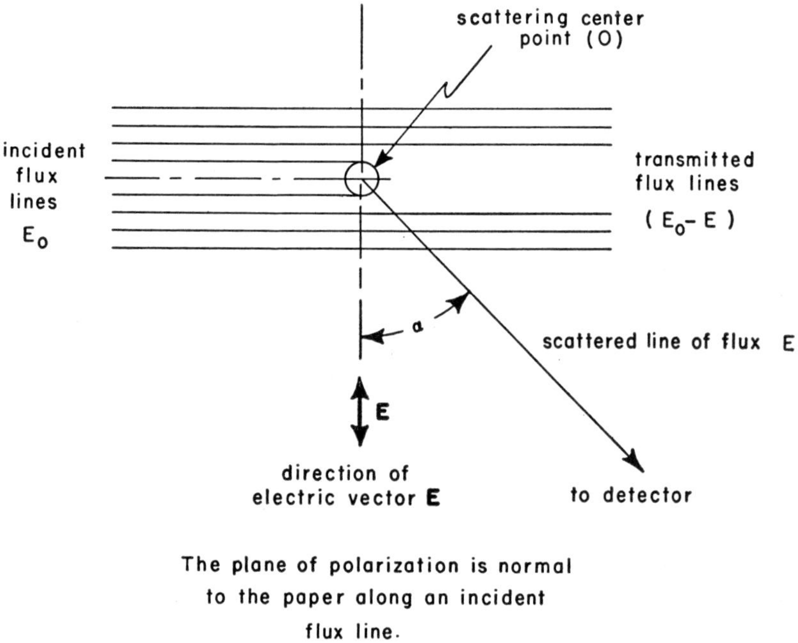

Fig. 2.1 The geometry of scattering of plane polarized electromagnetic energy by an object located at point O. The plane of polarization will always be oriented along a line of flux. If the plane of polarization is either in the plane of the diagram or normal to it, the scattered radiation will be plane polarized. If the plane of polarization makes an angle with the plane of the diagram, the scattered radiation will be elliptically polarized. The *total* intensity of flux detected at angle α will be unchanged by a rotation of the detector about an axis parallel to a line of incident flux and passing through the center of the scatterer.

coordinates in Fig. 2.1. Plane polarized radiation is used because then the wave front† is a plane surface normal to the direction of propagation, and the vibrations of the radiation are in one single plane oriented normal to the wave front. Such polarized radiation may be obtained by passing visible light through a crystal of tourmaline, by reflecting infrared rays from quartz at the polarizing angle, or by the use of a magnetron oscillator at microwave frequencies.

The section of the wave front impinging on a scatterer at O of

† A wave front is a surface of constant phase.

Fig. 2.1 is disturbed by the scatterer itself. This part of the wave front is deformed into a characteristic pattern in three-dimensional space, with the scatterer acting as a center of this axially symmetrical pattern, an example of which is shown in Fig. 2.5. The effect on the detector is the same as though the incident radiation were not present, and the scattered radiation emanated from the scatterer acting as a source.

Furthermore, because the energy in the original incremental wave front is conserved by being distributed over the entire wave front of the scattered wave, the scattered energy measured by any finite-sized detector located at any fixed angle from the scatterer is very small in comparison with the intensity of the incident radiation. However, theory predicts and experiment verifies that, in the absence of absorption, if all this scattered energy could be collected by an integrating sphere, the sum of the transmitted and scattered energy would be equal to the incident energy before scattering. For any measurable effect, one must usually measure the scattered energy from a large number of scatterers located in such a small volume about a point O, that the scattered energy may still be thought of as originating at this point in space. Fortunately, even in this small volume in space, each individual scatterer is so small that there are spaces many diameters in extent between neighboring scatterers. This fact plus the thermal motion of these particles masks any phase effects that may be present, e.g., effects such as to give interference between waves from neighboring particles. Because of *incoherent scattering*, the process just described, the total scattered energy is the sum of the individual energies scattered from each particle. There are examples, not to be treated here, when phase relationships are important in giving the correct scattered energy, characterized as *coherent scattering*. Crystals, wherein the relative positions of the atoms are fixed (in the mean), exhibit the latter process.

The preceding brief description of the scattering process can be placed on an ultimate quantitative basis by considering the scattering center to be an *oscillating electric dipole*. When the centers of electric charges of equal and opposite magnitudes are separated by a measurable distance, the product of one of the poles and the distance between the two is the *dipole moment*, and the entire system is the dipole. When the magnitude of dipole moment changes in a periodic fashion, the system is called an *oscillating* dipole. Because the dipole must be forced to oscillate by some external force, such as is present in radiation incident on the dipole, the frequency of oscillation of the dipole must be the same as the impressed frequency. The amplitude of oscillation of the dipole depends on the closeness with which the impressed frequency matches the frequency with which the dipole would tend to oscillate in the absence of this external force. Matching of these frequencies gives the maximum amplitude. Any oscillating dipole radiates energy at a frequency equal to its frequency of oscillation. The phase of the scattered energy is dependent on the distance from the dipole. This entire process means that the frequency of the scattered light is the same as that of the incident light. Furthermore, as both the incident and the scattered energy are in the same medium, there is also no change in wavelength on scattering.

In addition to electric dipoles, there are also magnetic dipoles, of which a bar magnet with a north and south magnetic pole is an example. Experiments with iron filings to indicate the magnetic lines of force are familiar to every student of general physics. Because of this, the qualitative aspects of the energy pattern together with the radiation of energy from an oscillating dipole will be illustrated through considering the following experiment using a magnetic instead of an electric dipole.

A small solenoid (a coil of wire with an iron core that will act like a small magnet) has an electric current passed through it. The ensuing magnetic field, which can be indicated by iron filings, causes the lines of flux (constant power) to take the form of two circles symmetrical about the axis of the solenoid (see Fig. 2.2). This magnetic field is of a characteristic form shown by all dipoles, whether electric or magnetic. As a more quantitative measure of the magnetic field, a small test magnet of small mass can be suspended at point P in Fig. 2.2 a distance r† from the central solenoid. Now, if the current is varied in a periodic fashion in the central solenoid, the strength of the field at P will change in the same periodic fashion with a time delay of r/c, the time taken for the energy to travel from the source to P at the speed of light. The changing magnetic field at P will be indicated by the small test magnet suspended at this point, which will be observed to oscillate at the same frequency at which the current varies through the solenoid. The use of an oscillating dipole is shown to be a way of transferring energy through

† As there is a distance of separation between the two poles of a dipole, there are actually two distances r_1 and r_2, to the point P. These distances are measured from each pole respectively. However, the measurements of scattered energy are made at such a great distance from the dipole that the interpolar distance is so small compared to r_1 or r_2 that $r_1 = r_2 = r$ for all practical purposes. This assumption treats the dipole as a point source.

space, with the energy detected at P by the test magnet being converted into a mechanical torque. This experiment illustrating the principle of dipole radiation shows the need for a constantly changing field in the continuous propagation of radiant energy. If the magnetic

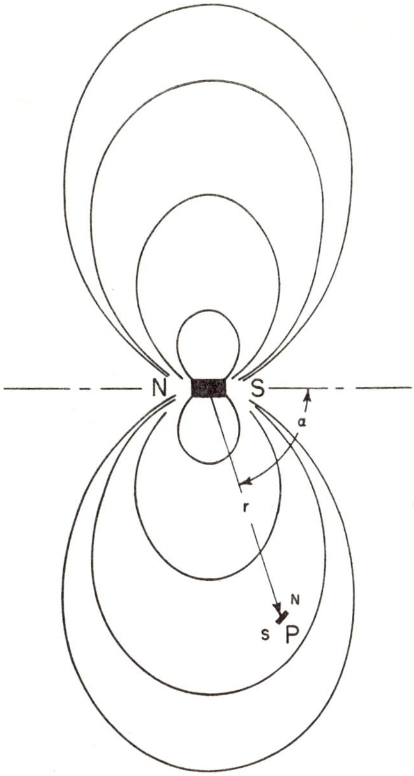

Fig. 2.2 Typical flux patterns associated with a dipole. The case illustrated is for a magnetic dipole. The solid lines are lines of constant flux, indicating that a constant amount of power will be detected by a receiver of finite area if its detecting surface remains normal to the radiating dipole and the distance from the dipole varies as one of the solid lines. Conversely, if the detector is rotated about the dipole as a center with the distance from dipole to detector remaining constant, the power received by the detector will vary continuously from zero to a maximum in rotating from $\alpha = 0$ to $\alpha = \pi/2$.

dipole does not oscillate, the test magnet takes a position of equilibrium with the resultant field and no energy is transferred.

In the theory of the scattering of electromagnetic radiation, an electric dipole replaces the magnetic dipole of the illustrative experiment, and the current in the solenoid is replaced by plane-polarized electromagnetic radiation of intensity I. We shall need to symbolize the following quantities for a quantitative description of scattering.

F = flux = total power passing through or impinging upon a surface perpendicular to the power flow [watts].
E = flux density = flux divided by the area [watts/meter^2].
r = radius of sphere with the source of flux at its center [meters].
I = intensity of radiation = flux per unit solid angle [watts/steradian].
ω = solid angle = any area projected on a sphere of radius r divided by the square of this radius [steradians].
Q = magnitude of the electric charge of the dipole [coulombs].
x = instantaneous separation of the two charges in the dipole whose pole strengths are Q and −Q [meters].
P = electric dipole moment whose magnitude is Qx [meter coulombs].
E = magnitude of the electric vector whose square is proportional to the flux density. The direction of the vector is normal to the flux vector. [Volts/meter.]
m = index of refraction [dimensionless].
c = speed of light [meters per second].
V = volume of scatterer, small compared to the wavelength λ [meters^3].
t = time [seconds].

These quantities, which may be in any set of consistent units such as the ones suggested above, are related in the following way.

If an emitter or scatterer of power is surrounded by a sphere of radius r (this sphere need only be any imaginary spherical surface in space), all the power that leaves the source will in the absence of an absorbing medium pass through the spherical surface with the speed of light c. r is so large that the source may be considered a point. As energy is conserved in passing from source to spherical surface

$$F = 4\pi r^2 E. \tag{2.1}$$

The number of steradians in a sphere of solid angle ω is by definition

$$\omega = \frac{4\pi r^2}{r^2} = 4\pi. \tag{2.2}$$

The intensity of radiation becomes

$$I = \frac{F}{4\pi} = Er^2, \tag{2.3}$$

which in differential form is

$$I = \frac{dF}{d\omega} \cdot \tag{2.4}$$

Writing Eq. 2.4 in the differential form has an advantage over Eq. 2.3 in that it admits of a variation of flux with solid angle. This condition is met in directional emitters, such as dipole scatterers or

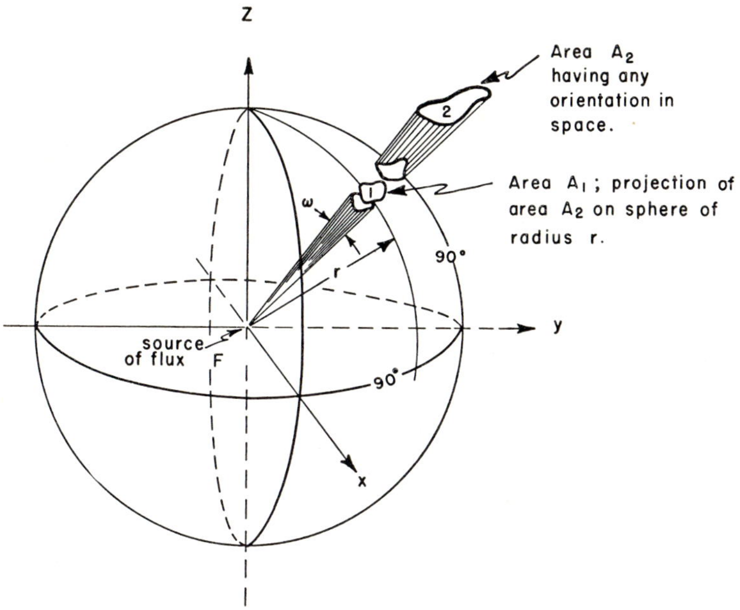

Properties not common to A_1 and A_2	Properties common to A_1 and A_2
$\text{flux density}_1 > \text{flux density}_2$	$\text{flux}_1 = \text{flux}_2$
$\text{radius}_1 < \text{radius}_2$	$\text{intensity}_1 = \text{intensity}_2$
$\text{area}_1 < \text{area}_2$	$\text{solid angle}_1 = \text{solid angle}_2$

The source has the same brightness when viewed from either A_1 or A_2

Fig. 2.3 The geometric interpretation of Eqs. 2.1 through 2.4.

dipole antennas. Directional emitters are called *anisotropic* sources, as contrasted with nondirectional or *isotropic* sources. The isotropic sources give the same illumination or flux density on every unit area of a reference sphere whose radius is r. Quantities indicated by Eqs. 2.1 through 2.4 are shown in Fig. 2.3. The instantaneous flux density can also be written as the magnitude of the power flow of an electromagnetic wave traveling with the speed of light whose electric vector is E. In this notation, E is the magnitude of the Poynting vector, which vector is equal to the power flow. This makes

$$†E = \frac{c}{4\pi} \mid \mathrm{E}^2 \mid . \tag{2.5}$$

It is this latter form with which we shall be most concerned in our scattering problem. The electrical units used are Gaussian. Equation 2.5 as it now stands indicates that E is the instantaneous flux density of the scattered electromagnetic wave and E is the instantaneous magnitude of the electric vector of this scattered wave.

In accordance with our understanding of the propagation of radiant energy, we have found that in the simplest case we must have an *oscillating* dipole as the source of scattered flux. Since periodically changing motion is implied and E is proportional to a force, it turns out to no one's surprise that E is proportional to the acceleration of the dipole moment. In order to specify the plane of vibration of the radiation from the oscillating dipole, we shall consider the dipole to be polarized, i.e., restricted to vibrating in one plane. The flux density falls off as the square of the distance from the source, so that E itself must be proportioned to $1/r$. Since both the scattered flux and the polarization of the dipole are directional in nature (vectors), the component of the radiation from the oscillating dipole along a given radius vector of magnitude r varies as sin α. α is the angle between the electric vector E of the incident light and the direction along which the scattered flux is detected. Expressing E as a

† It should be re-emphasized at this point in the discussion that E is an instantaneous value whose value lies between 0 and $\frac{c}{4\pi} \mid \mathrm{E}^2_{\mathrm{max}} \mid$. The measured average value of E is found by replacing $\mid \mathrm{E}^2 \mid$ by the square of the root mean square value of the electric vector. The latter is

$$\mathrm{E}_{\mathrm{rms}} = \frac{\mathrm{E}_{\mathrm{max}}}{\sqrt{2}}.$$

The average value of E becomes

$$\bar{E} = \frac{c}{8\pi} \mid \mathrm{E}^2_{\mathrm{max}} \mid .$$

We have assumed $\mathrm{E} = \mathrm{E}_{\mathrm{max}} \sin \frac{2\pi c}{\lambda} t$, where $\mathrm{E}_{\mathrm{max}}$ is the amplitude of the electric vector.

function of the variables mentioned above gives

$$E = \frac{1}{c^2}\frac{d^2P}{dt^2}\frac{\sin\alpha}{r}. \tag{2.6}$$

d^2P/dt^2 is the acceleration of the dipole and $1/c^2$ is the constant of proportionality in Gaussian units. The induced dipole moment P is related to the index of refraction through the Lorenz-Lorentz law. The strength of this induced dipole is directly proportional to the amplitude of the incident electric vector E_0. For our purposes, the expression for the amplitude of the induced dipole moment may be given as

$$P_0 = \frac{3E_0}{4\pi}\left(\frac{m^2-1}{m^2+2}\right)V. \tag{2.7}$$

The direction of E_0 and hence the direction of P is fixed in space because the incident radiant energy inducing the dipole to vibrate is assumed to be polarized. Assuming a simple sinusoidal variation for P, the instantaneous dipole moment, we have

$$P = P_0 \sin\frac{2\pi c}{\lambda}t. \tag{2.8}$$

The second derivative of this expression is

$$\frac{d^2P}{dt^2} = -\left(\frac{2\pi c}{\lambda}\right)^2 P_0 \sin\frac{2\pi c}{\lambda}t. \tag{2.9}$$

Substituting Eq. 2.7 into Eq. 2.9 to eliminate P_0 and substituting the resulting equation for d^2P/dt^2 into Eq. 2.6 makes

$$E = -3\pi\frac{E_0}{r}\left(\frac{m^2-1}{m^2+2}\right)\frac{V}{\lambda^2}\sin\frac{2\pi c}{\lambda}t\sin\alpha. \tag{2.10}$$

From Eq. 2.10, it is the average rather than the instantaneous value of E that we desire, because most detectors which convert radiant energy to heat or other types of energy, do an integration over the instantaneous power to indicate an average power. Even with a detector such as radar or radio which uses a dipole for detection, this integration has been done by the time the results are displayed on an indicator for use. Mathematically, we want

$$E_{\text{rms}}^2 = 9\pi^2\frac{E_0{}^2}{r^2}\left(\frac{m^2-1}{m^2+2}\right)^2\frac{V^2}{\lambda^4}\sin^2\alpha\,\overline{\sin^2\frac{2\pi c}{\lambda}t}, \tag{2.11}$$

where the bar indicates the average value of the quantity. The average value of the $\sin^2\theta$ is

$$\overline{\sin^2\theta} = \frac{\int_0^{2\pi} \sin^2\theta\, d\theta}{\int_0^{2\pi} d\theta}$$

$$= \frac{1}{2\pi}\left[\frac{1}{2}\int_0^{2\pi} d\theta - \frac{1}{4}\int_0^{4\pi} \cos 2\theta\, d(2\theta)\right] = \frac{1}{2}\cdot \quad (2.12)$$

We have let

$$\sin^2\theta = \tfrac{1}{2}(1 - \cos 2\theta) \quad (2.13)$$

where $\theta = (2\pi c/\lambda)\, t$ and $d\theta = (2\pi c/\lambda)\, dt$. The averaging has been done over one complete revolution, $\theta = 2\pi$. Averaging over additional periods does not change the value as we multiply and divide by the same number of periods over which the averaging was done.

Substituting the value of $\frac{1}{2}$ from Eq. 2.12 into 2.11, which in turn is substituted into Eq. 2.5, gives

$$\bar{E} = \frac{9\pi^2}{r^2}\left(\frac{c}{8\pi}\,\mathrm{E_0}^2\right)\left(\frac{m^2 - 1}{m^2 + 2}\right)^2 \frac{V^2}{\lambda^4}\sin^2\alpha. \quad (2.14)$$

(See the footnote to Eq. 2.5.) Analogous to Eq. 2.5,

$$\frac{c}{8\pi}\,\mathrm{E_0}^2 = \frac{c}{4\pi}\,|\,\mathrm{E}^2_{\mathrm{rms}}\,| = \bar{E}_0, \quad (2.15)$$

the average flux density of the incident radiation. When Eq. 2.15 is substituted into Eq. 2.14, we have a quantitative expression for the scattering of polarized radiation by nonabsorbing particles of small size, which is

$$\bar{E} = 9\pi^2\frac{\bar{E}_0}{r^2}\left(\frac{m^2 - 1}{m^2 + 2}\right)^2 \frac{V^2}{\lambda^4}\sin^2\alpha. \quad (2.16)$$

The total scattered flux can be found by multiplying both sides of Eq. 2.16 by the area of a sphere, $r^2\, d\omega$. The flux is

$$\bar{F} = 9\pi^2\,\bar{E}_0\left(\frac{m^2 - 1}{m^2 + 2}\right)^2 \frac{V^2}{\lambda^4}\int_0^{4\pi}\sin^2\alpha\, d\omega. \quad (2.17)$$

In spherical co-ordinates†

$$d\omega = 2\pi\sin\alpha\, d\alpha. \quad (2.18)$$

† In spherical co-ordinates, the differential area of a zone on the surface of a sphere of radius r and central angle α is

$$dA = 2\pi(r\sin\alpha)r\, d\alpha = 2\pi r^2\sin\alpha\, d\alpha = r^2\, d\omega.$$

When the limits of integration of ω are 0 and 4π, the corresponding limits on α are 0 and π. With the aid of Eq. 2.18, Eq. 2.17 becomes

$$\bar{F} = 18\pi^3 \bar{E}_0 \left(\frac{m^2 - 1}{m^2 + 2}\right)^2 \frac{V^2}{\lambda^4} \int_0^\pi \sin^3 \alpha \, d\alpha. \tag{2.19}$$

The integral on the right of Eq. 2.19 is a standard integral whose value is

$$\int_0^\pi \sin^3 \alpha \, d\alpha = \int_0^\pi \sin \alpha \, d\alpha - \int_0^\pi \cos^2 \alpha \sin \alpha \, d\alpha = \frac{4}{3}. \tag{2.20}$$

The substitution of $\frac{4}{3}$ for the integral in Eq. 2.19 gives

$$\frac{\bar{F}}{\bar{E}_0} = 24\pi^3 \left(\frac{m^2 - 1}{m^2 + 2}\right)^2 \frac{V^2}{\lambda^4}. \tag{2.21}$$

$\bar{F}/\bar{E}_0$ is the flux scattered out of a beam of radiant energy of unit flux density by a single scatterer. If there are n scatters per unit volume of space, the energy scattered out of the beam per unit length, called the scattering coefficient, k_s, is simply Eq. 2.21 multiplied by n. By this definition

$$k_s = 24\pi^3 n \left(\frac{m^2 - 1}{m^2 + 2}\right)^2 \frac{V^2}{\lambda^4}. \tag{2.22}$$

Equation 2.22, although derived for polarized incident energy, is equally true for unpolarized incident energy. If the scatterer is a sphere of diameter a

$$\frac{V^2}{\lambda^4} = \frac{\pi^2}{36} \frac{a^6}{\lambda^4}, \tag{2.23}$$

and the restriction on the use of Eq. 2.22 is that $a/\lambda \ll 1$. The inverse fourth power dependence of the scattering coefficient denotes Rayleigh scattering, named in honor of Lord Rayleigh,[11] the first man to develop a successful theory of scattering by small particles.

Once the equations for the scattering of polarized light have been developed, it is simple to derive the scattering law for unpolarized light. The secret of the derivation lies in the physical fact that unpolarized light can be treated as two beams of incoherent light of equal

---

The solid angle subtended by the zone is

$$d\omega = 2\pi \sin \alpha \, d\alpha.$$

Integration over the area of all zones gives the surface area of the sphere. This is a standard integral calculus problem.

intensity, plane polarized at right angles to one another.[5,B1] The intensity of unpolarized light becomes

$$\bar{I} = \tfrac{1}{2}(\bar{I}_\perp + \bar{I}_\parallel) \tag{2.24}$$

where $\bar{I}_\perp$ and $\bar{I}_\parallel$ are unit intensities of each of the beams polarized at right angles to one another. $\bar{I}$ is an unpolarized beam of unit intensity. The bar notation indicates root-mean-square values for the intensities; i.e., these quantities have been averaged over time in the fashion of Eq. 2.12.

If we let α_1 and α_2 be the angles associated respectively with polarized intensities $\bar{I}_\perp$ and $\bar{I}_\parallel$ in Eq. 2.16, Eq. 2.24 becomes

$$\bar{I} = \frac{9}{2}\pi^2 \bar{E}_0 \left(\frac{m^2 - 1}{m^2 + 2}\right)^2 \frac{V^2}{\lambda^4} (\sin^2 \alpha_1 + \sin^2 \alpha_2) \tag{2.25}$$

because $\bar{I} = \bar{E}r^2$ from Eq. 2.3.

The angular dependence of Eq. 2.25 can be simplified. Let ϕ be the angle between the incident unpolarized ray and the scattered ray measured at the scatterer, point O of Fig. 2.4. α_1 and α_2 are angles measured from the scattered ray to the two equal and mutually perpendicular polarized electric vectors. Each of these polarized vectors is also perpendicular to the unpolarized ray incident upon the scatterer, forming with the latter ray a mutually orthogonal set of vectors. From the geometry of Fig. 2.4

$$\cos^2 \alpha_1 + \cos^2 \alpha_2 = \sin^2 \phi.$$

Writing sines in terms of cosines and vice versa gives

$$1 - \sin^2 \alpha_1 + 1 - \sin^2 \alpha_2 = 1 - \cos^2 \phi.$$

This latter is simplified to the expression we want, namely,

$$\sin^2 \alpha_1 + \sin^2 \alpha_2 = 1 + \cos^2 \phi. \tag{2.26}$$

Multiplying Eq. 2.25 by $n/\bar{E}_0$ after substituting Eq. 2.26 gives the flux scattered from the beam per unit solid angle per unit length. This quantity will be denoted by $dk_s/d\omega$, because when this quantity is integrated over 4π steradians, it gives the scattering coefficient.

$$\frac{dk_s}{d\omega} = \frac{9}{2}\pi^2 n \left(\frac{m^2 - 1}{m^2 + 2}\right)^2 \frac{V^2}{\lambda^4} (1 + \cos^2 \phi). \tag{2.27}$$

Integration over $d\omega$, given by Eq. 2.18 with ϕ substituted for α, gives

$$k_s = 9\pi^3 n \left(\frac{m^2 - 1}{m^2 + 2}\right)^2 \frac{V^2}{\lambda^4} \int_0^\pi (1 + \cos^2 \phi) \sin \phi \, d\phi.$$

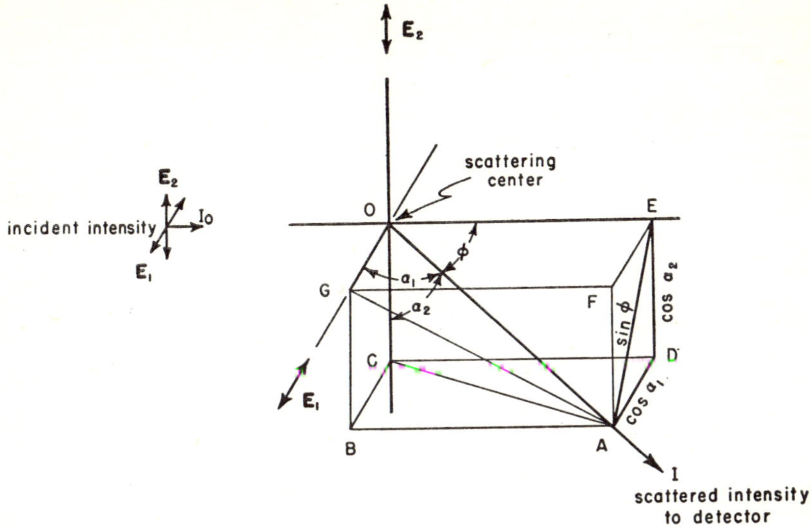

Fig. 2.4 The geometry associated with the angles α_1, α_2, and ϕ. Let the diagonal *OA* of the right parallelepiped *ABCDEFGO* be of unit length. The lengths of the sides become

$$OG = EF = BC = AD = \cos\,\alpha_1$$
$$BG = OC = ED = AF = \cos\,\alpha_2.$$

The diagonal

$$AE = \sin\phi.$$

From the right triangle *AED*

$$\cos^2\alpha_1 + \cos^2\alpha_2 = \sin^2\phi.$$

The value of the integral is $\frac{8}{3}$ so that

$$k_s = 24\pi^3 n\left(\frac{m^2 - 1}{m^2 + 2}\right)^2 \frac{V^2}{\lambda^4}. \tag{2.28}$$

Equations 2.28 and 2.22 are identical, although Eq. 2.28 was derived for a unit intensity of unpolarized light and Eq. 2.22 for a unit intensity of polarized light. From the development of Eqs. 2.28 and 2.22 we have implied a definition for k_s, which is

$$k_s = n\frac{\bar{F}}{\bar{E}_0} = nA\left(\frac{\bar{F}}{\bar{F}_0}\right) = nAK_s. \tag{2.29}$$

Stated literally, k_s is the ratio of the total flux scattered out of a beam of radiation in a unit length of n scatterers per unit volume, to the flux intercepted by the total geometric cross-sectional area of these n scatterers. K_s is called the scattering-area coefficient, or more simply the *scattering cross section*. This coefficient is very important in determining the attenuation of energy by scattering. A is the geometric cross-sectional area of a single scatterer.

In this chapter, discussion is devoted to the properties of the scattering function. In subsequent chapters, the scattering function will be used along with an analogous coefficient for absorption to predict quantitative decreases in intensity of beams of radiation because of scattering and absorption processes.

Let us rewrite the three equations that we shall use in the discussion, using n scatterers per unit volume and unit intensity of incident light, whether it be polarized or unpolarized. $\bar{I}$ now denotes the intensity from n particles. The equations are

$$\frac{\bar{I}}{\bar{I}_0} = \frac{1}{r^2}\frac{dk_s}{d\omega} = \frac{9\pi^2}{r^2} n \left(\frac{m^2-1}{m^2+2}\right)^2 \frac{V^2}{\lambda^4} \sin^2\alpha, \text{ for polarized } \bar{I}_0 \text{ (2.16).}$$

$$\frac{\bar{I}}{\bar{I}_0} = \frac{1}{r^2}\frac{dk_s}{d\omega} = \frac{9}{2}\frac{\pi^2}{r^2} n \left(\frac{m^2-1}{m^2+2}\right)^2 \frac{V^2}{\lambda^4} (1+\cos^2\phi), \text{for unpolarized } \bar{I}_0 \text{ (2.27).}$$

$$k_s = 24\pi^3 n \left(\frac{m^2-1}{m^2+2}\right)^2 \frac{V^2}{\lambda^4}, \text{ for either polarized or unpolarized } \bar{I}_0 \text{ (2.28).}$$

First, let us dispose of the common factors.

The evaluation of the nV^2 factor

The amount of scattered flux varies directly as the square of the volume and as the number of scatterers per unit volume. The factor nV^2 has very important applications, especially in research on precipitation processes in cloud physics. At this point it is sufficient to say that raindrops may be considered as spheres whose volume is $(\pi/6)a^3$. a is the diameter of the drop. The ratio a/λ is in the proper range in most cases for Rayleigh scattering when raindrops are illuminated by microwaves. nV^2 may be written as $n\pi^2a^6/36$ only when all of the scatterers are of the same size. When the scatterers are of different sizes, the proper way to express nV^2 is

$$nV^2 = \frac{\pi^2}{36}\sum_i n_i a_i^6 = \frac{\pi^2}{36}[n_1a_1^6 + n_2a_2^6 + n_3a_3^6 + \cdots + n_ia_i^6 + \cdots] \quad (2.30)$$

where a_i is the average diameter of a size class in which there are n_i†

† Let $n_i(a_i)$ be the number of particles per unit volume in an interval of diameter Δa_i. Then a_i is the value of the mean diameter in the interval Δa_i. For uniform intervals of Δa_i, we interpret

$$n = \sum_i n_i = \sum_i n_i(a_i)\,\Delta a_i = \Delta a \sum_i n_i(a_i).$$

members. Thus if there are 10 drops per 100 cubic centimeters whose average diameter is 0.5 millimeter, 65 whose diameter is 1.0 millimeter, and 18 of 1.5 millimeters diameter,

$$nV^2 = \frac{\pi^2}{36} (10 \times 0.5^6 + 65 \times 1.0^6 + 18 \times 1.5^6) \text{ mm}^6 \text{ per 100 cc.}$$

Equation 2.30, when allowed to go to an infinite number of classes (never done experimentally), is

$$\frac{\pi^2}{36} \sum_{i=1}^{p} n_i a_i^6 = \frac{\pi^2}{36} \sum_{i=1}^{p} n_i(a_i) a_i^6 \, \Delta a_i \rightarrow \frac{\pi^2}{36} \int_0^\infty n(a) a^6 \, da. \qquad (2.31)$$

Limit as $p \rightarrow \infty$.

Index of refraction

The index of refraction has an important bearing on the amount of scattered flux, especially in microwave scattering by water droplets and ice particles. Table 2.1 gives some values for $(m^2 - 1/m^2 + 2)^2$ at optical and microwave frequencies.

Table 2.1

Values for the Index of Refraction Factor $\left(\frac{m^2 - 1}{m^2 + 2}\right)^2$ for Water and Ice

The values for water at microwave frequencies are only approximate. Water absorbs energy at microwave frequencies, making the index of refraction m, a complex number, dependent in magnitude on the particular frequency chosen. Because of the variation in m, the magnitude of $\left(\frac{m^2 - 1}{m^2 + 2}\right)^2$ is correspondingly affected.

Substance	Spectral Region	m	$\left(\frac{m^2 - 1}{m^2 + 2}\right)^2$
Water	visible	1.333	0.042
Ice	visible	1.309	0.037
Water	microwave 3–10 cm	8.7–i	0.95
Ice	microwave 1–10 cm	1.75	0.17

The index of refraction term by itself can account for a stronger power return of microwave radiation from water droplets than is experienced from the same distribution of ice globules. The table shows

---

This latter form shows

$$n = \sum_i n_i(a_i) \, \Delta a_i \rightarrow \int_0^\infty n(a) \, da. \quad \text{Limit as } \Delta a_i \rightarrow 0.$$

n_i has the dimensions of length^{-3}, whereas $n_i(a_i)$ has the dimensions of length^{-4}.

their scattering coefficients under these conditions to be in the ratio of 5.5 to 1. Usually, there is not too much difference in the scattered power returned to the radar by water or ice. This condition usually signifies that n, the number of particles per unit volume, for ice is much larger than the n for raindrops. Cloud droplets do not enter into the power return as a significant factor. Their volume is so small that even though Rayleigh scattering occurs, the power return is so small compared with the rain echo (power return) that their contribution usually can be neglected.

We are now in a position to demonstrate how an extremely large number of particles per unit volume can be considered a continuous medium for refraction effects, yet discontinuous for scattering phenomena. Special reference will be made to visible light scattered by the enormously large number of air molecules. The factor

$$n\left(\frac{m^2-1}{m^2+2}\right)^2 V^2$$

which enters into the scattering equations may be written as

$$\frac{1}{n}\left[n\left(\frac{m^2-1}{m^2+2}\right)V\right]^2 = \frac{1}{n}\left[n\,\frac{(m-1)(m+1)}{(m^2+2)}\,V\right]^2 \to \frac{4}{9n}\,[n(m-1)V]^2.$$

Limit as $m \to 1$

For a gas, $(m-1)V$ refers to an individual molecule. This product is a constant for a given gas, albeit unmeasurable. Multiplying by n gives the number of this same product per unit volume. For a gas we define

$$(m-1)_{\text{gas}} = n(m-1)V = \text{constant} \times \rho \qquad (2.32)$$

because ρ, the gas density, is proportional to the number of particles per unit volume of space n. Equation 2.32 is the same expression as Eq. 1.3 in Chapter 1. Thus, Eq. 2.28 becomes

$$k_s = \frac{32\pi^3}{3n\lambda^4}\,(m-1)^2_{\text{gas}}. \qquad (2.33)$$

Equation 2.33 is the equation upon which Lord Rayleigh based his explanation for the blue of the sky.

At sea level, $(m-1)^2_{\text{gas}} = 293^2 \times 10^{-12}$, nearly independent of wavelength. n is the number of molecules per cubic centimeter of air at sea level, being of the order of 2.66×10^{19} centimeters^{-3}. The total number of molecules in this column from sea level to outer space† is

† Outer space is that region beyond the limits of the earth's atmosphere. For the scattering of light, no serious error is committed if 63 kilometers is selected as the lower limit of outer space.

the same as the number of molecules in a homogeneous atmosphere 7.991×10^5 centimeters in height. The flux scattered out of a beam of unit intensity by the molecules of the air is of the order of

$$k_s z \approx \frac{32\pi^3}{3 \times 2.66 \times 10^{19}} \times \frac{293^2 \times 10^{-12} \times 7.99 \times 10^5}{\lambda^4}$$
$$= \frac{8.56 \times 10^{-19}}{\lambda^4}. \quad (2.34)$$

This expression is valid only in the neighborhood of sea level. The numerator decreases with elevation because of the decrease of $\frac{(m-1)^2_{\text{gas}}}{n}$ with decreasing densities. λ is in centimeters making $k_s z$, a dimensionless quantity, of the order of 0.10 for visible light. The consequences of this number will be discussed under sky brightness and the attenuation of energy from the sun.

To explain the blue of the sky, it is necessary only to realize that at any level

$$k_s \sim \frac{1}{\lambda^4}.$$

Therefore, since blue light has a shorter wavelength than red light, $\lambda_{\text{blue}} \approx 425$ mμ and $\lambda_{\text{red}} \approx 650$ mμ, the ratio of the scattering coefficients of blue and red light is of the order $(650/425)^4 = 5.5$ because to a high degree of approximation white light from the sun gives the same incident energy at all wavelengths. The λ^{-4} law causes more of the blue light to be scattered than the red. Thus, the sun at the earth's surface appears yellowish or reddish, while the sky, which is viewed away from the sun's disk, appears blue. These results are in accord with observation and Eq. 2.28.

A few further inferences from Eqs. 2.33 and 2.34 may be drawn. The first is that the sky should gradually darken to become entirely black in outer space in directions away from the sun. This effect arises because of the decrease and finally absence of air molecules to scatter light. The sun itself should appear whiter and brighter with elevation.

Conversely, as the sun, moon, or stars approach the horizon, light reaching the earth's surface from them travels through more and more air. Hence, more and more blue is scattered out of the beam and the luminary appears a deeper yellow or red than at the zenith. This particular color effect is especially true of the sun or moon. Their angular subtense from the earth is so great that dispersion effects from atmospheric refraction are not discernible and their color is uniform over their entire surface. Stars, being point sources, undergo large color changes when low on the horizon. These color changes are a combination of the refractive effects described in Chapter 1 with scattering by the molecules in the long air path. Larger particles than air molecules also scatter but their effect is more complicated than Rayleigh scattering. It has been shown by the Mie theory for spheres that Rayleigh scattering is just the leading term in an expansion valid for all values of a/λ. The series[3] starts as

$$k_s = 24\pi^3 n \left(\frac{m^2 - 1}{m^2 + 2}\right)^2 \frac{V^2}{\lambda^4} \left[1 + \frac{6\pi^4}{5}\left(\frac{m^2 - 2}{m^2 + 2}\right)\frac{a^4}{\lambda^4} + \cdots\right]. \tag{2.35}$$

This series converges so slowly that it can be used with any facility only for values of a/λ slightly larger than at the Rayleigh limit. For larger values, the series can be expressed in a more compact but equivalent form.[5] Equation 2.35 does illustrate how scattering from particles only slightly larger than the Rayleigh limit tends to deviate quite rapidly from the inverse fourth power law for wavelength. The result in general is to scatter energy less selectively by wavelength. This means that the scattering from particles larger than air molecules, such as fine dust, tends to scatter white instead of blue light. As dust is always present to a greater or less degree in the lower atmosphere, the sky appears blue diluted with white scattered light, resulting in a less-pure blue sky than might be reasoned from pure Rayleigh scattering.

For microwaves, the wavelength of the radiation is usually invariant. This fact makes the wavelength a constant in the scattering equation for a selected high-frequency radio or radar set.

Polarization and the angular variation of scattered radiation (Rayleigh case)

Rayleigh scattering, in addition to explaining the blue of the sky and being a necessary tool for explaining the scattering of microwaves by rain, is very successful in explaining the general features of the polarization of scattered sunlight by air molecules. The explanation lies essentially in a discussion of the angular factors in Eq. 2.16 and especially Eq. 2.27. Equation 2.16 has some bearing on microwave scattering and will be discussed first.

Equation 2.16 says that the scattered intensity of a *polarized* beam of unit flux density is

$$\frac{\bar{I}}{\bar{E}_0} = \text{constant} \times \sin^2 \alpha. \tag{2.36}$$

At right angles to the beam $\alpha = 0°$, so that there is no scattered energy in this direction. There is a maximum of energy scattered, both in the direction of the beam, *forward scattering,* and directly backward toward the beam, *back scattering,* when α becomes either 90° or 270°, sin α becoming 1 and −1, respectively.

Even though the *incident* energy is *polarized* in a single plane, a plot of Eq. 2.36 is but a section of a figure of revolution in intensity space. The axis of this solid is parallel to the incident beam passing through the center of the scattering particle. As depicted by Fig. 2.1, the scattered radiation is in general *elliptically polarized.* This scattering pattern is entirely consistent with what is observed in our experiment with the magnetic dipole discussed earlier in the chapter. This dipole could be rotated through a complete revolution about the axis containing the two poles without changing the pattern of iron filings in a chosen horizontal plane. The experiment shows symmetry about this pole-to-pole axis.

These results are very important and in fact make such systems as radar possible. Since one of the directions of maximum scattering is directly backwards, a system can be designed to emit and detect scattered radiation at the same spot, whether the process be scattering or true reflection. This system using polarized microwaves emitted in pulses is radar.

From the point of view of meteorological optics, a discussion of the angular distribution of scattered, *initially unpolarized,* light is very important. Equation 2.27 sums up in concise mathematical form essentially the entire argument. Writing this equation as

$$\frac{\bar{I}}{\bar{E}_0} = \text{constant} \times \left(\frac{1 + \cos^2 \phi}{2}\right) \tag{2.37}$$

makes the constant the same as in the previous equation. Equation 2.37 also has the same axis of symmetry as Eq. 2.36. Thus it is necessary to discuss the pattern of scattered light in one plane only. The total scattered energy is the same in Eqs. 2.36 and 2.37, as attested by Eq. 2.28, but the angular distribution is different. This point has need of emphasis.

In accord with the definitions and development of the mathematics, forward scattering and back scattering occur when ϕ equals 0° and 180° respectively. This fact shows that for unpolarized as well as polarized light, the maximum scattered intensity occurs in the forward and backward directions, the factor $(1 + \cos^2 \phi)/2$ being unity in both cases. When $\phi = 90°$ (or 270°), i.e. scattering normal to the beam, the value of $(1 + \cos^2 90°)/2$ becomes $\frac{1}{2}$, showing that for the same incremental solid angle there is one half as much energy scattered to the side as is scattered forward. This fact contrasts markedly

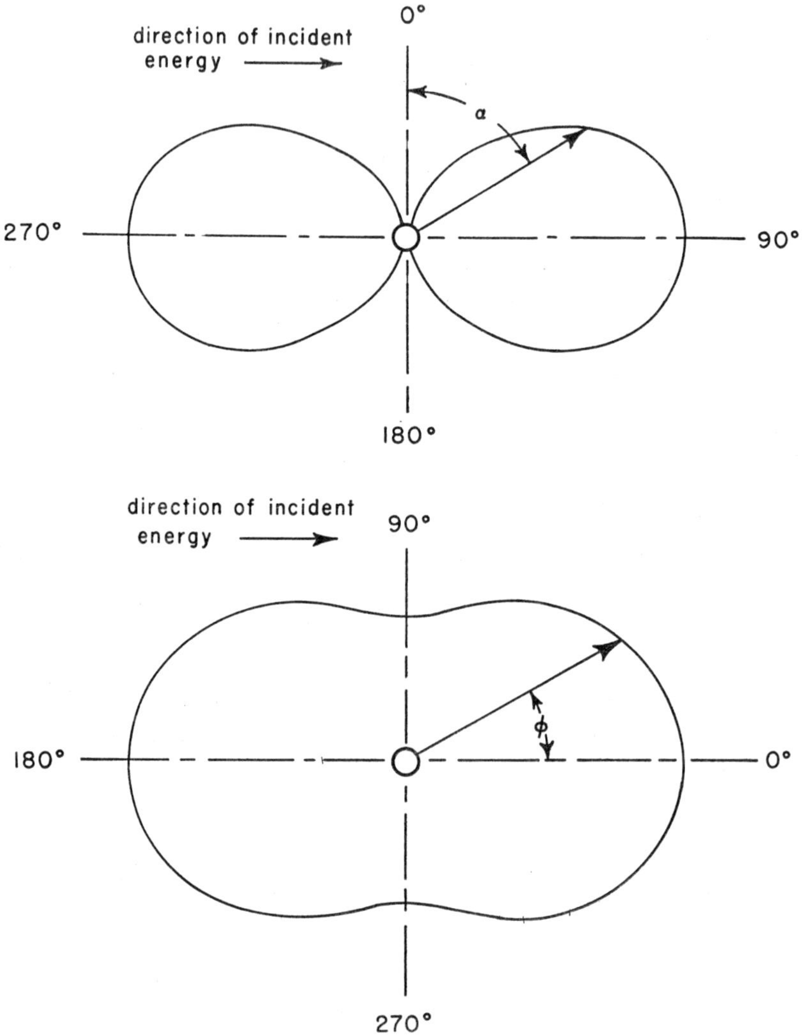

Fig. 2.5 The angular distribution of scattered energy from an initially polarized beam of light (upper figure) and from an initially unpolarized beam of light (lower figure).

with the zero scattered energy normal to the beam observed with polarized incident light. The angular distribution of scattered light illustrating Eqs. 2.36 and 2.37 is given in Fig. 2.5.

The sideward scattered light is also completely plane polarized, with the electric vector of the light being normal to a plane which by convention is called the plane of polarization. In this case, the plane of polarization is the plane containing the source, scatterer, and detector. This right angle scattering, first noticed and studied by Tyndall and later to be explained theoretically by Rayleigh is called *Tyndall scattering.* The plane of polarization of Tyndall scattering will be called the *positive plane of polarization.*

The observed distribution of polarization resulting from Rayleigh scattering can be explained by considering the factor $\frac{1}{2}$ to be the intensity of positive polarization. The intensity of this positive polarization is constant in all directions in space, being a sphere in intensity space of radius $\frac{1}{2}$. The $\cos^2 \phi/2$ term is the polarization normal to a plane at right angles to the positive plane of polarization called the *negative plane of polarization.* The polarization is called negative. Both planes contain the central ray of the incident unpolarized light as their intersection. The addition of the $\cos^2 \phi/2$ term to the constant term $\frac{1}{2}$ by the principle of the addition of polarized light at right angles, Eq. 2.24, gives for the scattered light (1) unpolarized light both for forward and back scattering; (2) plane polarized light at right angles to the beam; and (3) a variation of *plane polarized light* according to Eq. 2.37 for other angles, the amount of polarized light over the unpolarized light being

$$\frac{1 - \cos^2 \phi}{2} = \frac{\sin^2 \phi}{2}.$$

2 ($\cos^2 \phi/2$) is the amount of unpolarized light at any angle. The ratio of scattered light polarized in the positive plane to the total scattered light is

$$\Gamma = \frac{\sin^2 \phi}{1 + \cos^2 \phi}. \tag{2.38}$$

Γ is, by definition, the polarization.

Measurements of the polarization of sunlight show that even air molecules do not follow exactly the requirements of the Rayleigh theory. The effect is most noticeable at $\phi = 90°$, where Tyndall scattering occurs. Equation 2.38 predicts that at this point $\Gamma = 1$, i.e., complete polarization. In reality, the polarization is not complete, differing from unity by an amount related to δ, called the *coefficient of depolarization* or sometimes the *polarization defect.* A theory developed by Cabannes[1] takes into account the anisotropy of molecules in the scattering process, leading to this modification of Eq. 2.38

$$\Gamma = \frac{(1 - \delta)(\sin^2 \phi)}{1 + \cos^2 \phi + \delta \sin^2 \phi}. \tag{2.39}$$

In pure molecular scattering $\delta = 0.043$, leading to a maximum polarization at $\phi = 90°$ of 0.922 instead of 1. The angular position of the maximum and minimum polarizations is not changed in this modification of the Rayleigh theory.

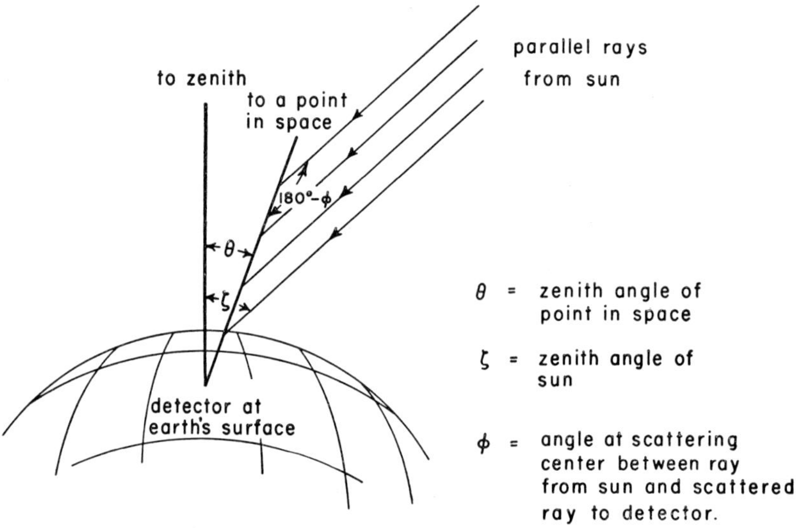

Fig. 2.6 The geometry of the polarization of molecularly scattered sunlight. This light, predominately blue in color, is called sky light.

Polarization of the daylight sky

The polarization of a cloudless daylight sky at altitudes high enough for only molecular scattering to be operative follows Rayleigh scattering very well. The sun is the source of parallel unpolarized light. The eye looking through a polarizing crystal used to analyze the sky light is the detector. When the eye glances at a certain point in the sky, the scattering by air molecules along the path joining the eye to the point in space is accomplished at a definite angle ϕ to the sun's parallel rays (see Fig. 2.6). Thus we have set up the conditions described in Eqs. 2.37 and 2.38. When the sun is on the horizon, the position of maximum polarization is at the zenith. As the sun rises, this point of maximum polarization moves toward the horizon, keeping at all times an angle of 90° between the sun, point in space under observation, and detector. Points of no polarization would be observed on looking into the sun, the *solar point* (not observed because of the intensity of the sun), and at the *antisolar point.* The latter point is found 180° away from the sun on a line which includes the center of the earth.

A closer investigation of the polarization of sky light shows that under the most ideal conditions for pure molecular scattering, complete polarization occurs nowhere in the sky. The point of maximum polarization agrees with theory, except for the polarization defect. The polarization defect may arise from radiation being scattered more than once, *secondary* and *multiple scattering,* as well as the anisotropy of the molecules. Consideration of both factors is omitted from the Rayleigh theory. Table 2.2 shows a calculation of the polarization using Eqs. 2.38 and 2.39, and results calculated from a more elaborate theory[14] taking secondary scattering of ground- or cloud-reflected light into consideration.

Table 2.2

Values of the Polarization Γ, Computed for the Following Conditions

Column 2	Rayleigh scattering, Eq. 2.38
Column 3	Modified Rayleigh scattering, Eq. 2.39
Column 4 through column 9	From a theory by Tousey and Hulburt.[14] The effect on the polarization of scattered sunlight by a reflecting surface (clouds, earth's surface) whose albedo is A is considered.

Angle $\phi = 0°$ indicates forward scattering. $\phi = 90°$ indicates normal or Tyndall scattering.

1	2	3	4	5	6	7	8	9
ϕ	Eq. 2.38	Eq. 2.39	$A=0$	$A=0.2$	$A=0.4$	$A=0.6$	$A=0.8$	$A=1.0$
0°	0	0	0	0	0	0	0	0
30°	0.143	0.136	–	–	–	–	–	–
60°	0.600	0.560	0.527	0.481	0.439	0.408	0.379	0.360
90°	1.000	0.922	0.852	0.775	0.666	0.612	0.550	0.492

Reflectivities for various surfaces are given in Table 2.3, Chapter 2.

More serious than the above considerations is the deviation from pure Rayleigh scattering when particles larger than molecules exist in the air. These particles, such as smoke and dust, always occur in the lower layers of the atmosphere. Finer ones, still large in respect to a wavelength of light (volcanic dust for example) exist to high levels and persist for long periods of time—months and even years. The distribution may be world-wide, thanks to the winds. The complete explanation for the complex polarization patterns is contained in the Mie theory. This theory has been worked out for a number of cases with simple geometric shapes for the scattering particles. Suffice it to say here that the Mie theory predicts that the point of maximum polar-

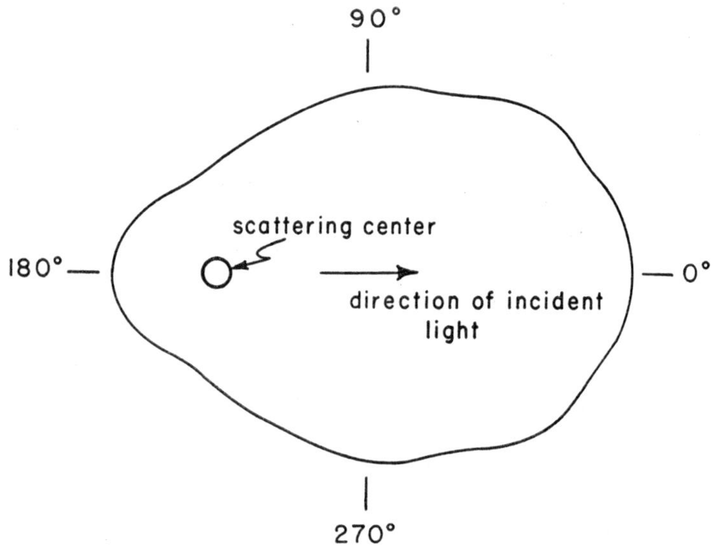

Fig. 2.7 The angular intensity pattern for a sphere where $a/\lambda = 1/2.6$. The index of refraction is $m = 1.33$.

ization will move to angles other than 90° as the particles become larger, the pattern of intensity changes may have several maxima as ϕ varies, and the scattered light will have a component that is *plane polarized.*† An example of the pattern is shown for a typical case in Fig. 2.7.

In addition, sky light consists to a small degree of sunlight that has been reflected from the earth (especially from water or snow) to the sky and scattered back to earth again. Simple physical theory,

† Scattered light from an *unpolarized source* is always *plane polarized* plus a component of unpolarized light. The scattered light from an unpolarized source is never elliptically polarized. *Elliptical polarization* of scattered light arises only when the *incident* light is *plane polarized.* In general, elliptical polarization is never encountered in the primary scattering of sunlight, but may arise as a component of the secondary scattering of sunlight that has first been plane polarized by primary scattering or reflection. Elliptical polarization is, therefore, only a second- or third-order effect.

available in any textbook on general physics, shows that light is polarized by reflection. In fact, for every plane surface there is an angle given by *Brewster's law* where unpolarized light can be completely plane-polarized by reflection. This polarizing angle depends on the index of refraction of the substance. The unreflected energy may be either transmitted or absorbed at the reflecting surface. Table 2.3 shows the per cent reflection from various surfaces. Both the incoming and reflected energy are measured at the reflecting surface itself, so that absorption effects by the atmosphere can be neglected.

Table 2.3

The Per Cent Reflection and Absorption of Visible Light† by Common Surfaces over the Earth

Type of Surface	Per Cent Reflection		Per Cent Absorption
Forest	4–10		96–90
Green fields	10–15		90–85
Dry grass fields	15–25		85–75
Dry plowed fields	20–25		80–75
Bay and rivers	6–10		94–90
Water (normal incidence)	2	Very deep plus bottom absorption	98
Water (grazing incidence)	80–100		20–0
Snow or ice	46–86		54–14
Dense clouds (Cu, Sc)	56–81	(1500 ft thick)	5–9
		(20,000 ft thick)	13–27
Dense clouds (nearly opaque) (As)	39–59		——
Thin clouds (Ci)	36–40		<5

Cloud types: Cu, cumulus; Sc, stratocumulus; As, altostratus; Ci, cirrus.

† Including the photographic region of the infrared ($\lambda < 2.5$ microns).

Large-particle scattering, secondary scattering, and reflection from the surface of the earth lead to polarization anomalies in the sky. The most noticeable of these irregularities are the neutral points. *Neutral points* are positions in the sky where no polarization is observed. The antisolar point is a neutral point given in the theory of Rayleigh scattering. In addition, there are three other common neutral points, and on occasion many others. Usually the neutral points lie in the vicinity of the solar or antisolar point. The three common neutral points are the Arago, 15° to 25° above the antisolar point; the Babinet, 15° to 25° above the sun; and the Brewster, 15° to 20° below the sun. A schematic drawing of these points is given in Fig. 2.8.

The angular values above are approximate and vary with the position of the sun. Arago's neutral point in particular has been the object of systematic study of late by H. Neuberger(9) and co-workers. He finds that the angular migration of the Arago point as measured by the sun is affected by many factors, the principal effect being from refraction as the sun dips below the horizon. At any given position

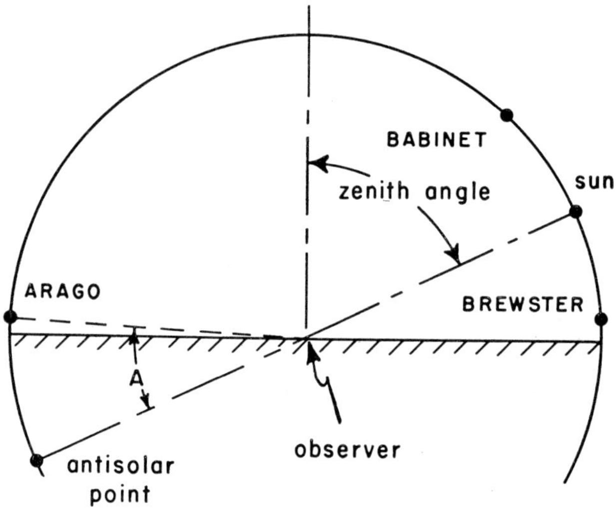

Fig. 2.8 (*After Neuberger.*) The position of the neutral points in the meridian of the sun.

of the sun, the neutral point always appears at a greater distance from the antisolar point when observed through a blue filter than when viewed through either no filter or a red filter. The red filter gives the minimum distance. The Arago point also varies very rapidly as clouds are forming, indicating the presence of particles larger than molecules minutes before the cloud becomes concentrated enough to be detected by the unaided eye.

Finally, two figures, Figs. 2.9 and 2.10, are presented to show the persistence of neutral points and the effect of reflection from surfaces on the position of the neutral point. Figure 2.9 shows data taken in Hamburg by C. Jensen and analyzed by H. Neuberger for the years 1911 through 1914. The data correlates very well with the eruption of the Alaskan volcano Katmai in June, 1912. The effects of the volcano dust were world-wide. Figure 2.9 shows that it took until 1914 for the turbid atmosphere from the eruption to reach the normal pure-air state of 1911. On the other hand, Fig. 2.10 shows the difference in the neutral point because of the difference in reflection from a land surface (State College, Pennsylvania) and a sea surface (the island of Sylt in the North Sea). The difference disappears for elevation angles less than 4.5 degrees.

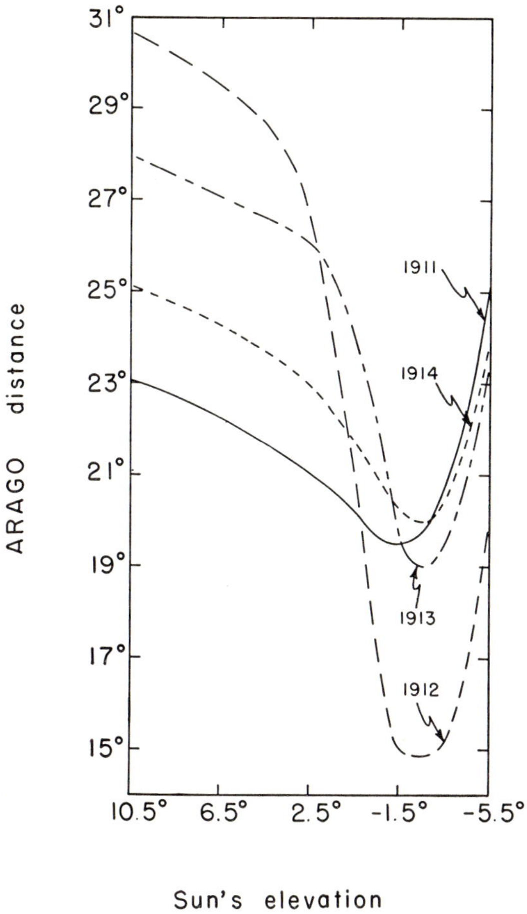

Fig. 2.9 (*After Neuberger.*) The effect of atmospheric turbidity on the position of Arago's neutral point.

Miscellaneous effects of scattering

Astronomical twilight is that period of the day between the times when the upper edge of the sun is on the horizon and the true position of its center is 18° below the horizon. Civil twilight, on the other hand, only extends until the sun's center is 6° below the horizon.

During the twilight period, the sky can exhibit its most colorful and variegated self, mainly through the process of scattering.

For example, with crenellated (turreted) clouds on the horizon, streaks of light may appear across the sky, apparently emanating from behind the clouds. These streaks are called *crepuscular rays* and are caused by scattered light from dust particles in the air. The same effect may be noticed during the middle of the day, when the skies are

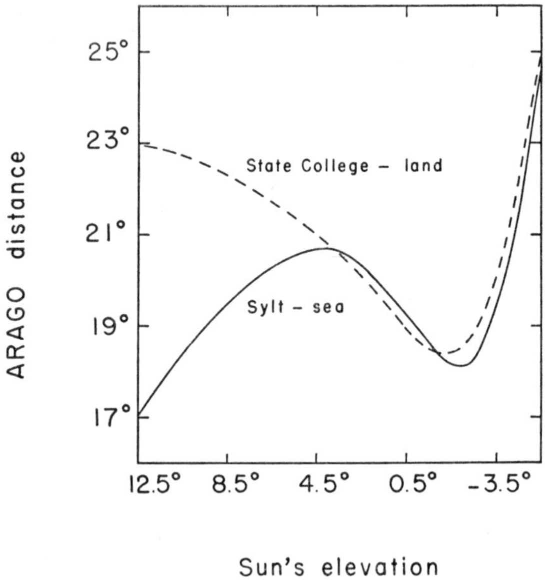

Fig. 2.10 (*After Neuberger.*) The effects of a water and of a land surface on Arago's neutral point.

partly cloudy. The shadows cast by the clouds are necessary, otherwise the dust particles increase the illumination uniformly instead of causing the appearance of light scattered from a beam.

The twilight sky goes through a variety of colors, the range and intensity of color being a function of the dust content of the air and the number and type of clouds in the air. Before sunset, the dust content is indicated by the reddish western sky. After sunset, if clouds are about, the red transmitted light will be reflected from the clouds.

At sunset and in the absence of clouds, scattering produces the gray-blue *dark segment* in the eastern sky rimmed by the *bright segment* or *twilight arch*. Since the earth's surface and atmosphere are curved, the path of the sun's rays nearest the earth is attenuated by scattering much more than the rays of the sun higher in the atmosphere. Strong attenuation of the lower rays plus the earth's shadow as the sun sinks low below the horizon cause the dark segment. The bright segment immediately above is the scattered light from the rays of the sun not blocked by the earth's bulk. For still higher rays, the density of the atmosphere has decreased so much that little light is scattered. The entire effect causes a bright segment of scattered light above the dark segment. The system moves across the zenith sky as the sun sets, being terminated only by darkness.

Simultaneously with the rise of the dark segment and twilight arch, *purple light* may appear in the western sky about half way between zenith and horizon. The purple light moves steadily toward the western horizon as the sun sets, disappearing at about a 6° depression of the sun's disk below the horizon. The purple light is caused by dust particles favorably located to scatter at a small angle, a beam of sunlight made progressively more red on passing through a dust layer in the atmosphere. This red transmitted light is reinforced by blue light scattered from clearer air higher in the atmosphere. Because of the integrating effect of the eye, the blue and red together are perceived as purple light.

The optical effects just described are only an indication of the variety of optical effects that may be observed in the twilight sky. At sunrise, the same effects are noticeable, but in a reversed sequence.

References

1. Cabannes, J., "Sur la diffusion de la lumière par les molécules des gaz transparents," *Ann. Phys.*, Paris (9), *15,* 5–149 (1921).
2. Fritz, S., "The albedo of the planet Earth and of clouds," *J. Meteorol.*, *6,* 277–282 (1949).
3. Hart, R. W., and Montroll, E. W., "On the Scattering of Plane Waves by Soft Obstacles. I. Spherical Particles," *J. Appl. Phys.*, *22,* 376–389 (1951).
4. Hulburt, E. O., "Explanation of the Brightness and Color of the Sky, Particularly the Twilight Sky," *J. Opt. Soc. Amer.*, *43,* 113–118 (1953).
5. Mie, G., "A Contribution to the Optics of Turbid Media, Especially Colloidal Metallic Suspensions," *Ann. Physik,* fourth series, *25,* 377–445 (1908) (in German; translation by H. B. Pickard and F. T. Gucker, Jr., Indiana University, 1943).
6. Moon, P., "A System of Photometric Concepts," *J. Opt. Soc. Amer.*, *32,* 348–362 (1942).
7. "Tables of Scattering Functions for Spherical Particles," *Nat. Bur. Standards,* Applied Math. Series, *4,* Washington (1949).
8. Neuberger, H., "General Meteorological Optics," *Compendium of Meteorology,* American Meteorological Society, Boston, 61–78 (1951).
9. Neuberger, H., "Arago's Neutral Point: A Neglected Tool in Meteorological Research," *Bull. Am. Meteorol. Soc.*, *31,* 119–125 (1950).

10. Packer, D. M., and Lock, C., "Brightness and Polarization of the Daylight Sky at Altitudes of 18,000 and 38,000 Feet above Sea Level," *Naval Research Laboratory, Report 3713,* Washington (1950).
11. Rayleigh, Lord, "On the Light from the Sky, Its Polarization and Colour," *Phil. Mag., 41,* 107–120, 274–279 (1871).
12. Sekera, Z., "Polarization of Sky Light," *Compendium of Meteorology,* American Meteorological Society, Boston, 79–90 (1951).
13. Sinclair, D., "Light Scattering by Spherical Particles," *J. Opt. Soc. Amer., 37,* 475–480 (1947).
14. Tousey, R., and Hulburt, E. O., "Brightness and Polarization of the Daylight Sky at Various Altitudes Above Sea Level," *J. Opt. Soc. Amer., 37,* 78–92 (1947).
15. Van der Hulst, H. C., "Optics of Spherical Particles," *Recherches Astronomiques de l'Observatoire d'Utrecht, XI,* Part 1 (1946), Part 2 (1949) (in English).

Source Books

B1. Chandrasekhar, S., *Radiative Transfer,* Oxford, Clarendon Press (1950), Chapters IX and X.
B2. Humphreys, W. J., *Physics of the Air,* McGraw-Hill Book Co., New York (1940), Part IV, Chapters VII and VIII.
B3. Stratton, J., *Electromagnetic Theory,* McGraw-Hill Book Co., New York (1941), Chapter IX, Sections 9.25, 9.26, and 9.27.

Problems

2.1 The complex index of refraction shall be defined as the complex number

$$m^* = m - m\bar{k}i,$$

where m is called the real index of refraction and $\bar{k}$, a dimensionless number, is called the electromagnetic absorption coefficient or index. If λ is the wavelength of the incident energy in vacuo (approximately the same as the wavelength in air), we find the diminution of plane parallel radiation in passing through x units of absorbing material is given by

$$\frac{E}{E_0} = e^{-k_a x} = e^{-4\pi \frac{m\bar{k}}{\lambda} x}.$$

The absorption coefficient per unit length is k_a. λ/m is the wavelength in the medium. (This definition of $\bar{k}$ fits that of Stratton,(B3) Ryde,(VIII–27) Aden,(VI–1) and the National Bureau of Standards,(7) but not that of Van der Hulst(15) and others.)

a. Water absorbs in the infrared. Measurements show

λ (microns)	k_a (cm^{-1})	m
1.90	31.5	1.309
2.40	40.3	1.276
4.70	545.0	1.338
6.0	2140.0	1.304

Find the complex index of refraction for these wavelengths (note: λ, x and k_a must be in the same units of length when finding $\bar{k}$ or $m\bar{k}$).

2.2 When considering Rayleigh scattering by a sphere whose absorption index is κ (see Prob. 2.1 for definition), the factor containing the index of refraction in Eq. 2.10 should be written as $\left(\frac{m^{*2}-1}{m^{*2}+2}\right)$, a complex number. When the scattered power (flux) is found, as in Eq. 2.16 for initially polarized radiation or in Eq. 2.27 for initially unpolarized radiation, this factor becomes $\left|\left(\frac{m^{*2}-1}{m^{*2}+2}\right)^2\right|$, a real number.

As a complex number, we can write

$$\frac{m^{*2}-1}{m^{*2}+2} = \frac{a+bi}{c+di},$$

where a and c are the real parts and b and d the imaginary parts of two complex numbers. By operating with complex numbers, it can be shown that

$$\left|\left(\frac{a+bi}{c+di}\right)^2\right| = \frac{a^2+b^2}{c^2+d^2}.$$

a. Show that for the complex index of refraction of Prob. 2.1, $m^* = m(1-\kappa i)$, Eq. 2.16 transforms into Eq. 8.8*a* of Chapter 8 when $\alpha = 90°$. Spherical droplets of diameter *a* are assumed.

b. The complex indices of refraction for microwaves having the following free space wavelengths at 0° C are:

λ(cm)	m^*
0.62	3.45–2.04i
1.24	4.75–2.77i
3.21	7.14–2.89i
10.00	8.99–1.47i

Calculate the factor $\left|\left(\frac{m^{*2}-1}{m^{*2}+2}\right)^2\right|$ and compare its magnitude with that of $\left(\frac{m^2-1}{m^2+2}\right)^2$, i.e., the index of refraction when the absorption term is neglected. What is the significance of the difference?

2.3 The attenuation by Rayleigh scattering of parallel light transmitted through the entire atmosphere may be expressed very nearly by

$$\tau = e^{\frac{-8.56\times10^{-3}m}{\lambda^4}}$$

when λ is expressed in microns (see Eq. 2.33). m is called the optical air mass and is nearly equal to the secant of the sun's zenith angle for angles less than 70°. Because of the curvature of the atmosphere, m has a finite value when the sun is on the horizon. τ, the transmissivity, is the ratio of the flux density of the light at the surface of the earth to the flux density at the top of the atmosphere. Normal incidence of flux on the detector is assumed. The flux density incident on the top of the atmosphere will be assumed independent of wavelength.

a. Find the transmissivity of solar energy at the following wavelengths: 0.4; 0.5; 0.555; 0.6; and 0.7 micron, when m has the following values.

Zenith angle	0°	35°	70°	89°
Air mass (m)	1	1.22	2.90	26.96

b. Plot the results as transmissivity versus wavelength for each value of air mass. Before plotting, normalize all transmissivities to unity at a wavelength of 0.7 microns.

c. What is the significance of the curves?

2.4 The zenith angle of the sun is expressed as

$$\cos \theta = \sin \varphi \sin \Delta + \cos \varphi \cos \Delta \cos h$$

where

θ = zenith distance of the sun;
φ = latitude of the observer;
Δ = declination of sun (a constant for a given day); and
h = hour angle of the sun (angular distance from the meridian of the observer).

If ζ is the zenith distance of the position of the maximum polarization of the sky (assuming Rayleigh scattering), find ζ as a function of φ, Δ, and h. Does ζ define a point in the sky, a curve, or an area?

2.5 Searchlight beams have been used to investigate the density of the upper atmosphere. Visualize a beam of parallel unpolarized light projected vertically into a clear sky. A detector of small numerical aperture located 20 km from the base of the searchlight is used to receive light scattered from the searchlight beam by air molecules. On the assumption that the intensity of the searchlight beam remains constant from 20 to 30 km, find the relative energy at a single wavelength scattered toward the receiver from the following heights.

Height, km	20	25	30
Density of air, g/cm^3	8.82×10^{-5}	4.03×10^{-5}	1.84×10^{-5}

Take the molecular weight of air to be 28.97. Rayleigh scattering only is assumed. At sea level, the density of air is 1.29×10^{-3} g/cm^3 and the index of refraction is 1.000293.

CHAPTER 3

Theory of Atmospheric Visibility

One of the meteorological parameters that are reported in synoptic weather reports is visibility. For these purposes, "Visibility is a term that denotes the greatest distance an object of specified characteristics can be seen and identified."†

Although specific instructions are given for estimating the visibility, observations made under the above rules may suffer from errors of judgment and lack of visual acuity on the part of observers, lack of suitable range markers, a bias toward low visibility both through instruction and through a feeling that underestimation is safer than overestimation, and a host of other defects. The theory of visibility is well advanced, but the instrumental techniques that would put visibility measurements on an exact quantitative basis are expensive enough so that general instrumental techniques are not practiced.

The need for visibility measurements may be divided into two classes. The first class, which answers the question of how far one can see, is usually thought of when visibility is mentioned. In line with modern terminology, visibility when used in this sense will be called the *visual range.* The visual range is of importance in air, sea, and land transportation, when visual ranges are low, impairing safe operation. Visual range may be of potential importance in characterizing the origin of air masses in meteorology, especially if the maximum visual range in any direction is used. The second class of visibility measurements might be characterized as those answering the question of how far away an unfamiliar object can be and still not escape detection. The second case, therefore, is a matter of positive recognition and has far more problems connected with its solution than the first case. Many of the recognition problems are psychological and physiological in nature, and depend on the alertness of the observer, the state of adaptation of his eyes, his familiarity with the object for which he is searching, and other

† Quoted from page 15 of *Circular N*, entitled "WBAN Manual of Surface Observations," published by United States Department of Commerce, Washington, D.C., January, 1949.

personal factors. By excluding or at least recognizing the existence of human factors, a great deal of progress has been made theoretically in evaluating the factors influencing visibility. As a practical application of visibility theory, one may cite the problem of air search at sea for a life raft carrying survivors from an air or sea disaster. This problem always reduces to searching a maximum area at sea consistent with adequate thoroughness. Evidently, the visibility is a determining factor on the size of the search pattern. Many other less spectacular events require positive recognition to insure maximum safety in daily pursuits.

In meteorology, the greatest direct interest is in the estimation of the visual range, but never to the total exclusion of problems in recognition.

The basic concept of visibility determinations

In developing a theory of the visual range, it was early recognized that the eye, or in fact any detector, is not capable to an infinitely fine degree of determining which of two quantities is the larger. This concept is of very general physical application, and is in general terms a null type of measurement, meaning that one of two quantities is considered a standard to which the other quantity is brought to a match. Examples of null measurements are equalizing electric potentials in potentiometers, currents in galvanometers, forces in chemical balances, and in fact all matching processes. But the match can never be perfect. There is always a small difference of magnitude in the matched quantities. This threshold value, in the language of the day, is a measure of the "noise" level.

For visibility work, the eye is stimulated by an optical quantity called brightness. The eye can determine the difference in the brightness of two objects down to a point called the threshold of brightness contrast, denoted by ϵ and defined as

$$\epsilon = \left| \frac{B_B - B_0}{B_B} \right| . \tag{3.1}$$

B_B is the brightness of the background and B_0 the brightness of the object under consideration. The background brightness is the standard. For the normal eye, a value of 0.02 for ϵ has been adopted as a standard.

Brightness as used here has a precise mathematical definition. The definition agrees, in its general aspect, with what we mean when we say that one object is brighter than another. In general, when the comparison is made, we are looking at rough surfaces giving diffusely reflected light. As an example, reflection from ground glass meets the requirements for giving diffuse reflection, whereas a mirror does not, giving as it does specular (mirrorlike) reflection. Brightness being a measure of diffuse reflection is independent of the angle of view. Mathematically, the *brightness* or its synonym *specific intensity* is the in-

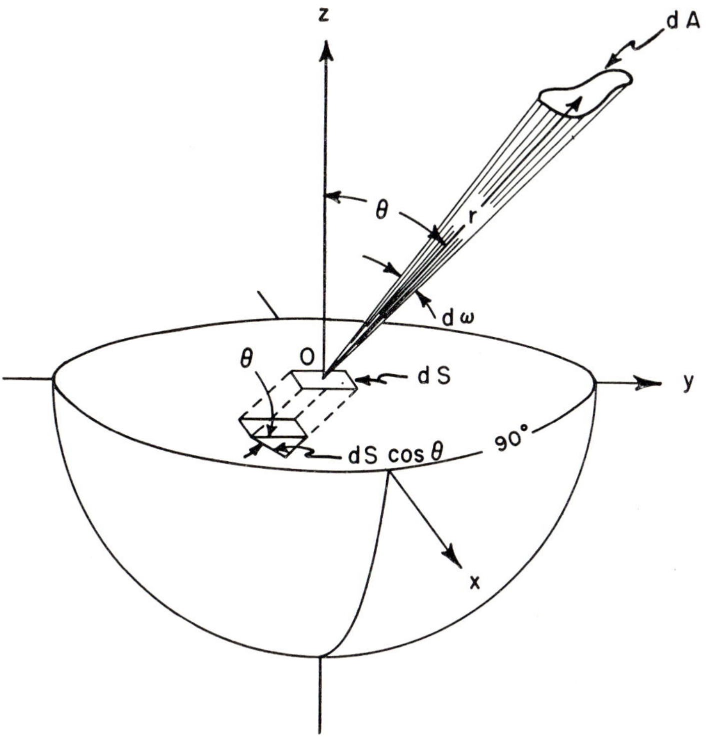

Fig. 3.1 This figure shows the relationship between the area of the source dA and the area of the detector dS. This figure is an elaboration of Fig. 2.3 and should be used in conjunction with it. The areas and the solid angle are expressed here as differential quantities for greater generality. $dS \cos \theta$ is the projection of the surface dS normal to the radiant energy emitted from dA in the direction of dS. $dS \cos \theta$ is positioned at dS, the apex of the solid angle $d\omega$.

tensity of radiant energy per unit area normal to the line of sight. To illustrate the definition, let us consider Fig. 3.1. In this figure, dS is a plane elementary area of the detector. Oz is the normal to dS. Radiation from an elementary area dA located a distance r from O at angle θ from Oz impinges on area dS.† From Eq. 2.4, the intensity

† Area dS is so small and is located at such a great distance from dA that the solid angle subtended by dA is the same, no matter from which point on dS dA is observed.

received at O from area dA is

$$I = \frac{dF}{d\omega}. \tag{3.2}$$

The brightness B from the definition given above becomes

$$B = \frac{dI}{\cos\theta\, dS}. \tag{3.3}$$

The total flux incident on all elements dS of the detector from all elements dA of the source (or conversely the total flux on dA from dS if the net radiation is in that direction) from Eqs. 3.2 and 3.3 is

$$F = \int_0^S dS \int_0^\omega B \cos\theta\, d\omega. \tag{3.4}$$

The flux density measured at dS is

$$E = \frac{dF}{dS} = \int_0^\omega B \cos\theta\, d\omega. \tag{3.5}$$

Equations 3.4 and 3.5 restrict $\int dS$ to a flat plate, which can also mean a tangent plane at the detector whose area is $\int dS$. Practically all elementary treatments of radiation make this assumption. Furthermore, all these equations as stated assume monochromatic radiation and no attenuation of the beam by scattering or absorption. The quantity B has the dimensions of brightness per unit wavelength and if dA is a substantial distance from dS B is identifiable with the intrinsic brightness $B^*(\lambda)$ to be introduced later in the chapter. The effect on the detector is the same as though measurements of brightness were made at the object itself. Let us digress a moment from a further development of the equations defining visibility to a discussion of the sensation of brightness that the eye or an instrument might detect. This next section is necessary background for a fuller appreciation of visibility measurements but can be omitted if continuity in the development of the visibility equations is desired.

The Measurement of Brightness. Colorimetry.[11,B1] The eye and in fact all detectors do not have a uniform frequency response over the entire spectrum. The quantitative expression for this selectivity of response by wavelength can most easily be expressed as the ratio of the energy detected to the energy impinging on the detector. This ratio is a fraction less than or in the limit equal to unity and will be symbolized by ψ_λ. When the frequency response characteristics of the eye are referred to, ψ_λ is called the *visibility* (not to be confused with meteorological visibility). For other detectors, ψ_λ is an efficiency factor. The visibility of the eye goes from zero at 400 millimicrons (mμ), to unity at 555 millimicrons, and down to zero at 700 millimicrons wavelength. Filters also have a visibility factor but much more narrow than that of the eye, and will cause the eye to perceive colored instead of white light. Curves for the visibility of the eye and for a typical reflector are given in Fig. 3.2.

The source of the radiation is also important in determining the level of brightness and the impression of color. In arriving at a measure of ψ_λ, the visibility can also be considered to be the actual energy utilized by the eye from a source that is emitting or reflecting radiation of unit brightness at every wavelength. When the eye views such a body, the impression is the color white, and the radiant source is called a white body. The definition of a white body has been generalized to include any region of the electromagnetic spectrum where the radiant energy emitted or reflected from a source is a maximum and independent of wavelength. For the visible part of the spectrum, $400\ \text{m}\mu < \lambda < 700\ \text{m}\mu$, the sun is a good approximation to a white body (see Fig. 3.3). Similarly a black body is one from which no energy is *reflected* or *transmitted* for any wavelength. A gray body has uniform spectral reflectivity at some level between a black and white body. It follows that some bodies can be white in the visible and black in the infrared. Snow is such a body, since it reflects in the visible but absorbs most of the energy in the infrared instead of reflecting it.

Finally, a body may have selective reflection even when illuminated by a source of white light. The reflection factor $\mathcal{R}_\lambda$, analogous to ψ_λ, is the fraction of light reflected at wavelength λ, and is therefore a function of wavelength. The illuminant, the source of the radiation, usually is not a white body even over the range considered. This source has a brightness $B^*(\lambda)$, expressed in energy units. The combination of the source and the reflection factor explains why an object that appears green in white light will appear black when illuminated by red light. The same goes for transmitted light. White light passing through a green filter appears green. None of this green light is transmitted if a red filter is placed in the system.

The total brightness B detected by a receiver whose visibility is ψ_λ is the brightness $B^*(\lambda)$ of the source reflected with an efficiency $\mathcal{R}_\lambda$ from the object under scrutiny after passing through any filters of

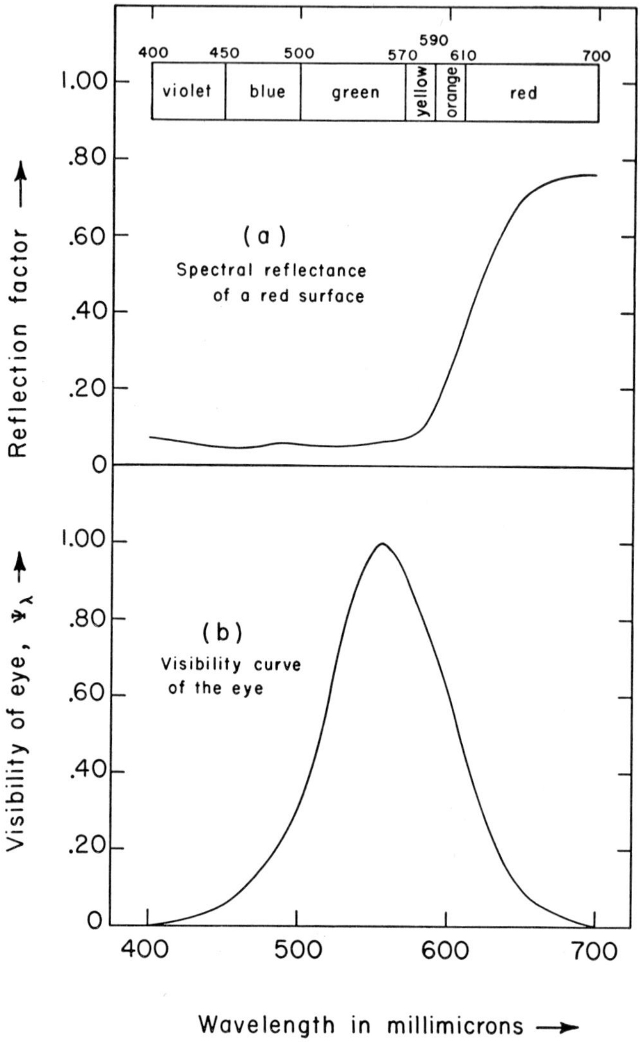

Fig. 3.2 The spectral reflectance of a typical red surface and the visibility curve of the eye. The product of the ordinates of the two curves wavelength by wavelength causes the physiological sensation of the color red.

transmission efficiency τ_λ (the transmissivity). In symbols,

$$B = \int_0^\infty B^*(\lambda)\mathcal{R}_\lambda \tau_\lambda \psi_\lambda \, d\lambda. \tag{3.6}$$

In any practical case many of these factors may be unity. For example, in the absence of absorption and scattering a red object of

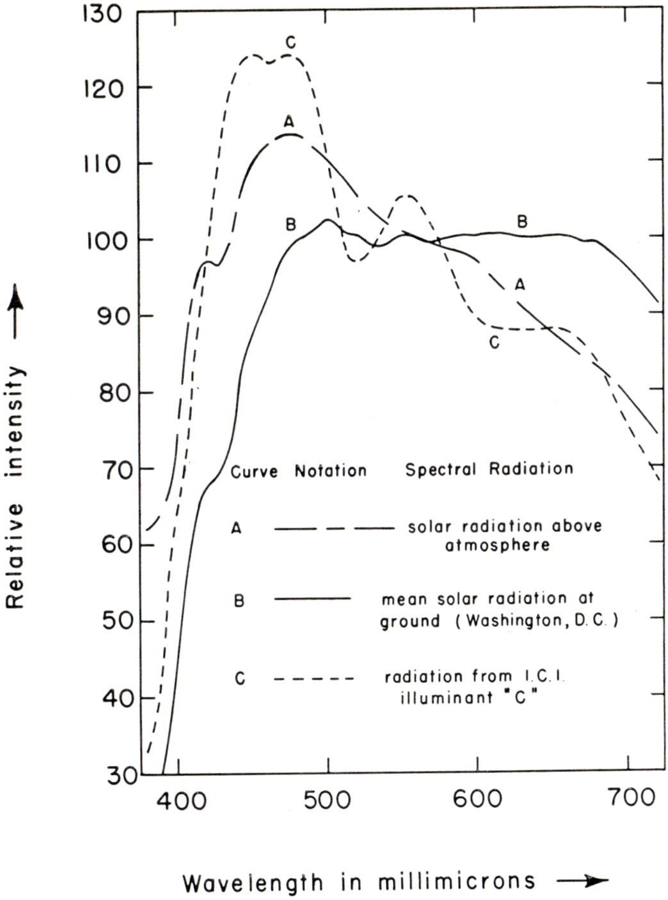

Fig. 3.3 The distribution of intensity in the visible region of the spectrum from three "white" sources.

reflectance $\mathcal{R}_\lambda$ illuminated by the sun appears to the eye to have a brightness equal to

$$B = \overline{B^*(\lambda)} \int_{400\ \mathrm{m}\mu}^{700\ \mathrm{m}\mu} \mathcal{R}_\lambda \psi_\lambda \, d\lambda. \tag{3.7}$$

As ψ_λ is zero for $\lambda < 400\ \mathrm{m}\mu$ and for $\lambda > 700\ \mathrm{m}\mu$, the integral is also zero for all wavelengths where ψ_λ is zero. B^* is the brightness of the sun in the 400 to 700 mμ range of λ and is essentially a constant. Figure 3.4 illustrates Eq. 3.7.

The above analysis shows that the eye, unlike the ear, is a relatively nonselective detector of frequency in the visible. If two wavelengths, such as those corresponding to blue and yellow, are combined, the eye perceives one sensation, namely, the sensation of green. The eye is able to distinguish not only this color, identified with a certain wavelength called the *dominant wavelength,* but also differences in *brightness,* and in the dilution of the dominant wavelength with white light, a decrease in *purity.* These three parameters, dominant wavelength, brightness,

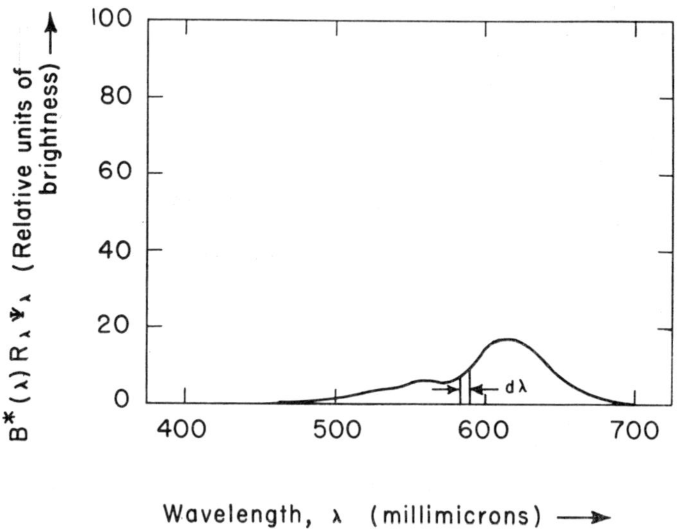

Fig. 3.4 The area under the curve represents the brightness B of Eq. 3.7 in relative energy units. The curve is the product of the ordinates of Fig. 3.2*a* and 3.2*b,* and the ordinate of curve B of Fig. 3.3.

and purity, form the subject of colorimetry. These quantities can be given a precise quantitative value that completely describes the physical as opposed to the aesthetic sensation of light on the eye.

A familiar experiment forms the basis of the quantitative aspect of colorimetry. It has long been known that any color can be duplicated in dominant wavelength and purity by the use of a Maxwell color disk. The color that is to be matched is uniformly illuminated by a standard light.† Three standard colors of high purity, usually a red, green, and blue, are displayed as sectors of a rotating disk. The ratios of these

† A very common standard light source is the International Commission on Illumination illuminant "C" (I.C.I. illuminant "C"). This artificial source duplicates the relative spectral brightness of natural sunlight and is plotted in Fig. 3.3.

three colors are adjusted until a color match between disk and sample is obtained. With purple, a match can be made only by adding one primary color to the sample and matching this result with the other two colors. Since a fixed ratio of red, green, and blue will give the sensation white, this ratio of color on the Maxwell disk can be replaced by white with no effect on the color matching. This substitution procedure, however, emphasizes the amount of white, or lack of purity, in virtually every color.

To prevent the need for subtracting colors to make a match, a mathematical transformation of the three colors described above is made. The resulting colors are of greater purity (or saturation) than is obtainable in nature and are used for standards. A proper mixture

Table 3.1

A Table Giving the Product of the Tristimulus Values $\bar{x}$, $\bar{y}$, and $\bar{z}$ with a Standard Light Source (I.C.I. Illuminant "C") Signified by $E_c(\lambda)$

(After Hardy)

Products $\bar{x}E_c(\lambda)$, $\bar{y}E_c(\lambda)$, and $\bar{z}E_c(\lambda)$ correspond to the general products $\bar{x}B^*(\lambda)$, $\bar{y}B^*(\lambda)$, and $\bar{z}B^*(\lambda)$ of Eq. 3.8. The products of the tristimulus values and the light source E_c are in relative units.

λ(mμ)	$\bar{x}E_c(\lambda)$	$\bar{y}E_c(\lambda)$	$\bar{z}E_c(\lambda)$
400	0.9	0	4.3
425	22.6	0.9	109.6
450	41.7	4.7	220.0
475	18.1	14.3	130.2
500	0.6	36.2	30.5
525	11.2	76.7	5.9
550	45.6	104.7	0.9
575	83.8	91.3	0.2
600	95.3	56.6	0.1
625	65.9	28.0	0
650	25.0	9.4	0
675	5.7	2.1	0
700	0.9	0.3	0

of these three standards can duplicate any color in the spectrum, giving the correct dominant wavelength, relative brightness, and purity. The relative magnitudes of the three colors are denoted respectively by $\bar{x}$, $\bar{y}$, and $\bar{z}$, called *tristimulus values*, that are a function of wavelength. The $\bar{y}$ distribution is purposely made identical with the visibility curve of the eye. The $\bar{x}$, $\bar{y}$, and $\bar{z}$ distributions are presented in Fig. 3.5. Tabular values of the product of the tristimulus values and I.C.I. illuminant C are listed in Table 3.1.

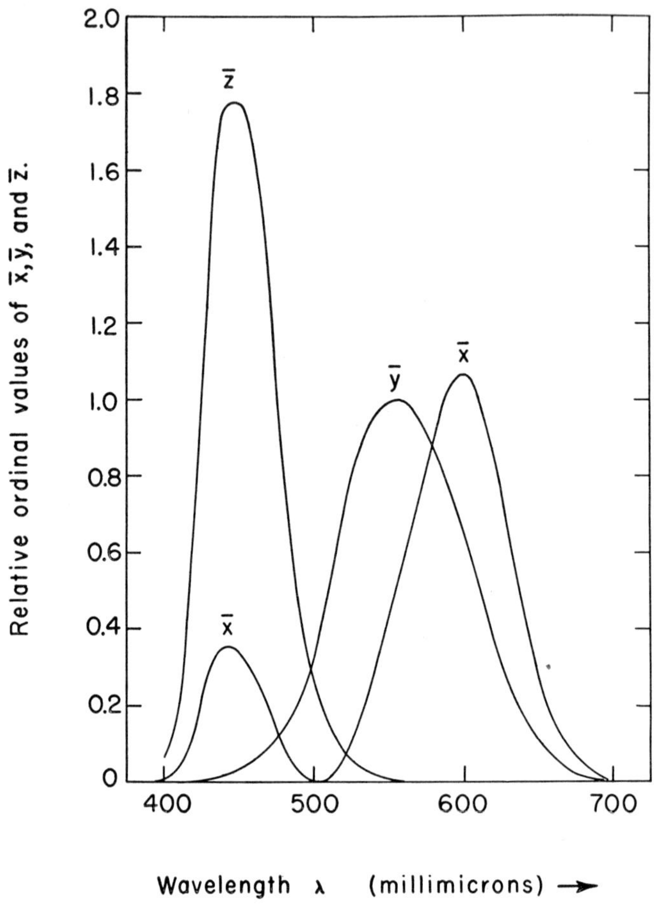

Fig. 3.5 The spectral distribution of the tristimulus values $\bar{x}$, $\bar{y}$, and $\bar{z}$.

Now, if an object has a reflectance $\mathcal{R}_\lambda$, we define the tristimulus values X, Y, Z as

$$
\begin{aligned}
X &= \int_0^\infty B^*(\lambda)\mathcal{R}_\lambda \bar{x}\, d\lambda, \\
Y &= \int_0^\infty B^*(\lambda)\mathcal{R}_\lambda \bar{y}\, d\lambda, \qquad (3.8) \\
Z &= \int_0^\infty B^*(\lambda)\mathcal{R}_\lambda \bar{z}\, d\lambda.
\end{aligned}
$$

The co-ordinates X, Y, and Z have the property of weighted means.

The system of equations expressed by Eq. 3.8 can be evaluated as a summation over a uniform interval $\Delta\lambda$, where the values of $\bar{x}$, $\bar{y}$, $\bar{z}$, and $\mathcal{R}_\lambda$ are chosen at the midpoints of the wavelength interval. For $B^*(\lambda) = E_c(\lambda)$, I.C.I. illuminant C, Eqs. 3.8 become

$$\begin{aligned} X &= \Delta\lambda \sum_{\lambda=400}^{\lambda=700} \mathcal{R}_\lambda \bar{x} E_c(\lambda), \\ Y &= \Delta\lambda \sum_{\lambda=400}^{\lambda=700} \mathcal{R}_\lambda \bar{y} E_c(\lambda), \\ Z &= \Delta\lambda \sum_{\lambda=400}^{\lambda=700} \mathcal{R}_\lambda \bar{z} E_c(\lambda). \end{aligned} \tag{3.9}$$

The trichromatic coefficients x, y, and z are defined as

$$\begin{aligned} x &= \frac{X}{X+Y+Z}, \\ y &= \frac{Y}{X+Y+Z}, \\ z &= \frac{Z}{X+Y+Z}. \end{aligned} \tag{3.10}$$

Only two of the three equations of (3.10) are independent. Because of this a color chart can be made whose ordinate is y and whose abscissa is x. The chart is shown schematically in Fig. 3.6. All possible colors lie within the enveloping curve. Point C represents white light. Points A, B, and C represent the chromatic co-ordinates of curves A, B, and C respectively of Fig. 3.3. The dashed line marked 25 per cent represents a line of constant purity whose value is 25 per cent. Any natural color whose co-ordinates are x and y will fall inside the enveloping curve. A line connecting the center and point (x, y) will terminate at point P on the curve. Point P determines the *dominant wavelength* of the object of reflectance $\mathcal{R}_\lambda$ when illuminated by the standard lamp. The ratio of the distance from the point (x, y) to C and the distance CP when multiplied by 100 is the *purity* in per cent. The relative *brightness* is the ratio of Y in Eq. 3.9 with reflectance $\mathcal{R}_\lambda$ for a given object to Y in the same equation with $\mathcal{R}_\lambda = 1$, typifying a perfect reflector. The color chart and its uses are described more fully in the *Handbook of Colorimetry*.†

Although the discussion of visibility will for the most part be carried out for monochromatic light, the equations developed can be utilized in

† Hardy, A. C., *Handbook of Colorimetry*, Technology Press, Cambridge, Mass., 1936.

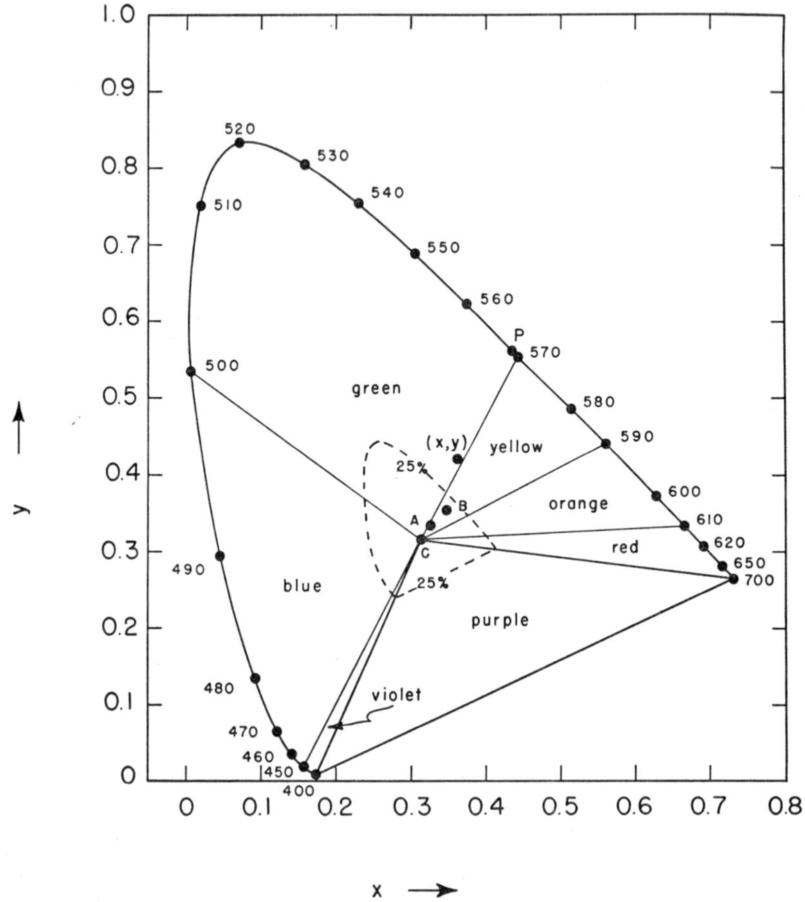

Fig. 3.6 A chromaticity diagram. The point (x, y) represents a color popularly called apple-green. Its spectral characteristics are:

Dominant wavelength	568 mμ
Purity	40%
Brightness	35%

theory only if the operation of integrating over all wavelengths as developed in this section is accomplished. One can often substitute mean values of certain individual quantities rather than take the true mean of all of these quantities as indicated. The approximate procedure, although useful in simplifying computations, needs to be guided by experience, because an indiscriminate use of averages can lead to erroneous results.†

† A drastic example of the improper use of mean values can be shown in the

The attenuation of radiant energy

Suppose an object of intrinsic brightness $B_0{}^*$ at a given wavelength is viewed through an atmosphere that attenuates this energy. The

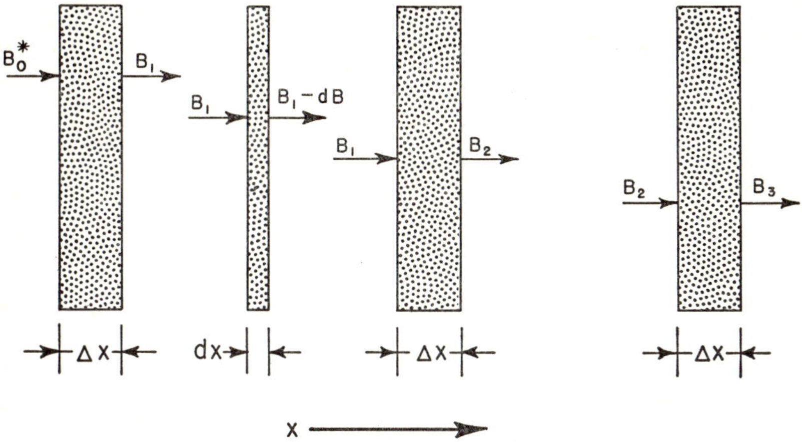

Fig. 3.7 Schematic drawing of the attenuation of energy by scattering or absorption. The fractional decrease of energy per unit length of the brightness B is given by Eq. 3.11. This equation is known variously as the Bouguer-Lambert law or Beers' law when absorption processes are considered, but with the proper coefficients also applies to scattering or a combination of scattering and absorption.

energy detected represents a brightness $B_1 < B_0{}^*$. The difference $B_0{}^* - B_1$ in Fig. 3.7 represents the energy lost to the detector over a distance Δx. The fractional change in brightness is $\dfrac{B_0{}^* - B_1}{B_0{}^* \, \Delta x}$ over the

---

difference between the square of the mean value of $\sin\theta$ over a wavelength ($\theta = 2\pi$) and the mean value of $\sin^2\theta$. These results are

$$\overline{(\sin\theta)}^2 = \left[\frac{\int_0^{2\pi} \sin\theta \, d\theta}{\int_0^{2\pi} d\theta} \right]^2 = 0,$$

$$\overline{\sin^2\theta} = \frac{\int_0^{2\pi} \sin^2\theta \, d\theta}{\int_0^{2\pi} d\theta} = \tfrac{1}{2} \neq \overline{(\sin\theta)}^2.$$

interval Δx. By repeating the same process over successive intervals experiment shows that

$$\frac{B_0{}^* - B_1}{B_0{}^* \, \Delta x} = \frac{B_1 - B_2}{B_1 \, \Delta x} = \frac{B_2 - B_3}{B_2 \, \Delta x} = \cdots = \frac{B_{n-1} - B_n}{B_{n-1} \, \Delta x} = \text{Constant.}$$

The above relationship implies a limiting process. In the mathematical limit, therefore, when Δx becomes so small as to approach the differential quantity dx, $\dfrac{B_{n-1} - B_n}{B_{n-1}} \to \dfrac{dB}{B}$. Letting the constant be symbolized by σ allows us to write

$$\frac{dB}{B} = -\sigma \, dx \tag{3.11}$$

where the negative sign is chosen to show a decrease in brightness for an increase in path length x. σ is called the extinction coefficient and is dependent on wavelength.† Attenuation can be caused by scattering, absorption, or reflection, alone or in any combination. Letting these coefficients be respectively k_s, k_a, and k_r

$$\sigma = k_s + k_a + k_r. \tag{3.12}$$

For example, if attenuation is caused only by pure Rayleigh scattering from air molecules

$$\sigma = k_s = \frac{32}{3n} \frac{\pi^3}{\lambda^4} (m - 1)^2_{\text{gas}} \tag{3.13}$$

where k_s is obtained from Eq. 2.33 in Chapter 2.

Equation 3.11 is a fundamental law in theoretical physics and is called the Bouguer-Lambert law. The same equation but with different physical quantities is the radioactive law of decay.

Integration between the intrinsic and apparent brightness B^* and B through a uniform atmosphere (σ, a constant independent of path x) is

$$\int_{B^*}^{B} \frac{dB}{B} = -\int_0^x \sigma \, dx$$

and gives

$$\tau = \frac{B}{B^*} = e^{-\int_0^x \sigma \, dx} = e^{-\sigma x}. \tag{3.14}$$

† All the quantities except distances and angles are dependent on wavelength. For ease of notation, the λ subscripts will be omitted. Comment will be made as to the importance of the wavelength dependency on the equations in the final form. As these equations now stand, the individual components are equal to the same components in other sections when the difference is in a λ subscript or (λ) to indicate wavelength dependency.

If σ is not constant along x, this fact must be taken account of in the integration. The ratio $B/B^* = \tau$ is called the transmissivity, a dimensionless number.

The visual range[7,8,10,B3]

Consider a volume of atmosphere dx thick along the horizontal. A cross section of this volume subtends a solid angle ω. Particulate matter in this elementary volume scatters light from the sky toward the

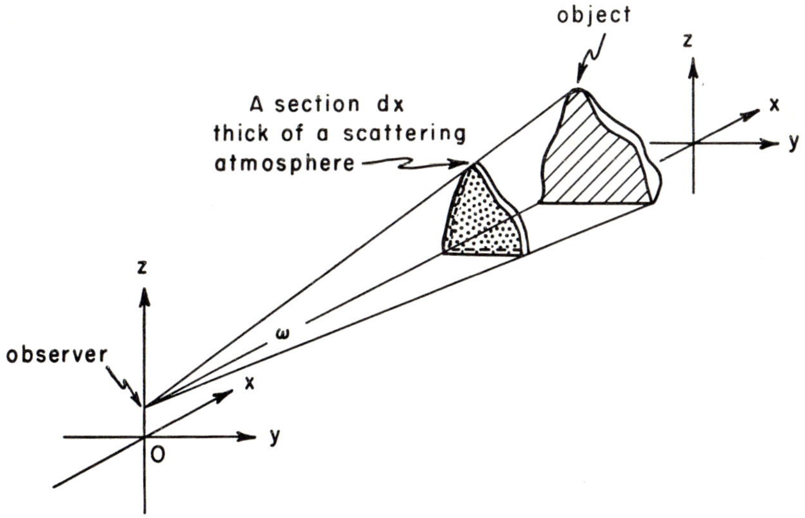

Fig. 3.8 A schematic diagram of an object of intrinsic brightness B_0^* that subtends an angle ω at the observer's eye. The contributions of scattered energy toward the eye from all sections dx in the solid angle ω increase the apparent brightness of the object to B_0 at the observer's position. The additional scattered energy from the atmosphere $B_0^* - B_0$ is called the airlight. The ultimate source of the airlight is the sun, but the sunlight may have been reflected from the ground or clouds, multiply scattered, or remained as direct sunlight before being scattered in the direction of the eye.

observer positioned at the apex of the solid angle. Figure 3.8 illustrates the geometry. Because of the scattering, the volume included in the solid angle has an *intrinsic* brightness B_H. This intrinsic brightness is diminished on passing through the remainder of the atmosphere between itself and the observer. As each individual element of volume in this solid angle behaves in the same manner, the *apparent* brightness at the observer's eye is the sum of all of this sunlight scattered toward the observer being integrated over a path from 0 to ∞.† The apparent brightness of the horizon is simply

$$B_B = B_H \int_0^\infty e^{-k_s x}\, d(k_s x) = B_H. \tag{3.15}$$

$k_s x$ or, in general, σx, is many times called the optical-path length u. The apparent brightness is proportional to u rather than to the geometric length in itself.

Now, let us view a black body situated a distance x from the observer. The contributions of the scattering particles to the apparent brightness in the path between observer and object are the same as in Eq. 3.15. However, the object itself terminates the length of atmosphere through which the observation is made. The object being black has zero intrinsic brightness. Therefore, the apparent brightness of the black object is

$$B_0 = B_H \int_0^x e^{-k_s x}\, d(k_s x) = B_H(1 - e^{-k_s x}). \tag{3.16}$$

It is clear from Eq. 3.16 that the apparent brightness of a black object varies from true black at zero distance from the observer to the brightness of the horizon sky when $x \to \infty$. A consideration of the threshold value of the eye in differentiating between two levels of brightness shows that the black object is indistinguishable from the horizon sky when from Eqs. 3.1, 3.15, and 3.16,

$$\epsilon = \frac{B_B - B_0}{B_B} = \frac{B_H - B_H(1 - e^{-k_s V_m})}{B_H} = e^{-k_s V_m}. \tag{3.17}$$

The particular value of x at this threshold limit is V_m, the meteorological visual range. Solving Eq. 3.17 for V_m by taking natural logarithms gives

$$V_m = \frac{1}{k_s} \ln \frac{1}{\epsilon} = \frac{3.912}{k_s}, \tag{3.18}$$

when the standard value of 0.02 for ϵ is used. Equation 3.18 is the defining equation for the visual range.

In the derivation of the visual range, it was assumed that a black object was viewed against a uniformly illuminated horizon sky (cloudless or completely cloudy sky, no shadows). The path was horizontal, monochromatic light was assumed, and the scattering particles were

† Infinity in its mathematical application is a relative thing. By the time the curvature of the earth becomes important in changing the density of the scattering particles, the function $e^{-\sigma x}$ has approached zero. A curved earth, therefore, need not be considered, unless the visibilities are greater than 40 or 50 miles.

uniformly distributed along this path. Scattering is the only process acting.

The use of Eq. 3.18 in practice is not so restricted as the derivation suggests. In most cases where Eq. 3.18 would be applied, the visual range is short, certainly under 10 miles, and often under 3 miles. For low visibilities, the particle size is so large, as in fogs, that the particles scatter light nonselectively by wavelength. The scattering process may be thought of as a sum of reflection and diffraction from particles greater than 5 microns in diameter. Thus, a ratio of $a/\lambda > 10$ marks a fair lower limit for geometric optics to begin to have meaning. In the geometric range the scattering coefficient k_s becomes independent of wavelength, so that monochromatic light need not be postulated. For these conditions of scattering, the background lighting becomes diffuse and uniform enough for the application of the theory. However, for visibility underneath a densely overcast sky, the brightness over the entire sky cannot be considered strictly uniform, as it may vary as much as 3 to 1 from zenith to horizon, the zenith sky being the brighter. On clear days, even in directions away from the sun, the variation in sky brightness may also be as large as a ratio of 3 to 1 on comparing horizon with zenith sky, where here the horizon sky is the brighter. The theory does not suffer much as long as the object has a low diffuse reflectance, approaching thereby a truly black object. Fortunately, a low reflectance is a characteristic of many natural objects (see Table 2.3). Equally true is the fact that black objects† in sizes sufficiently large for visibility measurements rarely, if ever, exist in nature.

The scattering coefficient k_s given by Eq. 2.29, valid for spherical particles (the usual assumption), is

$$k_s = \frac{n\pi}{4} K_s a^2 \tag{3.19}$$

where a is the diameter of the sphere and K_s is given by the Mie theory previously mentioned. A later chapter will discuss the consequences of the Mie theory in some detail, especially when diffraction effects are noted. At this time, it is only necessary to state the immediate consequences of K_s on the theory of visibility in general, and Eq. 3.19 in particular.

For large particles, the Mie theory predicts that $K_s \to 2$, appearing to make the scattering coefficient equal to twice the geometric cross section

† Black is a synonym for zero reflectance. It is not a color as used here. Many black-colored objects if highly polished are capable of reflecting relatively large amounts of light. A nephoscope for observing cloud motions uses such a reflecting surface.

of the spheres. This is not so, as long as a/λ is so large that virtually no diffraction pattern exists. At this latter limit, the spheres act as lenses of such short focal length that the energy intercepted by the geometrical cross section is spread so widely that little is intercepted by any detector. If we write that $K_s = 1$ unit of defocused light (including reflected and absorbed also) plus 1 unit of diffracted light and then realize that the

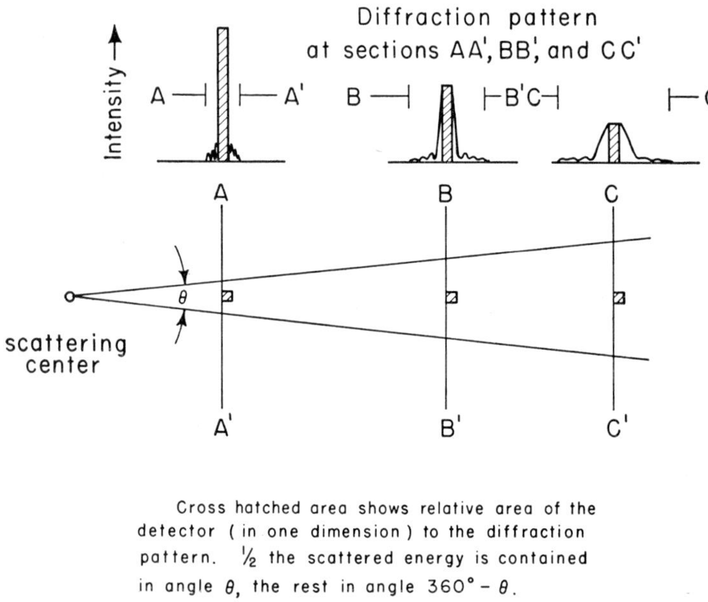

Fig. 3.9 The variation of the flux-collecting efficiency of a detector of finite area as a function of position. As the detector is moved away from the scattering center, less and less of the diffraction pattern is intercepted. Because of this the apparent value of K_s increases as the detector is moved from position A to C.

diffracted light has so little angular spread at $a/\lambda > 10$ that any detector collects nearly all of this diffracted light, then the attenuated light is only the one unit of the defocused light, making $K_s = 1$ instead of 2. Experiments have been devised to show this fact (Ref. 13, Chapter 2). It is in the range from $5 < a/\lambda < 10$ that K_s will vary from its theoretical value when measured by a detector of small but finite area. For $a/\lambda < 5$, the theoretical value prevails. Figure 3.9 indicates schematically the variation of K_s for a given geometry of the detecting system.

Hence, for water droplets > 5 to 10 microns when illuminated by visible light, the scattering coefficient may be written as

$$k_s = \frac{\pi}{4} \sum_i n_i a_i^2 \tag{3.20}$$

where the summation is used to indicate the effect of a distribution of drops. The meaning of the summation sign has been indicated in Chapter 2. Combining Eqs. 3.20 and 3.18 gives the visual range of a dark object in a uniformly distributed fog or cloud during daylight hours. k_s is virtually independent of wavelength so that k_s is taken to be the scattering coefficient for white light. Completing the substitution gives

$$\sum_i n_i a_i^2 = \frac{4.98}{V_m}. \tag{3.21}$$

V_m is in centimeters when n_i is in number of drops per cubic centimeter and a_i is in centimeters. Equation 3.21 is written as a formula for $\sum_i n_i a_i^2$ because at the present time cloud physics research is attempting to find the distribution of drop sizes in clouds, and this parameter can be obtained by the proper visibility measurements.

At this point it might be well to digress for a moment and discuss the meaning of $\sum_i n_i a_i^2$, or rather what it is not. Each cloud that contains n† drops per cm³ has an amount of liquid water equal to the sum of the masses of the individual drops in this unit volume of air. Taking into account the variation of size of the drops, the mass of liquid water per unit volume of air w is

$$w = \frac{\pi}{6} D \sum_i n_i a_i^3. \tag{3.22}$$

D, the density of water, is unity in the C.G.S. system and is many times omitted from formulas employing this system of units.

An assumption that is sometimes made is to write

$$w = \frac{\pi}{6} D\bar{a} \sum_i n_i a_i^2 = 2.61 D \frac{\bar{a}}{V_m},$$

so that a simultaneous measurement of the visibility V_m and liquid water content w gives a mean drop size $\bar{a}$. This assumption is strictly true only when all the drops in a unit volume of air are the same size, and this assumption increasingly becomes of doubtful value as the spread of drop sizes becomes large. This last statement is also true of methods that assume that cloud drops follow smooth distribution functions, one such being the Gaussian error curve. Evidence to date does not justify the use of such simplifying assumptions as have just been indicated, since the assumptions do not appear to be commensurate with the preciseness desired in measurements of visibility or cloud drop distributions.

† $n = \sum_i n_i$.

***Visibility Meters* (day).** Visibility meters for daytime use may be separated into two broad classes: those instruments that use Eq. 3.1 as their guiding principle and those that use Eq. 3.18 directly.

Visibility meters that compare brightnesses are in principle telephotometers. In accordance with the theory, the apparent brightness of the sky is compared with the apparent brightness of a dark target object. The comparison may be made either simultaneously, as is preferable, or through the use of a memory device when the measurement of target and sky are not made simultaneously. In the first case, a null measurement is usually made; in the second, two quantitative measurements of brightness.

To measure the visual range when the target is at a distance $x_0 < V_m$, we rewrite Eq. 3.17 in a more general form. For $x_0 < V_m$, $\epsilon \to |C|$, and $B_B \to B_H$ if the background is the horizon sky. Equation 3.17 becomes

$$k_s = \frac{1}{x_0} \ln \left| \frac{B_H}{B_H - B_0} \right| . \tag{3.23}$$

Since k_s is assumed the same over the entire visual range, Eq. 3.23 substituted into Eq. 3.18 gives

$$V_m = \frac{3.912 x_0}{\ln \dfrac{1}{|C|}} . \tag{3.24}$$

The apparent brightness contrast

$$C = \frac{B_0 - B_H}{B_H} \tag{3.25}$$

of a target at a known distance x_0 is measured. Calibration of the instrument in terms of Eq. 3.24 gives the visual range V_m. Figure 3.10 is a schematic drawing of an instrument using this principle. There are many variations of this basic principle, but no visibility meters are as yet in widespread use.

The second type of instrument measures k_s of Eq. 3.18 directly. Instruments of this type usually sample a small volume of air and assume that the remaining atmosphere through which the visual range would be determined has the same uniformity. This assumption leads to an instrument that can be used when no distant target objects are available (as at sea), but may give a severe sampling error if used over land in the vicinity of industrial areas.

k_s may be measured by a transmission measurement using Eq. 3.14 in the following form

$$k_s = \frac{1}{x_0} \ln \frac{1}{\tau} \tag{3.26}$$

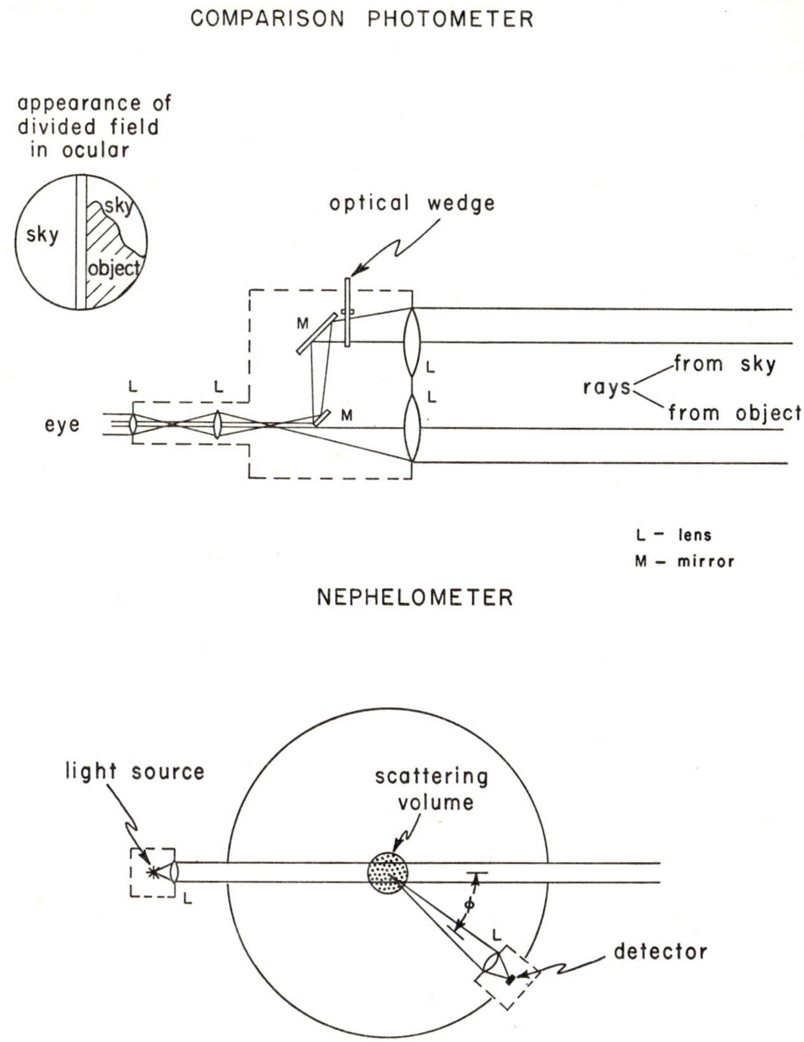

Fig. 3.10 Schematic drawing of a comparison photometer (upper drawing) with a divided field in the ocular. The optical wedge is rotated until the field in the left side is the same brightness as the object in the right side. The wedge is calibrated in distance. The nephelometer measures the scattered flux from the scattering volume. Measurements are made at enough angles so that the angular distribution of flux is obtained. Summation of this flux over the sphere enclosing the scattering volume gives the scattering coefficient.

through measuring τ, the transmission of a beam of light over a known distance x_0 and substituting for k_s in Eq. 3.18 to give an equation analogous in form to Eq. 3.24. k_s may also be measured through observing the angular scattering from an intense beam of light as it falls on a detector. The measurements are summed over the entire spherical area surrounding the illuminated sample of air. The ratio of the total scattered flux from a unit volume of air to the incident flux density is by definition k_s.

The general theory of the visual range

The theory of the visual range so far presented, although the most widely used and in practice not so restrictive as the assumptions might indicate, is nevertheless inadequate for predicting the inherent contrast or the visual range of objects seen against a background other than the sky. Furthermore, Eq. 3.18 cannot be used for determining the visual range when the extinction coefficient varies with the path length, unless k_s be interpreted as being an average value of the extinction coefficient over the entire path entering into the visual range determination. Both considerations enter into the recognition of objects on the ground from an aircraft.

In the general theory of visibility, the theoretical approach is an extension of the ideas expressed in the beginning of this chapter. We shall assume an extinction coefficient σ so that absorption as well as scattering can be considered. Let the intrinsic brightness of a target object at a specified wavelength be $B_0^* > 0$, instead of the zero brightness demanded by the more elementary theory. The intrinsic brightness B_0^* of this object will be attenuated to $B_0^* e^{-\sigma_0 \bar{r}}$ on passing through the atmosphere to the detector. The term $B_0^* e^{-\sigma_0 \bar{r}}$, being greater than zero, represents energy directed toward and received by the detector as an important addition to the air light contribution already considered in Eq. 3.16. The apparent brightness of the object at the detector becomes

$$B_0 = B_H(1 - e^{-\sigma_0 \bar{r}}) + B_0^* e^{-\sigma_0 \bar{r}}. \tag{3.27}$$

It is assumed that only air light is present with no specular (mirror-like) reflection present from the objects observed. We further assume that $\bar{r}$, the *optical slant range*, is given by

$$\bar{r} = \int_0^r f(r)\, dr \tag{3.28}$$

where r is the distance along the line of sight, not necessarily horizontal. Equation 3.28 becomes a mathematical device to do the following.

The slant visual range r is shown in Fig. 3.11. This figure represents the greatest distance at which an object on the ground is visible from an airplane flying at height h. $\bar{r}$ is defined so that

$$\sigma_0 \bar{r} = \sigma_0 x, \tag{3.29}$$

making $\bar{r}$ represent that horizontal distance in a uniform atmosphere at the ground for which the attenuation is the same as that actually

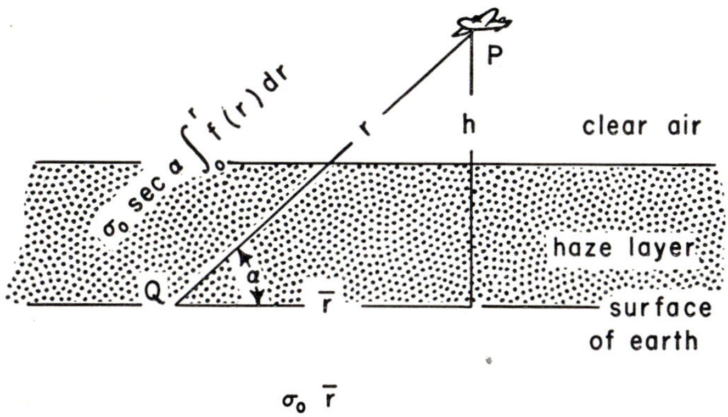

Fig. 3.11 The geometrical meaning of the optical slant range $\bar{r}$. The geometry shows that $\sigma_0 \bar{r} = \sigma_0 \int_0^r f(r)\, dr$.

encountered along the true line of sight of path length r. This device also means that the actual attenuation coefficient $\sigma(r)$ must be expressed as

$$\sigma(r) = \sigma_0 f(r) \tag{3.30}$$

where σ_0 is a constant extinction coefficient in the horizontal plane presently existing at ground level.

Let the target object be compared with its background, where the intrinsic brightness is B_B^*. By analogy with Eq. 3.27, the apparent brightness of the background as seen by the detector is

$$B_B = B_H(1 - e^{-\sigma_0 \bar{r}}) + B_B^* e^{-\sigma_0 \bar{r}}. \tag{3.31}$$

The apparent contrast C of object to its background will be defined as

$$C = \frac{B_0 - B_B}{B_B} = \frac{(B_0^* - B_B^*)e^{-\sigma_0 \bar{r}}}{B_H(1 - e^{-\sigma_0 \bar{r}}) + B_B^* e^{-\sigma_0 \bar{r}}} \tag{3.32}$$

with the help of Eqs. 3.27 and 3.31. Algebraic manipulation of Eq. 3.32 makes

$$C = C^* \left[1 - \frac{B_H}{B_B^*}(1 - e^{-\sigma_0 \bar{r}})\right]^{-1}, \tag{3.33}$$

provided that we define an intrinsic contrast C^* to be

$$C^* = \frac{B_0^* - B_B^*}{B_B^*}. \tag{3.34}$$

Solving Eq. 3.33 for $\bar{r}$ gives

$$\bar{r} = \frac{1}{\sigma_0} \ln \left[\frac{B_B^*}{B_H} \left(\frac{C^*}{C} - 1 \right) + 1 \right]. \tag{3.35}$$

As Eq. 3.35 can be used in air-to-ground visibility measurements, B_B^* is often the intrinsic brightness of the ground. B_H/B_B^* is therefore called the *sky-ground ratio*. Some typical values of the sky-ground ratio are given in Table 3.2.

Table 3.2
(After Duntley)

Sky Condition	Ground Condition	Sky-Ground Ratio
Overcast	Fresh snow	1
Overcast	Desert	7
Overcast	Forest	25
Clear	Fresh snow	0.2
Clear	Desert	1.4
Clear	Forest	5

It is implicit in the derivation of Eq. 3.35 that the visibility in direction P to Q may not be the same as in direction Q to P, even though the measurements are made in the same atmosphere. The difference arises because the contrast between an object and its background at Q need not be the same as the contrast between another object and its background at P as observed from Q. An example of this fact may be found in simultaneous measurements of air-to-ground visibility versus ground-to-air visibility over the same atmospheric path.

Special cases of the use of Eq. 3.35

I ***The Visual Range of a Black Object Viewed against the Horizon Sky.*** In this special case, $B_B^* = B_H$; $B_0^* = 0$. Because of this

$$\left| \frac{C^*}{C} \right| = \left| \frac{0 - B_H}{B_H} \cdot \frac{B_H}{B_0 - B_H} \right| = \frac{1}{\epsilon}.$$

Then Eq. 3.35 reduces to Eq. 3.18, the basic visibility equation.

II ***The Visual Range of an Object Having an Albedo A_0 against a Background Whose Albedo Is A_B.*** The albedo A will be considered to be the average reflectance in the visible region of the spectrum. The illumination from the sky is essentially white light so that we can apply Eq. 3.7 to the problem. If we define A as

$$A = \int_{400\ \mathrm{m}\mu}^{700\ \mathrm{m}\mu} \mathcal{R}_\lambda \psi_\lambda \, d\lambda, \tag{3.36}$$

Equation 3.7 becomes

$$B = B^* = A\, \overline{B^*(\lambda)}. \tag{3.37}$$

The source of illumination on this diffuse reflector is the flux from one half the sky provided that we are looking at a vertical target surface against a hill or wooded surface. The illumination on any element of air in the direction of the horizon arises from the flux contributions from the entire sky. The integrated effect of the flux from all volumes of air in a given solid angle of the sky is the horizon brightness B_H, but the target sees only one half the total volume of air. Therefore, the source intrinsic brightness should be one half the horizon brightness (neglecting any contributions from the ground), making

$$\overline{B^*(\lambda)} = \frac{B_H}{2}. \tag{3.38}$$

The intrinsic brightness of object and that of its background become respectively:

$$B_0^* = \frac{A_0}{2} B_H, \tag{3.39}$$

$$B_B^* = \frac{A_B}{2} B_H. \tag{3.40}$$

The ratio of the contrasts at the visual range is: $(C = -\epsilon)$,

$$\frac{C^*}{C} = -\frac{A_0 - A_B}{A_B \epsilon}, \tag{3.41}$$

and Eq. 3.35 in the horizontal becomes

$$\bar{r} = \frac{1}{\sigma_0} \ln \left[\frac{A_B - A_0}{2\epsilon} - \frac{A_B}{2} + 1 \right] \tag{3.42}$$

after Eqs. 3.40 and 3.41 have been used.

In case the background is the sky, $B_B^* = B_H$ and Eq. 3.35 becomes

$$\bar{r} = \frac{1}{\sigma_0} \ln \left[\frac{1}{\epsilon} - \frac{A_0}{2\epsilon} \right]. \tag{3.43}$$

Equations 3.42 and 3.43, although useful, contain approximations that call for some judgment. The distribution of the light is important in the calculation of B^* in Eq. 3.37, so that Eqs. 3.39 and 3.40, which lead to Eqs. 3.42 and 3.43, are approximations at best. For example, under actual conditions, the terrain immediately in front of the target object may have a reflectivity much different from the average reflectivity of the terrain contributing to the air light. Because of this, the source brightness may be enhanced over the air light. This effect makes the right-hand side of Eq. 3.38 a poor approximation to $\overline{B^*(\lambda)}$. Effects such as this one, plus specular reflections, are outside the scope of Eqs. 3.42 and 3.43.

The colors of distant objects

In the general problem of visibility one uses Eqs. 3.27 and 3.6, which indicate that

$$B_{0\text{ average}} = \int_{400\text{ m}\mu}^{700\text{ m}\mu} B_H(\lambda)(1 - e^{-\sigma_{0\lambda}\bar{r}})\psi_\lambda \, d\lambda + \int_{400\text{ m}\mu}^{700\text{ m}\mu} B^*(\lambda)e^{-\sigma_{0\lambda}\bar{r}}\mathcal{R}_\lambda\psi_\lambda \, d\lambda. \quad (3.44)$$

When one is interested in the colors of distant objects, at large distances but still not near the visual range, $\sigma_{0\lambda}$ is small, small particle scattering is the dominant factor, and the λ^{-4} law of Eq. 2.33 is closely followed. $\bar{r}$ is the horizontal distance and is of the order of 20 to 50 miles. Equation 3.44 can be analyzed by colorimetric techniques, using the color co-ordinates x, y, and z described in the section on colorimetry, for which the integral equation is approximated by a summation. However, Eq. 3.44, as it now stands, can guide us in describing the colors to be expected.

The first integral on the right of the equation predicts that the horizon brightness $B_H(\lambda)$, which for molecular scattering and haze (particles $< 1\mu$ in diameter) produces a blue color, is effectively a constant for large $\bar{r}$. The second integral represents a colored object whose color is a function of its reflectivity $\mathcal{R}_\lambda$. For a dark object $\mathcal{R}_\lambda \rightarrow 0$, and the object which appears black at $\bar{r} = 0$ becomes a deep purple or blue as $\bar{r}$ increases. The color change occurs because of the increased effect of the air light with distance through the dependence of the air light on λ^{-4} scattering. Distant mountains, heavily wooded, take on this appearance. At the other end of the scale, a white object such as a snow field illuminated by the sun changes toward an orange color, faint to be sure, but still tinted. This phenomenon follows from the second integral on the right of Eq. 3.44. The $e^{-\sigma_{0\lambda}\bar{r}}$ factor predicts that the blue light will be attenuated more effectively than the red light. The first integral contributes little to the color because the ratio of the reflected light from the snow to the sky light is so great (see Table 3.2). The apparent color to the eye of a distant snow field with increasing range $\bar{r}$ is a change from white to pale yellow. For colored objects, the most noticeable effect is an increasing lack of saturation (purity) in the color, and an eventual blending with the horizon sky at the visual range. It is because of the achromatic effect at large distances that approximate equations such as (3.18), (3.42), and (3.43) express the visual range so well.

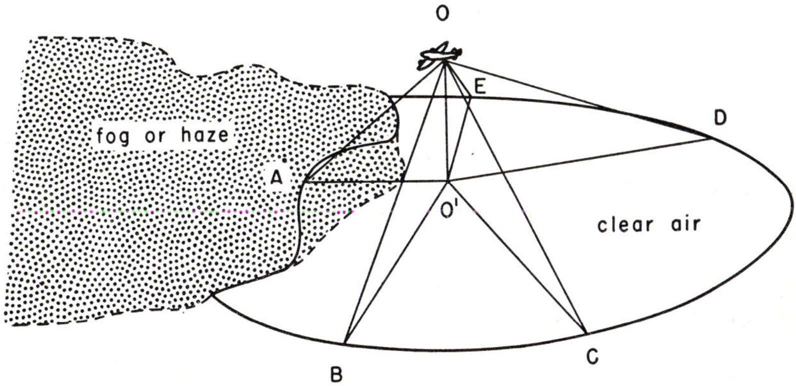

Fig. 3.12 Illustration of the aviator's "cone of visibility." The lines $O'A$, $O'B$, $O'C$, $O'D$, and $O'E$ represent the radii $\bar{r}$ of the "circle of visibility" measured at the ground. The airplane is at O. The dashed lines represent the edge of the fog or haze bank. The solid line is the limiting "circle of visibility."

The Air-to-Ground Visual Range. It is to be expected that an observer in an aircraft will see a distance $\bar{r}_{max}$ in all directions in a uniform atmosphere. The visual range would sweep out a circular area on the ground because objects at any azimuth would be visible for $\bar{r} < \bar{r}_{max}$ and invisible for $\bar{r} > \bar{r}_{max}$. This circle of visibility is the circle of radius $\bar{r}_{max}$ that encloses the base area of the aviator's "cone of visibility," the airplane being thought of as located at the apex of the cone. A schematic drawing of the cone of visibility is given as Fig. 3.12.

In practice, the cone of visibility may become somewhat biased from a true circle. Equation 3.35 governs as usual, but now $\bar{r}$ is related to the slant height and the variable factor $f(r)$ of the extinction coefficient in the definition of $\bar{r}$ (Eqs. 3.28 and 3.30).

First and foremost, $\bar{r}$ is a function of the elevation angle. Equations have been derived under suitable assumptions and nomographs constructed showing the optimum altitude to give the greatest visual range under varying contrast conditions of the target and its background. It is obvious that such information is needed in search operations.

At a given altitude, the value of $\bar{r}_{max}$ in the cone of visibility is affected by the difference in the attenuation coefficient at different azimuths. This effect can become very serious under extremely stable atmospheric conditions when smoke or fog can exist in a surface layer several hundred

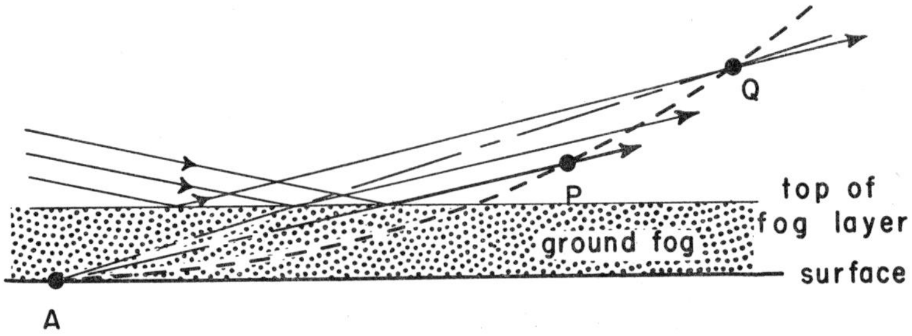

Fig. 3.13 Illustration of the decrease in visibility due to reflection from the top of a layer of ground fog. An airplane has a glide path Q, P, A along dashed curve. Pilot's eyes focus on A, the touchdown point, at all times. Pilot is blinded by sun's reflection when at P but not at Q.

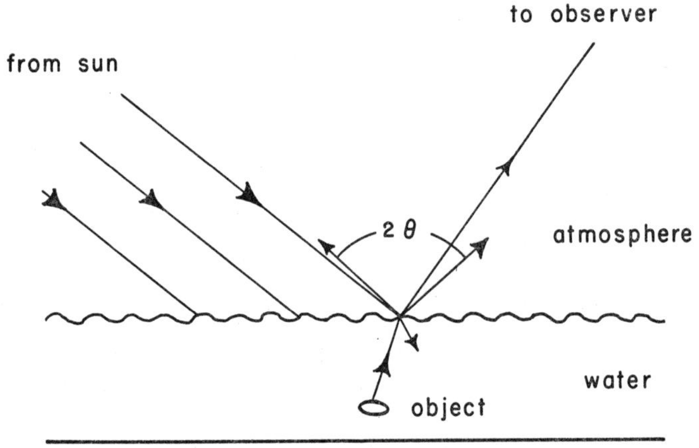

Fig. 3.14 Visibility of submerged objects is complicated by the presence of a water-air interface and reflection effects at the boundary. In addition, if small capillary waves are present on the water's surface, there is a fluctuating reflection through the angle 2θ at the apex of a cone. The additional reflections of the sun's rays toward the observer by the choppy waves reduce the visibility of the object.

feet thick, or even in an elevated stratum of air. Here, the problem of visibility not only includes the effect of different extinction coefficients in the various scattering media (see Table 3.3), but also the reflections at the interface (boundary) between media. A combination of low sun, ground fog, and an airplane making a flat approach for a landing into the sun can cut the forward visibility to practically nothing, because the glide angle coincides with the angle giving specular reflection from the upper surface of the fog. Figure 3.13 illustrates the effect just described, whereas Fig. 3.14 illustrates the more complicated problem of seeing a submerged object. In this latter problem, there is the additional complicating effect of a changing mirrorlike surface, because of the slope of the tiny capillary wavelets superimposed on the larger gravity waves of the ocean. Considerable progress has been made in the solution of these problems, much of it on the theoretical side.

Table 3.3

Some Typical Values of the Extinction Coefficient as a Function of Height from Nephelometer Measurements Made in England in 1942

(After Waldram)

The units of σ, k_s, and k_a are meters^{-1}. $\sigma = k_s + k_a$.

Height, feet	Rayleigh Scattering $\sigma = k_s$	Good Visibility $\sigma = k_s$
0	0.16×10^{-4}	0.6 to 2.0×10^{-4}
10,000	0.11	0.3 to 0.7
20,000	0.08	0.2 to 0.5
30,000	0.06	0.1 to 0.3

	Poor Visibility in Industrial Smoke		
Height, feet	σ	k_s	k_a
0	20 to 30×10^{-4}	5 to 27×10^{-4}	3 to 15×10^{-4}
1000	10 to 20	5 to 17	3 to 15
2000	$\sim$10	$\sim$10	0

Finally, the reflectivities of the targets and their contrasts affect the cone of visibility. In the ideal case of a pure white body viewed against the horizon sky, the object appears brighter when facing the sun, darker when viewed with the sun to the object's back. By theory, the object appears to blend in with the horizon sky at about 100° from the sun, independent of distance. In certain practical cases, this effect may lead to reduced visibilities in directions normal to the sun.

Variations in the threshold of brightness contrast

The standard value for the threshold of brightness contrast ϵ is chosen for the definition of the meteorological range to be 0.02. Experiments conducted during the last war have shown conclusively that this value is subject to a wide range of variation, depending principally on the general level of illumination and the angular subtense of the target. Of course, the experiments were conducted with observers whose uncorrected vision is normal (at least 20/20). Some of the data are produced as Fig. 3.15.

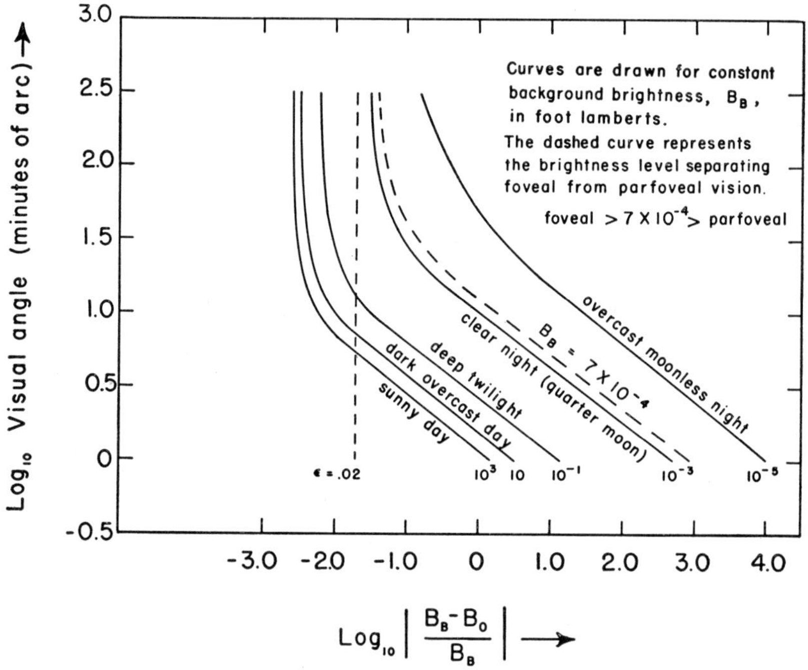

Fig. 3.15 (*After Blackwell.*) The threshold of brightness contrast as a function of the angular subtense of the object viewed. Note that for objects whose angular diameter is greater than about 30 to 100 minutes of arc, the brightness level is the determining parameter in fixing the magnitude of the threshold of brightness contrast.

The data of Fig. 3.15 are valid for a white-light background. A dotted curve is shown at a value of the background brightness of 7×10^{-4} foot-lambert, equivalent to a moonlight night. It was found in these tests that brightness contrasts normally were detected through looking directly at the object and its background (foveal vision), whereas at lower brightness levels the contrast between object and its background was detected through not looking directly at the object but seeing it out of the corner of one's eye (parfoveal vision).

When one changes from foveal to parfoveal vision, one gradually changes from photopic vision (the ability to distinguish colors) to scotopic vision (inability to distinguish colors). Because of the structure of the eye, considerable time is necessary to change from foveal to parfoveal vision, an adjustment the eye must make when the illumination of its field of vision is suddenly decreased from bright to dim. Once the eye has become dark adapted, a matter of 10 to 30 minutes, parfoveal vision of the eye is extremely sensitive: extremely low levels of brightness can be detected, especially if the source of light is flashing. As noted before, the color of the flashing light cannot be identified. However, for photopic vision, red and blue-green are the colors that can be positively identified to low brightness levels against a dark background.

The visual range at night

At night, the visual range is determined by the use of lights suitably spaced. There are objections to this procedure, and some doubt as to whether a complete equivalence between daylight and nighttime visual range can be established, because at night we make an estimation of the scattering coefficient by detecting the transmitted flux from a light source against a dark background, whereas during the day we make an estimation of the scattering coefficient by determining a contrast threshold induced by scattered air light. However the scattering coefficient is determined at night, the simple fact remains that even at night we distinguish objects by a brightness contrast, even when both object and its background are illuminated by lights. Unfortunately, objects of sufficient size to be used as targets cannot be illuminated artificially; instead, the lights themselves must substitute as targets.

Let us see how Eq. 3.35 can be adapted for nighttime use. First $\bar{r} = D$, the visual range at night. σ_0 is a constant equal to k_s for pure scattering and may be related to the daytime visual range by Eq. 3.18. At night, as the background and horizon brightnesses are at low intensities, $B_B{}^*/B_H \approx 1$. On the other hand, $B_0{}^*/B_H$, the intrinsic brightness of the source related to the horizon brightness, is a dimensionless ratio that is very large, because, according to Fig. 3.15, B_H is very small at night. However, when the candle power and dimensions of the source are known, the effective brightness of the source can be computed if not given. From this information, the desired brightness ratio can be determined. For example, an uncollimated tungsten filament lamp on a bright moonlight night would have a brightness ratio of the order 10,000 to 1. Uncollimated lights should be used, because in a collimated light, care must be exercised to make the observations of the beam in the region of its full rated beam candlepower, a region of small angular width at best.

Now let us examine the ratio of the contrasts C^*/C. C^*, the intrinsic contrast from Eq. 3.34, is

$$C^* = \frac{B_0^*}{B_H} - 1 \cong \frac{B_0^*}{B_H} \tag{3.45}$$

for all practical purposes because $B_0^*/B_H \gg 1$. C, the apparent contrast, is determined by the Tiffany data given as Fig. 3.15. Let us denote the threshold value of C by C_0. From Fig. 3.15, C_0 is found to have a strong dependence on the angular size of the light source. As the angular size decreases for increasing distance because the dimensions of the light remain constant C_0 is a function of the visual range D. Therefore, if we write Eq. 3.35 in terms of the quantities we have discussed, for $B_0^*/B_H = C^*$,

$$D = \frac{1}{k_s} \ln \frac{C^*}{C_0} = \frac{V_m}{3.912} \ln \frac{C^*}{C_0}. \tag{3.46}$$

There are two ways of using Eq. 3.46. The first way is to measure k_s. This measurement can be made by a comparison photometer or a nephelometer. The latter instrument has been described, and the former consists of the following.

A standard light of known intensity I_0 is viewed at a fixed distance x_0. The flux density E_0 reaching the detector is, in the absence of scattering particles,

$$E_0 = \frac{I_0}{x_0^2}. \tag{3.47}$$

Refer to Fig. 3.16 for the arrangement of components. A second standard light of known intensity I_1, and located a known distance x_1 from the detector, is viewed through a scattering atmosphere. The flux density on the detector from the second light is

$$E_1 = \frac{I_1}{x_1^2} e^{-k_s x_1}. \tag{3.48}$$

If a null method of comparison is used, $E_1 = E_0$, and, combining Eqs. 3.47 and 3.48, we obtain on solving for k_s

$$k_s = \frac{1}{x_1} \ln \left(\frac{I_1}{I_0}\right) \left(\frac{x_0}{x_1}\right)^2. \tag{3.49}$$

An instrument using the philosophy of Eq. 3.49 is a comparison photometer.

Once k_s has been determined either by this method or through assuming that V_m, the visual range, has been unchanged from day to night, the range D at which any lights can be seen at night can be determined from Eq. 3.46. For a given light on a given night, C^* is known and k_s has been measured. Only C_0 and D remain to be found. These quantities are not independent, so a graphical solution or "cut-and-try" procedure must be used. One proceeds as follows.

All constant values are substituted in Eq. 3.46, so that the equation reads

$$D = \text{constant} \times \ln \frac{\text{constant}_2}{C_0}.$$

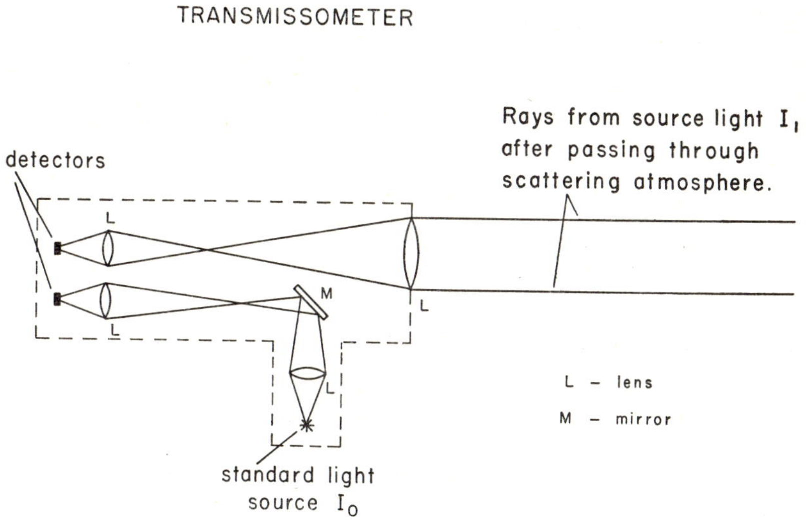

Fig. 3.16 Schematic drawing of a transmissometer for the measurement of the visual range at night. Null methods can be used, by the use of an optical wedge in the optical system or by controlling either the output from standard light I_0, or by suitable attenuation either optically or electrically at the detectors. The transmissometer is used as a comparison photometer.

A value of D is selected. From the size of the target (its largest dimension, called L), the angular size of the target in minutes can be determined by the formula

$$\frac{L}{D} \times \frac{10{,}800}{\pi} \tag{3.50}$$

where L and D are in the same units. The angular size of the target and the background brightness being known, C_0 can be found from Fig. 3.15. Knowing C_0, D can be computed. For Eq. 3.46 to be a true equation, the D used to find C_0 must be the same as the D found after the equation is solved. Sufficient trials must be made until the D's agree, so that Eq. 3.46 is a true equation. This type of solution is known as a "cut-and-try" solution.

The second way in which Eq. 3.46 can be used is as an aid in determining what intensity lights make suitable markers for determining the visual range at night. The problem is to make $D = V_m$ by a suitable choice of $B_0{}^*$, the intrinsic brightness of the lamps. This condition makes

$$\frac{1}{3.912} \ln \frac{C^*}{C_0} = 1 \tag{3.51}$$

from Eq. 3.46 or

$$C^* = C_0 e^{3.912}. \tag{3.52}$$

For example, suppose a certain light has an effective luminous circular area of 100 square inches and the background brightness is taken to be

Table 3.4

The Visual Range at Night Compared with the Visual Range by Day in the Same Atmosphere

(After Bennett)

Values greater than $1\frac{1}{4}$ miles for the visual range by day are extrapolated from the experimental data.

Daytime Visibility	Visual Range at Night for Lights of			
	1 c.p.	10^2 c.p.	10^4 c.p.	10^6 c.p.
27 yards	38 yards	53 yards	70 yards	87 yards
55 yards	69 yards	111 yards	139 yards	174 yards
110 yards	125 yards	204 yards	269 yards	347 yards
220 yards	226 yards	371 yards	525 yards	689 yards
550 yards	451 yards	829 yards	1240 yards	1690 yards
1100 yards	738 yards	1490 yards	$1\frac{1}{3}$ miles	2 miles
$1\frac{1}{4}$ miles	1140 yards	$1\frac{1}{2}$ miles	$2\frac{1}{2}$ miles	$3\frac{2}{3}$ miles
$2\frac{1}{2}$ miles	1650 yards	$2\frac{1}{2}$ miles	$4\frac{2}{3}$ miles	7 miles
$4\frac{1}{3}$ miles	$1\frac{1}{4}$ miles	$3\frac{3}{4}$ miles	$7\frac{1}{2}$ miles	12 miles
$6\frac{1}{4}$ miles	$1\frac{1}{3}$ miles	$4\frac{3}{4}$ miles	10 miles	16 miles
$12\frac{1}{2}$ miles	$1\frac{1}{2}$ miles	$7\frac{1}{2}$ miles	18 miles	30 miles
18 miles	$1\frac{2}{3}$ miles	9 miles	24 miles	44 miles
31 miles	$1\frac{3}{4}$ miles	11 miles	34 miles	67 miles

For orders of magnitude, an ordinary tungsten bulb rated at 40 watts has a candlepower of about 32, airport boundary lights have of the order of 100 candlepower, whereas airway beacons may be of the order of several million candlepower output.

7×10^{-4} foot-lambert. The visual range is to be 2 miles. From Eq. 3.50, the light subtends an angle of 0.31 minute. Extrapolation of Fig. 3.15 shows $C_0 = 1.59 \times 10^4$. Solving Eq. 3.52 makes $C^* =$

7.95×10^5 and the brightness of the source 556 foot-lamberts, as the background brightness has been taken to be 7×10^{-4} foot-lambert. From data taken from Table 8–37 of the *I.E.S. Lighting Handbook*,(B2) a 75-watt incandescent light source has an intrinsic brightness of about 556 foot-lamberts (537 to be exact) provided that the element efficiency is as low as 50 per cent. The results of this problem illustrate the large distances to which relatively dim lights can be seen, although there may be some doubt as to the validity of the extrapolation of the Tiffany data. Nomographs showing the relationship between intensity of source to give the correct daytime visual range under varying conditions of background brightness can be made.

In 1935, M. G. Bennett(2) made some empirical studies on the relationship between the visual range at night as estimated from lights of varying candlepower and the visual range during the day. The atmosphere was assumed not to change from day to night. His results are reproduced in Table 3.4.

Bennett's results indicate that ordinary lights may be used as visibility markers up to about 4 miles without serious error. These lights, if used as visibility markers beyond 4 miles, will lead to underestimation of the range. Unfortunately, it is seldom that suitable markers are available for these greater ranges.

Table 3.5

A Table of Photometric Units

The *lumen* is taken as the fundamental unit. The *lumen* is a unit of *flux* for visible light and is equal to $(1/685\psi_\lambda)$ watts of radiant energy. ψ_λ is the visibility factor of the eye. The *luminous efficiency* is the ratio in lumens/watt with which electrical power is converted into light.

Unit	Measured in	Equivalent Magnitudes
Diffuse brightness	foot-lamberts	1 foot-lambert = 1 lumen/ft^2 = $(1/\pi)$ candles†/ft^2
Illumination	lumens/ft^2	1 lumen/ft^2 = $(1/4\pi)$ candles‡/ft^2 = foot-candle
Intensity	lumen/steradian	1 lumen/steradian = 1 candle
Intensity (point source)	spherical candles	1 candle = 4π lumens
Diffuse intensity (extended source)	apparent candles	1 candle = π lumens

† The *apparent candlepower* of an extended source a specified distance away is meant. It is the candlepower of a point source that would produce the same illumination at that distance.

‡ The *mean spherical candlepower* is implied, as though the radiation were emitted from a point source.

References

1. Aufm Kampe, H. J., "Visibility and liquid water content in clouds in the free atmosphere," *J. Meteorol., 7,* 54–57 (1950).
2. Bennett, M. G., "Further Conclusions Concerning Visibility by Day and Night," *Quart. J. Roy. Meteorol. Soc., 61,* 179–186 (1935).
3. Beuttell, R. G., and Brewer, A. W., "Instruments for the Measurement of the Visual Range," *J. Sci. Instr.,* London, *26,* 357 (1949).
4. Blackwell, H. R., "Contrast Thresholds of the Human Eye," *J. Opt. Soc. Amer., 36,* 624–643 (1946).
5. Bricard, M. J., "Propagation of Visible and Infrared Radiation Through Fog," *Centenary Proc. Roy. Meteorol. Soc.,* 36 (1950).
6. Coleman, H. S., and Rosenberger, H. E., "A Comparison of Visual and Photoelectric Measurements of the Attenuation of Brightness Contrast by the Atmosphere," *J. Opt. Soc. Amer., 40,* 371–372 (1950).
7. Duntley, S. Q., "The Visibility of Objects Seen Through the Atmosphere," *J. Opt. Soc. Amer., 37,* 635–641 (1947).
8. Duntley, S. Q., "The Reduction of Apparent Contrast by the Atmosphere," *J. Opt. Soc. Amer., 38,* 179–191 (1948).
9. Foitzik, L., "Sichtbeobachtung–Sichtmessung," *Z. Meteorol. 5,* 1–14 (1951).
10. Koschmieder, H., "Theorie der horizontalen Sichtweite," *Beitr. Phys. frei. Atmos., 12,* 33–55, 171–181 (1924).
11. Lovell, D. J., "Principles of Colorimetry," *Am. J. Phys., 18,* 104 (1950).
12. Middleton, W. E. K., "The Colors of Distant Objects," *J. Opt. Soc. Amer., 40,* 373 (1950).
13. Middleton, W. E. K., "Visibility in Meteorology," *Compendium of Meteorology,* American Meteorol. Society, Boston, 91–97 (1951).
14. Pearson, C. A., Koomen, J. J., and Tousey, R., "Visual Measurements of Atmospheric Transmission of Light at Night," *Bull. Am. Meteorol. Soc., 33,* 117–121 (1952).
15. Sanderson, J. A., "The Transmission of Infrared Light Through Fog," *J. Opt. Soc. Amer., 30,* 405 (1940).
16. Sinclair, D., and LaMer, V. K., "Light scattering as a measure of particle size in aerosols," *Chem. Rev., 44,* 245–267 (1949).
17. Waldram, J. M., "Measurement of the Photometric Properties of the Upper Atmosphere," *Quart. J. Roy. Meteorol. Soc., 71,* 319–336 (1945).

Source Books

B1. Hardy, A. C., *Handbook of Colorimetry,* Technology Press, Massachusetts Institute of Technology, Cambridge, Mass. (1936).
B2. *I.E.S. Lighting Handbook,* first edition, Illuminating Engineering Society, New York (1947).
B3. Middleton, W. E. K., *Vision Through the Atmosphere,* University of Toronto Press (1952).

Problems

3.1 A comparison photometer is to be used to measure the brightness of the sky. Radiation from the sky is allowed to enter a blackened tube of radius r and illuminates a screen x_0 units away. Coincidentally, light from a standard point source of intensity I_0 and located on the axis of the tube is allowed to illuminate the reverse side of the screen. The standard light is moved until a position is found x units from the screen where the flux density from the standard source is the same as that from the diffuse source. Find the brightness of the sky as a function of I_0, x, x_0 and r. Assume $x_0 \gg r$ and that the sky has a uniform brightness.

3.2 On a typical sunny day, a white object is viewed against a green background. The spectral reflectances of the two objects are:

Wavelength (mμ)	400	450	500	550	600	650	700
$\mathcal{R}_\lambda$ (white object)	0.60	0.70	0.78	0.83	0.87	0.88	0.89
$\mathcal{R}_\lambda$ (green object)	0.11	0.16	0.39	0.27	0.10	0.06	0.11

a. Assuming that the sun can be approximated by an ICI Illuminant "C", find the relative brightness of both the white object and the green object. Find the intrinsic brightness contrast between the two objects.

b. Assuming that the white object is a rectangular patch of snow 2000 by 500 feet on a mountainside, find the visual range. The scattering coefficient k_s is independent of wavelength and is equal to 0.020 per mile.

c. Let the scattering coefficient for air be expressed as

$$k_s = \frac{A}{\lambda_D}.$$

What is the visual range of the object in part *b* on a day when $A = 11.2$ mμ per mile? k_s is in units of miles^{-1} when the dominant wavelength λ_D is expressed in millimicrons. For the white object take $\lambda_D = 560$ mμ.

3.3 The meteorological range is given by Eq. 3.18. A certain observer consistently reports 7 miles when the visual range is known to be 8 miles. What is the true visual range when this observer reports a visual range of $2\frac{3}{4}$ miles?

3.4 On a certain day, the meteorological visual range is 9 miles. Smoke from a fire reduces the visibility within the smoke pall to 2 miles. What is the maximum distance an object 1 mile behind the smoke pall can be seen if an observer must look through $\frac{3}{4}$ mile of smoke when viewing the object? Assume $\epsilon = 0.02$ and that the object is a black body against a horizon sky. Neglect any reflection effects.

3.5 A simultaneous measure of visual range and drop-size distribution in cloud shows the visual range to be 250 feet and the drop-size distribution to be as follows.

Drop diameter, microns	4	8	12	16	20	24	28	32
Number fraction	0.30	0.22	0.19	0.15	0.07	0.04	0.02	0.01

a. What is the liquid-water content of the air?

b. The visibility decreases to 100 feet with no measurable change in the liquid-water content. What is the apparent explanation? Will a simultaneous measurement of the drop-size distribution necessarily check the results obtained?

CHAPTER 4

Radiation Processes in the Earth's Atmosphere

In the last analysis, the sun is the source of all useful energy on this planet, whether stored for ages in coal and oil formed from vegetation nurtured by the sun or amassed daily in water power being constantly replenished by the evaporation and condensation cycles in the atmosphere. These sources have led indirectly even to the release of atomic energy, which once released is not then intimately connected with the sun. On a less grand scale, the study of the energy from the sun and the loss of energy from the earth and its component substances, both natural and man-made, has application to the solar heating of houses, to the production of favorable climatic conditions for farming, and to physiological effects on man, especially at high altitudes.

In meteorology, energy from the sun is the driving energy for the great atmospheric heat engine. Since energy from the sun is transmitted to the earth by radiation, some of which is ultimately reradiated to space from the earth-atmosphere system, the study of radiation from hot bodies (a relative term) is very important in many geophysical problems. Although by convention radiation processes are a concern of physical meteorology, and convection and conduction in the atmosphere itself are a concern of dynamic meteorology, such a neat division of effort occurs only rarely in the solution of actual problems. In accord with convention, however, in this chapter and especially in the chapter immediately following, the burden of the proofs and explanations will be directed toward the radiation problem. Information or concepts involved in conduction or convective transport of energy in the atmosphere will be borrowed as needed from dynamic meteorology.

General features of the sun

From astrophysical measurements, we know that the sun is a slowly rotating body of hot and highly condensed gas surrounded by an atmosphere of rarefied gas of varying degrees of temperature. When the sun is viewed by eye through smoked glass, it appears as a flat circular disk. This part of the sun, corresponding in form and having one-fourth the density of the earth, is called the *photosphere.* It is possible to obscure the sun's disk (photosphere) and distinguish three separate regions of the sun's gaseous atmosphere. The regions in order from the photosphere are (1) a thin *reversing layer* at high temperature but still cooler than the photosphere, responsible for the Fraunhofer lines in the solar spectrum; (2) the *chromosphere,* the upper, low-pressure portion of the solar atmosphere (the reversing layer plus the chromosphere is ordinarily considered the solar atmosphere, and it extends the order of 1.5×10^4 kilometers from the photosphere); and (3) a faintly luminous region called the *corona,* extending many solar diameters from the photosphere.

On the photosphere itself, the temperature is not uniform. There are regions of concentrated circulation and cooler temperature called *sun spots,* or hotter areas called *faculae* and *flocculi.* The hotter areas can be detected most readily only in monochromatic light; the sun spots are clearly visible when the sun is viewed through a dense photographic negative. The molten or concentrated gaseous as opposed to a solid state of the sun is indicated by the observation on the limb of the sun of eruptions of matter from the photosphere. These eruptions often lead to severe electrical and magnetic effects in the earth's atmosphere.†

Astronomical measurements show that the diameter of the sun (used synonymously with photosphere) is of the order of 1.42×10^6 kilometers (0.88×10^6 miles). The mean solar distance of the earth from the sun is 149×10^6 kilometers (92.6×10^6 miles). The earth is at the mean solar distance on or about April 3 and October 5. The orbit of the earth about the sun is an ellipse, causing the earth to be nearest the sun on January 1 (the perihelion distance) and farthest from the sun about July 1 (the aphelion distance). The solar distances are 98.3 per cent and 101.7 per cent of the mean solar distance at perihelion and aphelion respectively. The sun is at the zenith (normal to the earth's surface) over the earth's equator at the time of the equinoxes (March 21 and September 23), farthest north in the summer solstice (June 22) and farthest south in the winter solstice (December 22). In general, these are the astronomical data needed in radiation studies.[B3]

These data show that the mean angular diameter of the sun is of the order of one-half a degree, 0.545 degree to be exact. The earth

† See Chapter 10.

is so far away from the sun that the sun may be considered nearly a point source and the radiation received at the earth from the sun is parallel radiation. If the flux emitted from the sun is constant and none of this flux is absorbed in outer space, the flux incident at the top of the earth's atmosphere would vary between 0.97 $E_0 < E_0 < 1.04$ E_0. The calculation is made from Eq. 2.3, with substitutions of the maximum, mean, and minimum solar distances respectively. E_0, the flux density incident at the top of the atmosphere on a surface perpendicular to the beam at mean solar distance, is called the solar constant. Measurements of the solar constant made since 1902, principally by the Smithsonian Institution, place its value between 1.9 and 2 calories centimeters^{-2} minutes^{-1} with a generally accepted value of 1.94. Satisfactory direct measurements of the solar constant have not been made as of the date of this writing, but measurements from rockets may attain satisfactory accuracy in the near future. Until such time as a long series of accurate direct measurements can be made by rockets or other extraterrestrial satellites, it will be necessary as in the past to make measurements of solar radiation at the surface of the earth. From these data, the true solar constant can be deduced after the solar-radiation measurements have been corrected for geometry; for scattering, reflection, and absorption losses in the atmosphere; and for certain instrumental errors. During the 50 years of continuous measurements by the Smithsonian Institution and shorter periods of measurements by other investigators in the world, the technique of measurement has been refined to a high degree of precision. As errors have been discovered, much of the older information has been subsequently corrected to make the data comparable to later measurements. The invariance of the solar constant is continually questioned, especially as a result of large energy variations in certain narrow regions of the spectrum, but all the evidence to date indicates that, for the entire spectrum, the energy must lie between the limits quoted and probably very close to 1.94 calories centimeters^{-2} minutes^{-1}.

The solar spectra measured at the ground and corrected to the scattering and absorbing limit of the earth's upper atmosphere in the visible region of the spectrum were shown in Chapter 3, Fig. 3.3. The solar spectrum over a larger wavelength region is shown as Fig. 4.1 together with a theoretical curve of radiation versus wavelength for a body whose temperature is 6000° K. The agreement in the shape of the curves is so close that the studies of the emission properties of the sun, and in fact of any radiating body, are facilitated by a knowledge of the properties of an ideal radiator, called a black body. The curve of 6000° K in Fig. 4.1 is the radiation from a black body of that temperature. We have the curious situation of nomenclature that in the visible range 400 m$\mu < \lambda <$ 700 mμ a black body at 6000° K meets to a high degree of approximation the definition of a white body as given in Chapter 3.

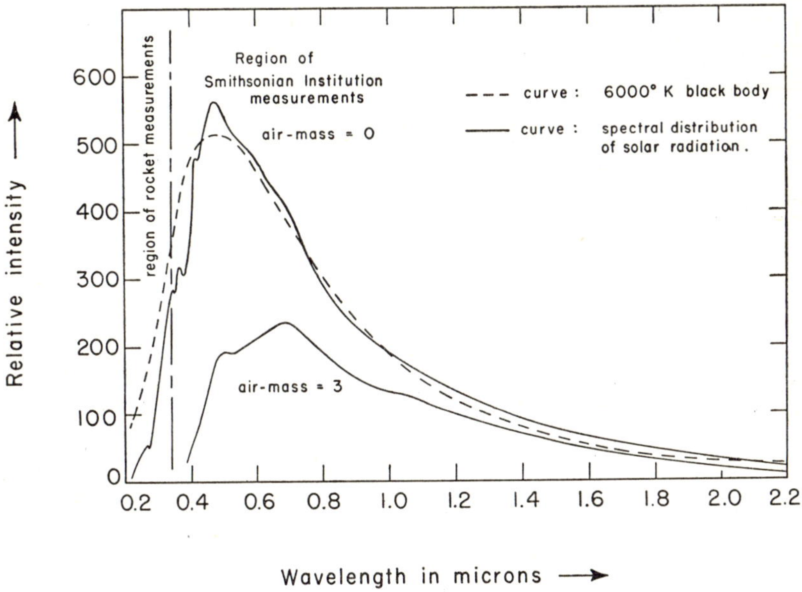

Fig. 4.1 Composite curve (air-mass = 0) of the spectral distribution of solar radiation at the top of the earth's atmosphere. The curve to the right of $\lambda =$ 0.34 micron is after Abbot, Fowle, and Aldrich,[1] whereas the curve to the left of $\lambda = 0.34\ \mu$ is from a rocket ascent to 55 kilometers (see Chapter 10). A curve of air-mass 3 corrected for absorption is also shown. Notice the shift of position of the wavelength of maximum energy because of scattering at shorter wavelengths.

Properties of Black Bodies. The development of black-body radiation in this section is based entirely on theory. No real black body exists in nature over the entire spectral region, although some solids and some liquids under high pressure approach the ideal quite closely. The explanation of the spectral distribution of black-body radiation is due to the work of Max Planck[20] in 1901. His theory tied together so many experimental facts and empirical laws of radiation that the theoretical distribution is now postulated as the flux distribution from an ideal radiator. In the derivation of the theory, it was necessary to postulate the quantization of energy in discrete units. Before

Planck's postulate, classical electromagnetic theory required a continuous distribution of energies to be emitted as a body radiated. The assumption of discrete energy levels and its subsequent power as a tool in correctly explaining phenomena for which classical theory gave an incorrect explanation, is the basis of the old quantum theory, now included in the newer (1925) wave mechanics.

We shall omit the derivation of the black-body spectral curve and merely postulate it. For parallel radiation, the flux of *specific intensity* (brightness) of a black-body emitter is

$$B^*(\lambda, T) = \frac{c_1 \lambda^{-5}}{e^{c_2/\lambda T} - 1} \cdot \quad \left[\frac{\text{watts}}{\text{steradian cm}^3}\right] \tag{4.1}$$

The constants in *Planck's law* are

$$c_1 = 1.177 \times 10^{-12} \text{ cm}^2 \text{ watts/steradians}$$

$$c_2 = 1.432 \text{ cm °K.}$$

λ is wavelength in centimeters and T is in degrees Kelvin. Curves of $B^*(\lambda, T)$ versus λ for constant temperatures are shown in Fig. 4.2.

Equation 4.1 may be simplified into two asymptotic expressions, both of which were derived originally from classical physics. For small values of λT such that $x = c_2/\lambda T > 50$, Eq. 4.1 can be expressed to an accuracy greater than 1 per cent, as

$$B^*(\lambda, T) = c_1 \lambda^{-5} e^{-c_2/\lambda T} \tag{4.2}$$

because $e^{c_2/\lambda T} - 1 \to e^{c_2/\lambda T}$. Equation 4.2 is known as the *Wien radiation law.*

For large values of λT, Eq. 4.1 can be reduced to

$$B^*(\lambda, T) = \frac{c_1}{c_2} \lambda^{-4} T \tag{4.3}$$

giving an accuracy of > 1 per cent only if $x = c_2/\lambda T < 0.019$. Equation 4.3 is an approximation to Eq. 4.1, because a series expansion of e^x is equal to $1 + x$ when x is sufficiently small. Equation 4.3 is called the *Rayleigh-Jeans law* of radiation.

From simple calculus considerations, the total radiation and the wavelength of maximum emission for a black body at constant temperature can be obtained through operating on Eq. 4.1. The calculus indicates that the total radiation is the area under a plot of $B^*(\lambda, T)$ versus λ at constant T; i.e.,

$$B^*(T) = \int_0^\infty B^*(\lambda, T)\, d\lambda = c_1 \int_0^\infty \frac{\lambda^{-5}\, d\lambda}{e^{c_2/\lambda T} - 1} \cdot \tag{4.4}$$

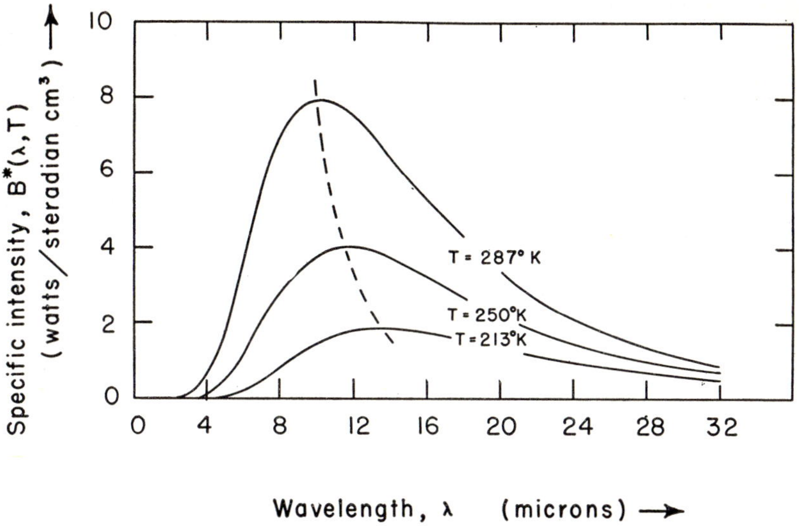

Fig. 4.2 Black-body curves for temperatures of 14° C, −23° C, and −60° C. These temperatures correspond to the mean temperature of the earth, the planetary temperature, and a characteristic temperature of the tropopause. The dashed line gives the locus of the maxima of the curves. The area under each curve is equal to $(10^4/\pi)\sigma T^4$ watts/steradian centimeter². Multiplication by π gives the radiant energy density emitted by a black body in units of watts/centimeter².

The integration of Eq. 4.4 is best accomplished through making a change of variable from wavelength λ to frequency ν through the well-known relation $\lambda = c/\nu$, c being the speed of light. Differentiation of λ gives $d\lambda = -(c/\nu^2)\,d\nu$. The limits of integration of Eq. 4.4 change to ∞ and 0 from 0 and ∞. Writing Eq. 4.4 in terms of frequencies makes

$$B^*(T) = \frac{c_1}{c^4}\int_0^\infty \frac{\nu^3\,d\nu}{e^{\frac{c_2\nu}{cT}} - 1}. \tag{4.5}$$

Remembering that T is a constant during this integration, we may put Eq. 4.5 into a form suitable for integration by letting $y = (c_2/cT)\nu$. Then

$$B^*(T) = \frac{c_1}{c_2^{\,4}}\,T^4 \int_0^\infty \frac{y^3\,dy}{e^y - 1}. \tag{4.6}$$

The integral of Eq. 4.6 may be integrated by parts. A series of terms in y results that converges to a value of $\pi^4/15$ when limits are substituted. If we let $b = (\pi^4/15)\ (c_1/c_2^{\,4})$, a lumped constant, we find from Eq. 4.6 that

$$B^*(T) = bT^4. \tag{4.7}$$

This last equation is the Stefan-Boltzmann law for parallel radiation. The value of $b = 1.826 \times 10^{-12} \dfrac{\text{watts cm}^{-2}\,{}^\circ\text{K}^{-4}}{\text{steradians}}$. Most black bodies like the sun are extensive sources and hence emit diffuse radiation. The Stefan-Boltzmann law for diffuse radiation gives the relationship between the flux density of the emitter and the brightness of the emitter. Equation 4.34*a* shows that for a body of uniform brightness, following Lambert's cosine law,

$$E = \pi B^*(T) = \pi b T^4 = \sigma T^4 \tag{4.8}$$

where $\sigma = \pi b = 5.735 \times 10^{-12}$ watt centimeter^{-2} $^\circ$K^{-4} $= 8.22 \times 10^{-11}$ langley minute^{-1}. σ is called the Stefan-Boltzmann constant. One langley (ly) equals 1 calorie centimeter^{-2} of flux.

Differential calculus indicates that the wavelength of the maximum energy of the radiation curve follows from the condition that at constant T, the rate of change of $B^*(\lambda, T)$ with λ at the maximum point is

$$\left[\frac{\partial B^*(\lambda, T)}{\partial \lambda}\right]_T = 0. \tag{4.9}$$

Differentiating Eq. 4.1 with respect to λ can be accomplished most easily by taking natural logarithms and then differentiating. These two operations give

$$\frac{1}{B^*(\lambda, T)} \frac{dB^*(\lambda, T)}{d\lambda} = 0 = \frac{-5}{\lambda} + \frac{c_2 e^{c_2/\lambda T}}{\lambda^2 T (e^{c_2/\lambda T} - 1)} \quad \text{[for constant } T\text{]}. \tag{4.10}$$

Equation 4.10 can be simplified to

$$\frac{5 - (5 - c_2/\lambda T)e^{c_2/\lambda T}}{\lambda(e^{c_2/\lambda T} - 1)} = 0. \tag{4.11}$$

Equation 4.11 is true, provided that the denominator remains finite and not equal to zero, if the numerator is set equal to zero. With this condition,

$$5e^{-c_2/\lambda T} = 5 - \frac{c_2}{\lambda T}. \tag{4.12}$$

Equation 4.12 is a transcendental equation that is satisfied only if the dimensionless ratio

$$\frac{c_2}{\lambda T} = 4.965. \tag{4.13}$$

Substituting for c_2 its value and solving for λT give the *Wien displacement law,*

$$\lambda_{max} T = 2884 \text{ [microns °K]} \tag{4.14}$$

provided that λ_{max} is in microns and T in degrees absolute. Figure 4.3 illustrates Eqs. 4.1, 4.7, and 4.14.

From measurements of the solar spectrum corrected to the top of the earth's atmosphere, we now have several means of estimating the temperature of the photosphere of the sun.

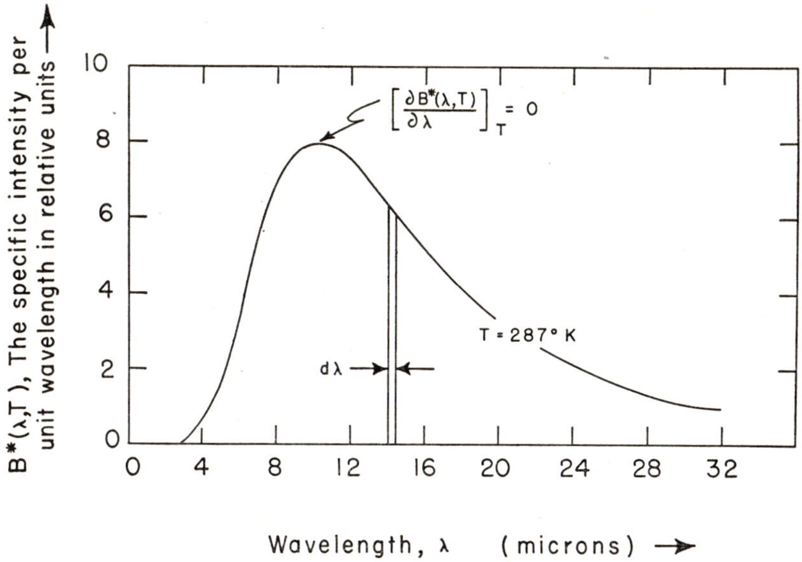

Fig. 4.3 The black-body curve is a plot of Eq. 4.1 for $T = 287°$ K. The maximum point is found from the condition set by Eq. 4.9, which leads to Eq. 4.14. The area under the curve is the graphical expression of Eq. 4.7. Units of the ordinate are watts/steradian centimeter³.

1. From Fig. 4.1 we find that the wavelength of maximum specific intensity is $\lambda_{max} = 0.474$ micron. Substituting of λ_{max} in Eq. 4.14 makes $T = 6080°$ K. A temperature found in this manner is called the color temperature of the sun.

2. The temperature of the sun can be estimated from measurements of the solar constant E_0. If a is the diameter of the sun, and D the mean solar distance, the black-body temperature of the sun is

$$4\pi D^2 E_0 = 4\pi \left(\frac{a}{2}\right)^2 \sigma T^4 \tag{4.15}$$

from the condition that in the absence of absorption, the same total flux that leaves the sun must pass through a spherical surface of radius D concentric with the sun. Putting in the values of σ, D, E_0, and a given earlier in the chapter makes $T = 5700°$ K. Temperatures arrived at by means of Eq. 4.15 are called radiative temperatures.

3. A third method of arriving at solar temperatures is to match the spectral curve of the measured radiation with Eq. 4.1. This procedure usually can be accomplished only in a relatively narrow spectral region. By this method, the sun's temperature is equivalent to a black body of 6000° K in the wavelength region of 0.45 to 2 microns, about 7000° K in the infrared region below 24 microns, and of the order of 10^{6}° K in the microwave region. In the ultraviolet, the black-body temperature of the sun is of the order of 4000 to 5000° K down to at least 0.22 micron. For shorter wavelengths, < 0.10 micron, the black-body temperature may be very high again, possibly to 10^{6}° K. Whatever these temperatures may be outside the region 0.34 micron $< \lambda <$ 2.5 microns, there will probably be a variation of not more than 1 or 2 per cent in the solar constant.

Equation 4.1 is also used in a relative way to specify temperatures. Thus, if the ratio of the relative brightness of several wavelengths in a spectral energy curve of a radiator, filter, or reflector is the same over a selected region (usually in the visible) as the brightness ratio of the same wavelengths for a black body of a certain temperature, the radiator, filter, or reflector is characterized as having this black-body temperature. This method of nomenclature is used for colored filters for cameras so that all colors will have the proper relative exposure, and is also the basis of naming the ICI standard lamps. The sky can also be given a color temperature that is extremely high because of its characteristic blueness. For example, a clear blue sky may have a color temperature of 20,000° K; a blue sky with thin white clouds a temperature of 12,000° K; and only an overcast sky will have the sun's relative spectral distribution of color, namely 6000° K. This method is satisfactory for specifying the ratio of the brightness of the blue to the red in any light source.

4. For gases and vapors, the use of the black-body temperature is not so obvious. Gases have emission or absorption only in narrow regions of the spectrum, called lines or bands, depending on the extent of the spectral region. The specific intensity of the lines is governed by the temperature of the gas, so that no line can emit energy greater than its black-body temperature at that wavelength. With metallic vapors, the black-body temperature can be extremely high. Gases in the corona of the sun have been estimated to be of the order of 10^{6}° K in the extremely narrow regions of the spectrum at which they emit.

Kirchhoff's Law. Before we make a detailed study of the radiation from the sun, earth, atmosphere, and clouds, it might be worth while to develop a few simple concepts on the transfer of radiation from one body to another. In particular, we shall be interested in the process of absorption of radiation by the earth and atmosphere. We have intimated before that energy incident on a surface is either absorbed, reflected (a general term which for the purposes of Eq. 4.16 includes scattering), or transmitted. By conservation principles, the total flux F_0 incident on a surface is distributed as

$$F_0 = F_a + F_{\mathcal{R}} + F_\tau \tag{4.16}$$

where F_a is the flux absorbed, $F_{\mathcal{R}}$ is the flux reflected, and F_τ is the transmitted flux. Dividing through by F_0 to obtain ratios, we obtain terms called the absorptivity, reflectivity, and transmissivity respectively; i.e.,

$$1 = \mathcal{A} + \mathcal{R} + \tau \tag{4.17}$$

where the order of terms in Eq. 4.17 is the same as in Eq. 4.16.

During a process of radiation transfer, it is the absorbed flux that goes into the internal energy of the absorbing medium. The internal energy increase is indicated principally as a temperature rise of the medium, although some pressure and volume changes may occur. After an interval, even in the absence of convection or conduction, no further changes in internal energy are noted even though the incident, reflected, and transmitted flux may remain virtually unchanged. The only explanation that makes sense is that radiative equilibrium has been attained, wherein as much radiation is being emitted as is being absorbed. However, the emitted radiation must not be of the same wavelength as the incident, reflected, or transmitted radiation; otherwise our measuring instruments would assign the emitted radiation to either the reflected or transmitted radiation. The emitted radiation is in thermal equilibrium with the source and is emitted at the temperature of the absorbing medium.

With these facts in mind, we define a quantity called the emissivity symbolized by ϵ. The emissivity is the ratio of the flux emitted by a body to the flux emitted by a black body at the same temperature. At radiative equilibrium, the flux absorbed and the flux emitted by the same body must be equal, as no temperature change occurs. This equivalence† in magnitude of emission and absorption,

$$\epsilon = \mathcal{A}, \tag{4.18}$$

† The only way Eq. 4.18 can be satisfied wavelength for wavelength by any real body is to have this body completely surrounded by the walls of a cavity that are maintained at a uniform temperature.

is called *Kirchhoff's law.* For a true black body, $\mathcal{R} = \tau = 0$ in Eq. 4.17, so that on combining Eqs. 4.17 and 4.18,

$$\epsilon = \mathcal{A} = 1. \tag{4.19}$$

Equations 4.16 through 4.19 are true in general only if $\mathcal{A}$, $\mathcal{R}$, τ, and ϵ are measured at the same wavelength. When $\mathcal{A}$ and ϵ are considered independent of wavelength, so that Eq. 4.8 may be expressed as $E = \epsilon\sigma T^4$, Eq. 4.8 defines a gray body.

Sky Brightness. In the study of visibility, we found that a fundamental concept in the development of the theory involved the comparison of the brightnesses of two adjacent media, an object and its background. In many practical cases, the background was a sky of brightness B_H. Many times it was not necessary to know the absolute value of the brightness, because the equations were written in such a fashion as to demand a knowledge of a brightness ratio only. In radiation studies, where the total flux from the sun, called the insolation, on a surface is desired, it is necessary to know the absolute brightness of the sky and sun in order to compute the flux. As the problem of the transfer of short-wave radiation by scattering which in turn leads to a theory of sky brightness is in many important respects formally analogous to the more important problem of the transfer of long-wave radiation by absorption and re-emission, a brief study of the theory of sky brightness will be undertaken.

As it is only the number and not the distribution of light-scattering molecules that enters into studies of sky brightness, the concept of the homogeneous atmosphere may advantageously be used again. The problem resolves itself into finding the brightness of the sky along a fixed direction in the sky, which brightness is brought about by the summation of the scattered light from the molecules in each volume of air along the line of sight.

Let us then consider a volume of light-scattering molecules located along a zenith angle θ and at a level in the atmosphere where the pressure is p. For the geometry see Fig. 4.4. Sunlight of brightness $B^*(\lambda)$ for a chosen wavelength originates at $p = 0$ (outer space), and is attenuated in its passage through the atmosphere to a particular scattering volume located at the level where the pressure is p. At this level, flux is scattered toward the eye, making an angle of $\pi - \phi$ between the direction of the direct sunlight and the scattered flux. Up to this point, we can say mathematically the following.

At a fixed wavelength, the optical path u_λ, extending from $p = 0$ to $p = p$, from the geometry is

$$u_\lambda = k_s\,(z_\infty - z) \sec \zeta \tag{4.20}$$

where $z_\infty - z$ is the vertical distance from outer space at z_∞, to the scattering level at z. $\zeta = (\pi/2) - \alpha$, where α is the elevation angle of the sun in radians. k_s is the total scattering coefficient given by Eq. 2.33 and u_λ has the same concept as in Chapter 3, Eq. 3.15.

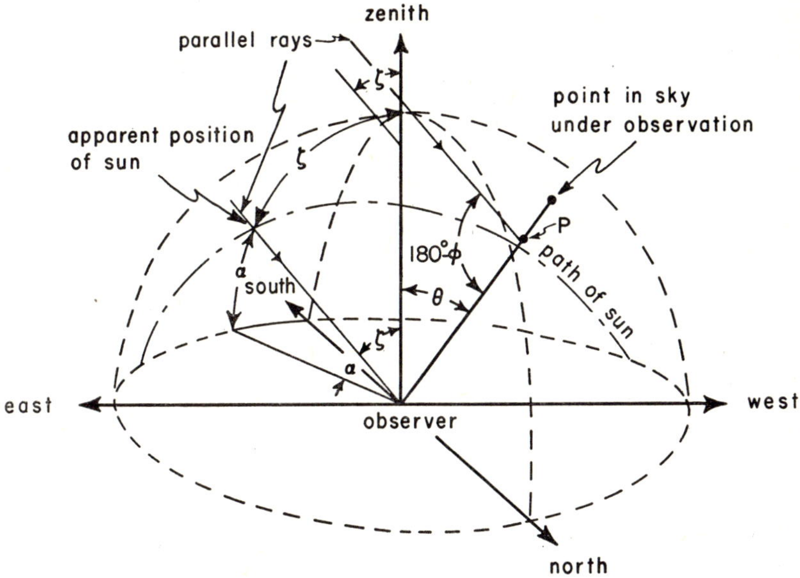

Fig. 4.4 The geometry of primary scattering used for determining the brightness of the sky. A particular ray from the sun is scattered from point P to an observer at the ground, making an angle $180° - \phi$ between the direct and scattered ray.

For a homogeneous atmosphere extending from p to 0,

$$z_\infty - z = \frac{p}{\rho_0 g} = \frac{p}{p_0}\frac{R_G}{g}T_0 = \frac{T_0}{\gamma_A p_0}p. \qquad (4.21)$$

Equation 4.21 was obtained with the help of definitions used for Eqs. 1.9 and 1.10. The null subscripts refer to ground level. $\gamma_A = 34.1°$ C kilometers^{-1}.

Therefore, because of attenuation of the flux by the atmosphere in traveling from outside the atmosphere to the level wherein the scattering volume under analysis is located, the apparent brightness of the sun viewed from this level is

$$B_p(\lambda) = B^*(\lambda)e^{-u_\lambda} \qquad (4.22)$$

where u_λ is given by Eq. 4.20. Equation 4.20 can be made a function of pressure through the use of Eq. 4.21, and a function of wavelength through the use of Eq. 2.33. With these substitutions,

$$u_\lambda = \frac{32}{3n_0}\frac{\pi^3}{\lambda^4}(m_0 - 1)^2_{\text{gas}}\,\gamma_A \frac{T_0}{p_0}\, p \sec \zeta = \beta\lambda^{-4}\, p \sec \zeta, \quad (4.23)$$

if we set $\beta = \dfrac{32}{3n_0}\pi^3(m_0 - 1)^2_{\text{gas}}\,\gamma_A \dfrac{T_0}{p_0}$, a constant.

A certain fraction of the flux, represented by the specific intensity $B(\lambda)$, is scattered toward the eye. If we consider only Rayleigh scattering, the amount of flux that is scattered in the direction of the eye is the angular scattering function multiplied by $B(\lambda)$. The angular scattering function is just Eq. 2.27, written for a gas through the use of Eq. 2.32. Thus, the flux scattered toward the eye is proportional to

$$\frac{2\pi^2}{n_0}(m_0 - 1)^2_{\text{gas}} \frac{(1 + \cos^2\phi)}{\lambda^4} B_p(\lambda) = \Omega \frac{(1 + \cos^2\phi)}{\lambda^4} B_p(\lambda) \quad (4.24)$$

if we let $\Omega = \dfrac{2\pi^2}{n_0}(m_0 - 1)^2_{\text{gas}}$, a constant. The flux starting toward the eye is attenuated by an amount proportional to $e^{-\beta\lambda^{-4}\sec\theta\,(p_0 - p)}$, on the basis of the same reasoning as was used to obtain Eqs. 4.22 and 4.23. Now, however, the observer is situated at the ground looking in a direction θ toward a spot in the sky of brightness $B_p(\lambda)$. Moreover, there is not just one scattering volume in the line of sight, but a continuum of scattering volumes each giving a contribution $\Omega\lambda^{-4}(1 + \cos^2\phi)B_p(\lambda)e^{-\beta\lambda^{-4}\sec\theta\,(p_0 - p)}$. Thus the apparent brightness of the sky in the direction θ from an observer on the ground is

$$B(\lambda) = \int_{p_0}^{0} \Omega\lambda^{-4}(1 + \cos^2\phi)B_p(\lambda)e^{-\beta\lambda^{-4}\sec\theta\,(p_0 - p)} \times d\,[\beta\lambda^{-4}\sec\theta\,(p_0 - p)]. \quad (4.25)$$

Equation 4.25 expresses the same philosophy used in developing Eqs. 3.15 and 3.16, namely that for a gas that scatters (or radiates) energy

$$\text{Apparent brightness} = \int_{\text{optical path}} \text{intrinsic brightness} \times e^{-\text{optical path}}\, d\,(\text{optical path}) \quad (4.26)$$

where $e^{-\text{optical path}}$ is the transmissivity.

$B_p(\lambda)$ in Eq. 4.25 can be found from Eq. 4.22 and is

$$B_p(\lambda) = B^*(\lambda)e^{-\beta\lambda^{-4}\,p \sec \zeta}. \quad (4.27)$$

We also find that

$$d[\beta\lambda^{-4}\sec\theta\,(p_0 - p)] = -\beta\lambda^{-4}\sec\theta\, dp. \quad (4.28)$$

When Eqs. 4.28 and 4.27 are substituted into 4.25, this latter equation for the apparent brightness of the sky at one wavelength becomes

$$B(\lambda) = \Omega\beta\lambda^{-8}(1 + \cos^2\phi)\sec\theta\, B^*(\lambda)$$
$$\times \int_0^{p_0} e^{-\beta\lambda^{-4} p \sec\zeta}\, e^{-\beta\lambda^{-4}\sec\theta\,(p_0 - p)}\, dp. \qquad (4.29)$$

The total apparent brightness of the sky as perceived by the eye is

$$B = \int_0^\infty B(\lambda)\psi_\lambda\, d\lambda \qquad (4.30)$$

from the discussion of colorimetry in Chapter 3.

The integration of equations corresponding to Eqs. 4.29 and 4.30 has been carried out by Hulburt.(28) He finds that

$$B = \bar{B}^*(1 + \cos^2\phi)\,\frac{e^{-\bar{\beta}p_0\sec\zeta} - e^{-\bar{\beta}p_0\sec\theta}}{1 - \sec\zeta\cos\theta} \qquad (4.31)$$

under the assumptions that an average brightness of the sun $\bar{B}^*$ over all wavelengths given by

$$\bar{B}^* = \frac{\int_0^\infty B^*(\lambda)\psi_\lambda\, d\lambda}{\int_0^\infty \psi_\lambda\, d\lambda} \qquad (4.32)$$

can be used and an average scattering coefficient $\bar{\beta}$ over all wavelengths in the visible can also be used. $\bar{\beta}$ is given by

$$\bar{\beta} = \frac{\beta\int_0^\infty \lambda^{-4}\psi_\lambda\, d\lambda}{\int_0^\infty \psi_\lambda\, d\lambda}\,. \qquad (4.33)$$

Some results from this somewhat lengthy development are tabulated in Table 4.1. The values from the theory are realistic up to $\zeta = \theta < 80°$, see Table 4.2, after which the assumption of a flat earth, inherent in the derivation, becomes increasingly unrealistic. It is interesting to note that in all the problems treated in this and previous chapters the curvature of the earth does not become important in problems involving extremely long path lengths until the zenith angle approaches 70 to 80 degrees.

Although the methods of this section have shown that theory can be utilized to predict the distribution of sky brightness, and hence the total scattered flux from the sky, theory plays its most important part in guiding the number and type of direct measurements of insolation that must be made. First of all, theory shows that as the flux falling on a unit horizontal area from the entire hemispherical sky is given by

$$E = 2\pi \int_0^{\frac{\pi}{2}} B \cos\theta \sin\theta \, d\theta \qquad (4.34)$$

in agreement with Eq. 3.5, B cannot be assumed to be a constant over the entire sky if Eq. 4.31 has any validity. The possible nonuni-

Table 4.1

Sky-Brightness Measurements at 38,000 Feet above Sea Level

Theoretical values from Eq. 4.31 by Hulbert. Data from Naval Research Laboratory Report 3713, by D. M. Packer and C. Lock, Washington, July 31, 1950.

Solar Altitude $\alpha = \frac{\pi}{2} - \zeta$	Azimuth from Sun	Zenith Angle, θ				
		Measured 75°	Theoretical 75°	Measured 15°	Theoretical 15°	
20°	0°	162	112	—	48	
	45°	150	107	76	46	
	90°	116	97	60	42	
	135°	105	93	49	39	
	180°	105	92	47	38	units candles/ft²
60°	0°	—	—	32	30	
	45°	180	122	35	31	
	90°	108	94	29	28	
	135°	135	98	25	25	
	180°	145	108	24	24	

formity of B was stressed in visibility measurements and is verified by theory and measurement (see Table 4.2). It is only when B is a constant that

$$E = 2\pi B \int_0^{\frac{\pi}{2}} \cos\theta \sin\theta \, d\theta = \pi B. \qquad (4.34a)$$

E is the flux impinging on a surface from a diffuse source of infinite extent, a condition *not* met when the brightness of the entire unclouded sky is considered.

Table 4.2

Averages of Sky-Brightness Measurements Expressed in Terms of Zenith Brightness for Clear Skies in Summer at Washington, D.C.

Table adapted from Table 2 in H. H. Kimball and I. F. Hand, *Monthly Weather Review*, *50*, 615, 1922.

Solar Altitude $\alpha = \frac{\pi}{2} - \zeta$	Azimuth from Sun	Zenith Angle, θ							Zenith Brightness, milli-lamberts
		88°	75°	60°	45°	30°	15°	0°	
20°	0°	21.10	—	10.85	4.06	2.34	1.40	1.00	400
	45°	7.72	5.90	3.91	2.74	1.94	1.31		
	90°	3.42	2.81	1.78	1.35	1.07	1.14		
	135°	2.48	1.86	1.23	0.83	0.72	0.82		
	180°	2.85	2.13	1.19	0.85	0.74	0.76		
40°	0°	7.35	6.19	8.43	—	2.75	1.58	1.00	803
	45°	4.15	3.13	2.69	2.45	1.88	1.42		
	90°	2.13	1.41	1.13	0.97	0.99	1.04		
	135°	1.74	1.13	0.74	0.60	0.62	0.74		
	180°	1.83	1.11	0.68	0.54	0.53	0.68		
60°	0°	2.08	1.74	1.84	2.51	—	1.65	1.00	1650
	45°	1.74	1.53	1.30	1.53	1.67	1.41		
	90°	1.16	0.88	0.72	0.74	0.88	0.99		
	135°	0.95	0.61	0.47	0.50	0.58	0.75		
	180°	1.00	0.64	0.47	0.50	0.54	0.72		
70°	0°	1.45	1.29	1.13	1.60	3.30	—	1.00	2300
	45°	1.25	0.90	0.89	1.07	1.33	1.53		
	90°	0.77	0.65	0.56	0.54	0.62	0.89		
	135°	0.62	0.42	0.37	0.38	0.44	0.66		
	180°	0.58	0.39	0.31	0.31	0.41	0.61		

Measurements of solar energy

In solar energy measurements, three primary instruments are used: the pyrheliometer, the pyranometer, and the spectrobolometer.

The pyrheliometer† is the most important single instrument of the three. This instrument is a carefully constructed differential bolometer or thermopile, evacuated and protected from convective and conductive heat losses. The heart of this instrument is a carefully

† The instrument described is of the type most likely to be encountered by the student. It is not an "absolute" instrument in the sense of the Smithsonian silver disk pyrheliometer. This latter instrument, used in solar constant work, measures radiant energy directly by a calorimetric technique.

measured surface usually placed parallel to the earth's surface, and blackened so as to absorb as much of the energy from the sun's rays as possible. Surrounding this absorber of radiation is another disk with a highly reflecting coating for the sun's rays. The absorbing area

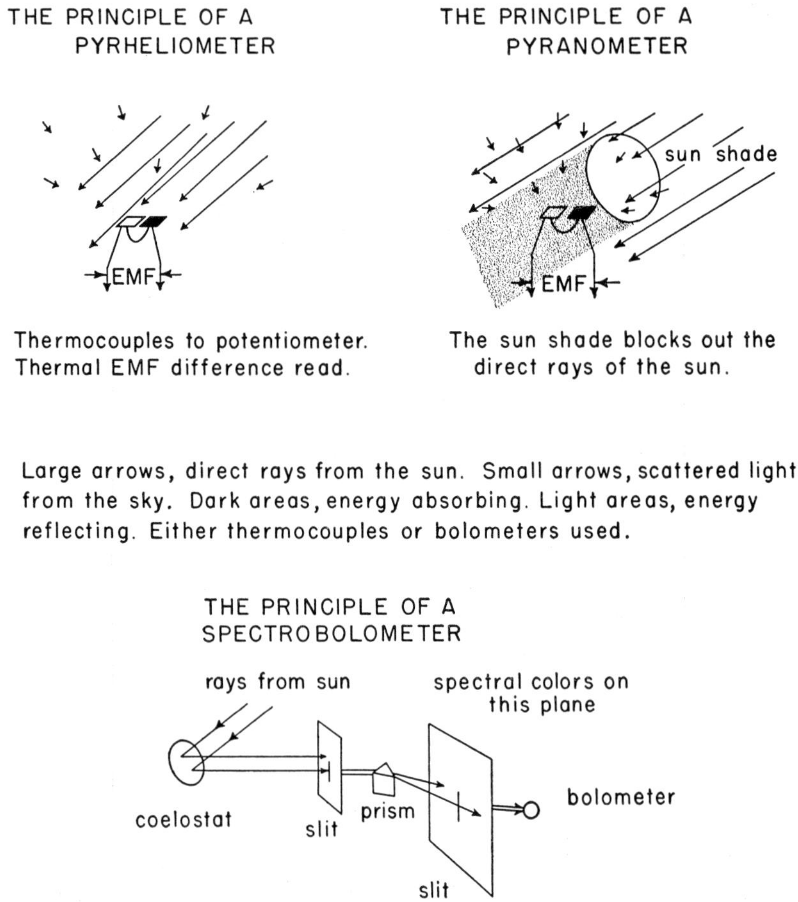

Fig. 4.5 Schematic drawings of radiation instruments. In a practical spectrobolometer there are lenses in the system to collect a large amount of energy and focus this energy on the slits and on the bolometer. The coelostat also has a focusing mirror.

heats up while the reflecting area remains at the initial temperature. A knowledge of the thermal mass of the components and the differential heating of the surfaces gives a measure of the total flux from the direct sunlight plus the scattered sunlight from the sky falling on the blackened target. In principle, the pyrheliometer described can be calibrated against an absolute instrument for flux in absolute units.

In solar constant work, only the direct sunlight is wanted. For this a second instrument called the pyranometer is used in conjunction with a pyrheliometer oriented normal to the solar rays. The pyranometer measures only the flux scattered by the sky on a unit area

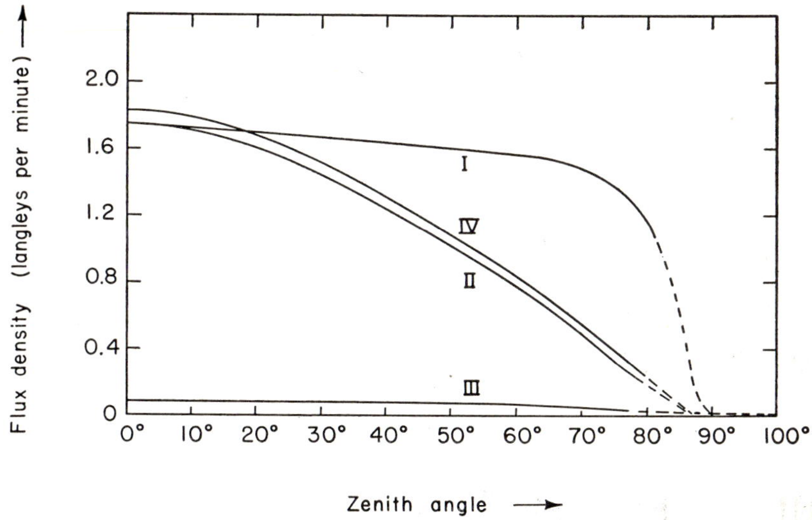

Fig. 4.6 (*After L. W. King.*) Total solar and sky radiation as measured at Mt. Wilson Observatory. Curve I approximates the solar constant.

Curve I. Total intensity of solar radiation normal to sun's rays.
Curve II. Total intensity of solar radiation on a horizontal plane.
Curve III. Total intensity of diffuse sky radiation on a horizontal plane.
Curve IV. Total intensity of solar plus diffuse sky radiation on a horizontal plane.

normal to the sun at the earth's surface.† A suitable shield prevents the direct rays of the sun from striking the instrument, so that the diffuse sky radiation E of Eq. 4.34 is measured. The direct sunlight is given then by the flux density measured by the pyrheliometer less the flux density measured by the pyranometer. The directly measured sunlight corrected for absorption and scattering in the atmosphere represents an amount of flux that is rather closely related to the Stefan-Boltzmann law discussed earlier in the chapter.

† For general solar energy measurements, a pyranometer with a horizontal area may be constructed for use in conjunction with the pyrheliometer mounted in a horizontal position. This arrangement is useful for solar heating studies

The spectrobolometer is essentially a spectrograph in combination with a coelostat, a mirror that follows the sun, focusing its rays continuously on the entrance slit of the spectrograph. The spectrograph disperses the sun's rays into their component wavelengths by means of a prism or diffraction grating, permitting a measurement of the energy wavelength for wavelength relative to a standard wavelength. In the Smithsonian solar constant measurements, some forty standard wavelengths between 0.34 micron $< \lambda <$ 2.5 microns are measured nearly simultaneously from the record, called a bologram, of the spectrograph trace.

Figure 4.5 illustrates the principle of these various instruments, and Fig. 4.6 shows some pyrheliometric and pyranometric measurements taken as a function of the zenith angle of the sun.

The solar constant. The Smithsonian long method

There are two techniques of measuring the solar constant, the "long" and the "short" methods of the Smithsonian Institution. The long method is the more nearly fundamental, and establishes the basis for the short method. Though the latter method is more widely used at present, the long method will be described first. For convenience, we shall define a quantity called the air-mass. It is a natural parameter to use if we recall that the atmospheric attenuation of specific intensity from the sun for any wavelength is given by

$$\frac{B(\lambda)}{B^*(\lambda)} = \frac{E(\lambda)}{E_0(\lambda)} = e^{-\sigma_\lambda z \sec \theta} = a_\lambda{}^m. \tag{4.35}$$

m, called the air-mass, for zenith angles $< 70°$ is equal to sec θ. m is really the ratio of the number of attenuating particles between the sun and observer at the surface of the earth to the number of attenuating particles at zenith distance. a_λ, a function of wavelength, is equal to $e^{-\sigma_\lambda z}$. Outside tropical latitudes, an air-mass of unity is never observed at sea level because the sun never reaches the zenith even at local noon. $B^*(\lambda)$ is the intrinsic brightness of the sun and $B(\lambda)$ is the apparent brightness of the sun at the earth's surface. Because the attenuating atmosphere is so thin compared to the distance of earth from the sun, $B(\lambda)/B^*(\lambda)$ may be replaced by $E(\lambda)/E_0(\lambda)$. $E(\lambda)$ is the flux density at a particular wavelength received at the surface of the earth while $E_0(\lambda)$ is the flux density at this same wavelength at the top of the atmosphere. In other words, there is negligible divergence of the beam between the top and bottom of the atmosphere. Now, let us take a number of observations of $E(\lambda)$ during the course of the day. If we consider that $E_0(\lambda)$ does not vary and that the atmospheric conditions are unchanged during the period of observation (both restrictive assumptions, especially the latter), we may draw a graph of m versus $\ln E(\lambda)$ for each of the forty wavelengths. By taking logarithms of Eq. 4.35 we see that

$$\ln E(\lambda) = m \ln a_\lambda + \ln E_0(\lambda)\,. \tag{4.36}$$

In general, for $1.2 < m < 3$, the curve of m versus $\ln E\ (\lambda)$ is a straight line which may be extrapolated to $m = 0$. The slope of the curve is

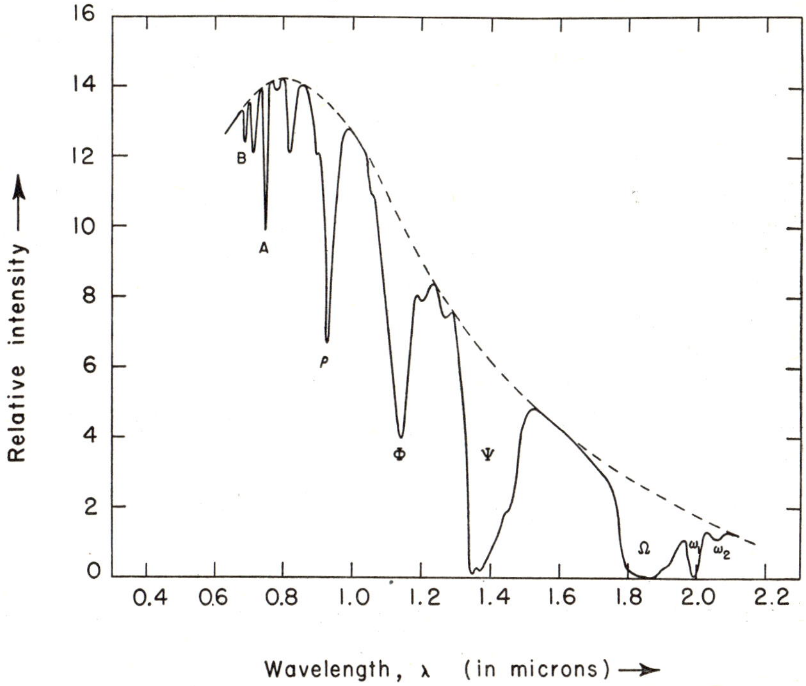

Fig. 4.7 Contour solar energy curve, adapted to a linear wavelength scale after a curve by F. E. Fowle.(10) See Fig. 14 of Ref. 10. Original measurements were made at an air-mass of 3. Assume the solid curve corrected for scattering.

$\ln a_\lambda$. At $m = 0$, $E(\lambda) = E_0(\lambda)$, and the flux density of the sun at the solar distance and in the absence of an atmosphere that scatters light is obtained. When the process described by Eq. 4.36 is repeated for each of the standard wavelengths, a curve of $E_0(\lambda)$ versus wavelength can be drawn. Such a curve would have the appearance of the dashed curve of Fig. 4.7.

Exhaustive experiments carried out in the laboratory have indicated that water vapor, carbon dioxide, and ozone are the principal absorbing gases in the atmosphere. The locations of these regions of absorption have been identified by wavelength and, as shown by Fig. 1.2, their absorption coefficients have been measured. Inspection shows that the regions of absorption in the sun's bologram (solid curve of Fig. 4.7) coincide with the absorption spectra of the known gases in the atmosphere. Therefore, the energy that was incident normal to the top of the earth's atmosphere must have been the smooth envelope given by the dashed line of Fig. 4.7 that joins the regions where the atmosphere is transparent (nonabsorbing) to the sun's rays.

Now, the smooth curve that is obtained has a marked similarity to a black-body curve given by Eq. 4.1. It only remains to convert the black-body curve to absolute units of flux. The pyrheliometer measurement corrected to direct sunlight is used for this operation.

In the development of the Stefan-Boltzmann law, it was stated that the area under the curve $E(\lambda)$ versus λ was equal to the total flux density E. The ratio of the area under the solid curve to the area under the dotted curve of Fig. 4.7 is taken to be the same as the flux in absolute units as measured by the pyrheliometer corrected to a zero air-mass is to the solar constant. A small amount of flux (of the order of 3.4 per cent of the total energy) for the regions < 0.34 micron and > 2.5 microns is added to the curves of relative flux to correct for the flux not measured in the ultraviolet (u.v.) and infrared (i.r.) regions of the spectrum. The solar constant E_0 becomes

$$E_0 = \frac{\text{Area (under dashed curve)} + (\text{u.v.} + \text{i.r. correction})}{\text{Area (under solid curve)} + (\text{u.v.} + \text{i.r. correction})} E_{\text{pyrhel.}} \tag{4.37}$$

From Eq. 4.37 we find that the magnitude of the solar constant is

$$E_0 \approx 1.94 \text{ ly minutes}^{-1}. \tag{4.38}$$

By Eq. 4.15, 1.94 ly minutes^{-1} is equivalent to a sun whose temperature is 5700° K.

It is found that a temperature for the sun of 6000° K fits the data very well over much of the measured spectrum. However, in the far infrared the temperature is about 7000° K, and the ultraviolet temperature is certainly much less than 6000° K, certainly down to $\lambda = 0.22$ micron. The magnitude of these discrepancies is a matter of concern in determining an unequivocal value for the solar constant, or the true variation of the solar constant if such a variation exists.

An objection to the long method of determining the solar constant concerns the length of time (2 to 3 hours) required to make one determination. During this time, there is no guarantee that either atmospheric or solar conditions will remain unchanged. Because of this uncertainty, the short method of measuring the solar constant was devised.

In the short method, use is made of the fact that the brightness of the sky in the neighborhood of the sun as measured by a pyranometer is primarily a function of the turbidity (number of scattering particles) of the sky. If over a long period of time, a mean brightness of sky has been determined for a given locality, an individual pyranometer reading will differ from the mean by an amount ϵ, called the pyranometer excess. Attenuation of the sun's energy by water vapor as well as scattering occurs. If we define a quantity called the total precipitable water,† denoted by w, an empirical relationship that is proportional to the total attenuation in the atmosphere by scattering and absorption can be found. The expression is of the form

$$F = w + Q\epsilon. \tag{4.39}$$

Q is a constant empirically determined. Thus at a given time, a bologram of the sun can be made, an observation of sky brightness taken by the pyranometer, and the position of the sun determined by a theodolite. These three measurements take 10 to 15 minutes.

From the measurement of the transmission in the ρ band of water vapor, the precipitable water can be found. The observation and substantiation of the relationship were done by Fowle, whose curves are reproduced as Fig. 4.8. The pyranometer excess ϵ is determined from the pyranometer reading. With Q known for each wavelength, the value of F for each wavelength can be found.

From a long series of observations of F, m, and a_λ previously made at the particular locality where the solar constant measurement is made, a graph of F versus the air-mass m for various values of a_λ has been constructed. For a particular measurement of F at a given wavelength through a known air-mass m, the graph can be entered to give the corresponding value of a . This value of a_λ when substituted with m into Eq. 4.35 yields the numerical ratio $E(\lambda)/E_0(\lambda)$. $E(\lambda)$ is determined from the calibrated bolometer trace which is the additional piece of information necessary to determine $E_0(\lambda)$. Once $E_0(\lambda)$ has been determined for the standard wavelengths, the evaluation

† The total precipitable water in the atmosphere is simply the height to which the vapor in a column of uniform cross section extending from the surface to outer space would extend if the vapor were all condensed to liquid water. For a column normal to the earth's surface, the total precipitable water lies between $1 \text{ cm} < w < 3 \text{ cm}$.

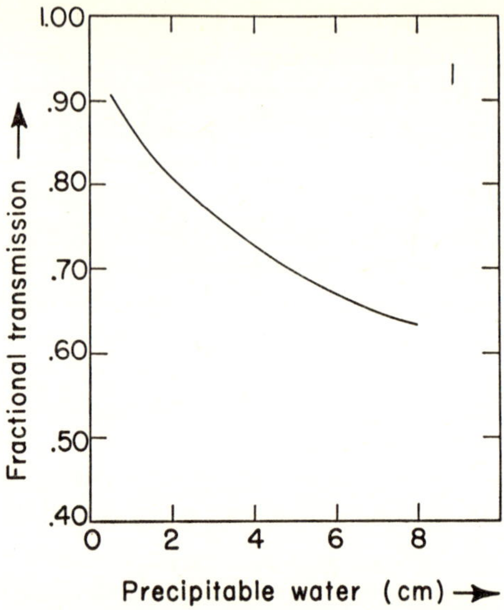

Fig. 4.8 (*After Fowle.*) Fractional transmission of solar energy by the ρ band of water vapor, centered at about 0.92 micron.

of the solar constant proceeds in the same manner as in the "long method."

The determination of long-wave radiation in the atmosphere

We shall defer our discussion of the ways in which the short-wave radiation enters into the heat budget of the earth until the following chapter. For the remainder of this chapter, we shall concern ourselves with the philosophy of measurement of the transference of radiation from a series of low-temperature sources, both gases and solids. Then, in the next chapter, we can consider the equilibrium between high-temperature radiation from the sun to the earth and atmosphere with the low-temperature radiation emitted back to space from our earth and atmosphere.

The study of low-temperature radiation is called the study of long-wave radiation (a strictly relative term). The terms long- and short-wave radiation arose because by Wien's displacement law (Eq. 4.14), the wavelength of maximum emission at the planetary temperature† is 11.5 microns, whereas the wavelength of maximum radiation from the sun has been measured at 0.474 micron. The planetary temperature just mentioned is an equilibrium temperature that the earth and

† To be derived.

atmosphere system must attain with the solar energy from the sun in order to keep this planet's internal energy constant. That such an equilibrium condition exists is attested by the fact that there has been no significant change in the mean temperature of the earth and atmosphere over recorded time. If A is the mean albedo of the earth, the amount of flux absorbed from the sun at a given instant by the earth is $(1 - A)E_0 \cdot \pi R^2$. E_0 is the solar constant and πR^2 is the cross-sectional area of the earth intercepting the sun's rays. The earth acts as a diffuse emitter, radiating in the mean a flux per unit area of σT^4 from the *entire* surface of the earth, equal to $4\pi R^2$ units of area. Thus, at equilibrium

$$\pi R^2(1 - A)E_0 = 4\pi R^2 \sigma T^4 \tag{4.40}$$

so that

$$T = \left[\frac{(1 - A)E_0}{4\sigma}\right]^{1/4}. \tag{4.41}$$

Taking $E_0 = 1.94$ langleys minute^{-1} and $A = 0.34$ makes

$$T = 250° \text{ K}(-23° \text{ C}).$$

Other estimations of the mean albedo of the earth range from $0.34 < A < 0.45$. No combination of the values of E_0 and A will affect the validity of the argument for the relative positions of the wavelengths of maximum emission of flux in the electromagnetic spectrum from the earth and sun.

The planetary temperature gives an insight into the problem of long-wave radiation. The planetary temperature is a mean or equilibrium temperature, so that one might expect to compare the mean temperature of the earth with the planetary temperature. Now, although the surface temperature of the earth may vary from reasonable extremes of −40° C in polar latitudes to 40° C in deserts in tropical latitudes, the mean temperature of the earth's surface is 14.3° C, much higher than the planetary temperature of −23° C. On the other hand, the planetary temperature of the earth is much greater than the coldest temperatures experienced in the atmosphere below 70 kilometers, where a mean of −60° C at the tropopause is reasonable. It appears, and theory will show, that the planetary radiation is composed of contributions of energy both from the earth's surface and lower atmosphere at higher than planetary temperatures together with less energy flux from the upper troposphere at lower than planetary temperatures. The combination of fluxes from different sites of origin emits to space, in the mean, a flux equal to an equivalent black body at the planetary temperature. Thus, there must be some radiation process acting in the earth's atmosphere that will keep the surface of the earth warmer than the upper troposphere, and thus sustain life.

Suppose we consider what happens when two plane parallel sheets of material of semi-infinite extent are illuminated by the sun (see Fig. 4.9). Let the sheet nearer the sun be a pane of glass, transparent to the sun's rays (out to about 2.5 microns), but opaque to long-wave radiation. The lower sheet can be the surface of the ground, warmed by the sun's rays until an equilibrium temperature has been reached. For true radiative equilibrium, the experiment should be done in a vacuum so that heat cannot be transported by convection of the air or radiated by the atmosphere itself.

If the glass is truly transparent to the sun's rays, there will be black-body equilibrium between the plate of glass and the earth's surface on one side and the plate of glass and outer space on the other side. Thus, at equilibrium, in this ideal case of no atmosphere, radiation in the amount $\sigma T_0{}^4$ is received on each unit area of the glass facing the earth whose temperature is T_0. No flux will be absorbed by the opposite side of the glass which faces the sun. But the glass plate at temperature T radiates $2\sigma T^4$ units of flux for each unit area facing the earth. At equilibrium

$$2\sigma T^4 = \sigma T_0{}^4 \tag{4.42}$$

or

$$T = (\tfrac{1}{2})^{1/4}\, T_0 < T_0. \tag{4.43}$$

Now, if a number of parallel glass plates are used between the earth and sun and only radiation transfer with outer space occurs, it can be shown that the temperature of the nth glass plate would be

$$T_n = \left(\frac{1}{n+1}\right)^{1/4} T_0 \tag{4.44}$$

and the energy lost to space would be

$$E = \sigma T_n{}^4 \tag{4.45}$$

for every unit area of the earth's surface. Equation 4.45 expresses the well-known fact that when we have perfect radiators (and absorbers), it is only the temperature of the last radiator that determines how much flux will be exchanged with the surroundings. Even in the presence of air, radiation effects such as have just been described are working to keep the temperature of the earth T_0 greater than the temperature of the glass plates. A greenhouse is governed by this principle, and since the atmosphere acts in a somewhat analogous manner through absorption and reradiation of long-wave flux to keep the

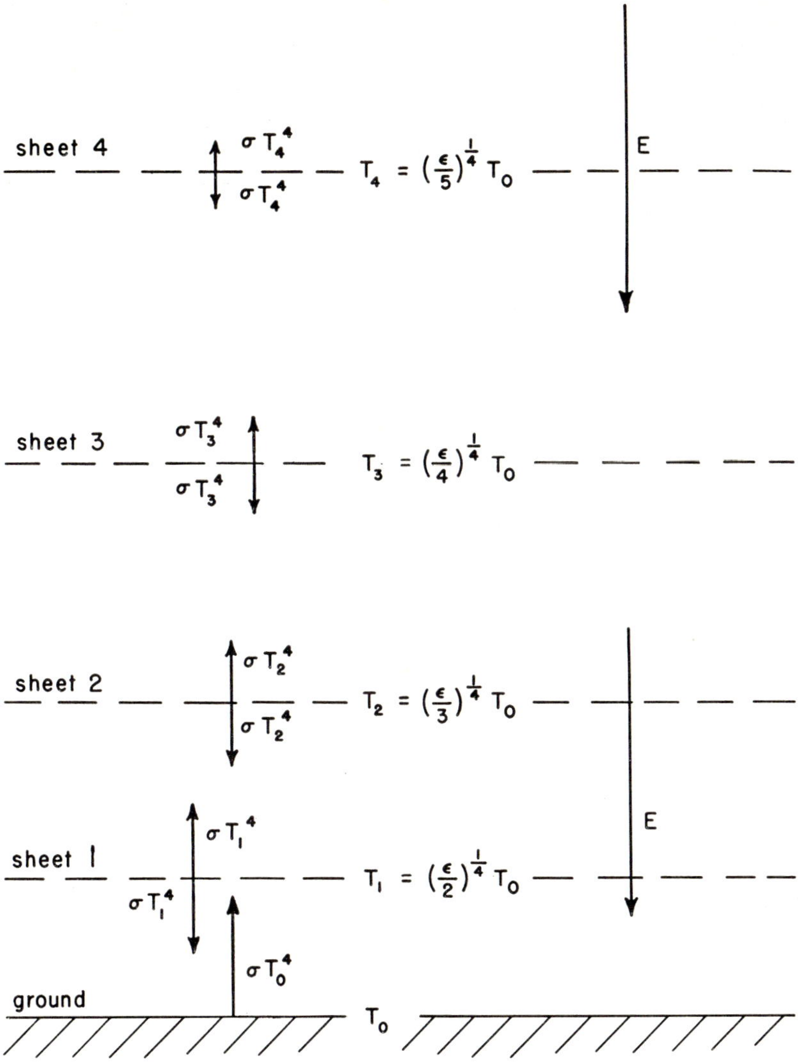

Fig. 4.9 Schematic of the "greenhouse" effect. E represents the short-wave flux from sun and sky, whereas $\sigma T_n{}^4$ represents the long-wave flux emitted by successive layers of the atmosphere acting as a gray-body radiator. In the atmosphere, E is not independent of height.

surface temperature of the earth greater than atmospheric temperatures, this effect is referred to as the "greenhouse" effect of the atmosphere.

However, if the model of the perfectly absorbing plates is taken over in toto and applied to the atmosphere, one immediate trouble presents itself; namely, the atmosphere is not a perfect black body. Measurements have shown that the water vapor and carbon dioxide in the atmosphere absorb in bands (see Fig. 4.10). At normal atmospheric temperatures, a layer of the atmosphere containing 0.3 millimeter of precipitable water absorbs and re-emits about 20 per cent of black-body radiation at the mean temperature of the layer. The

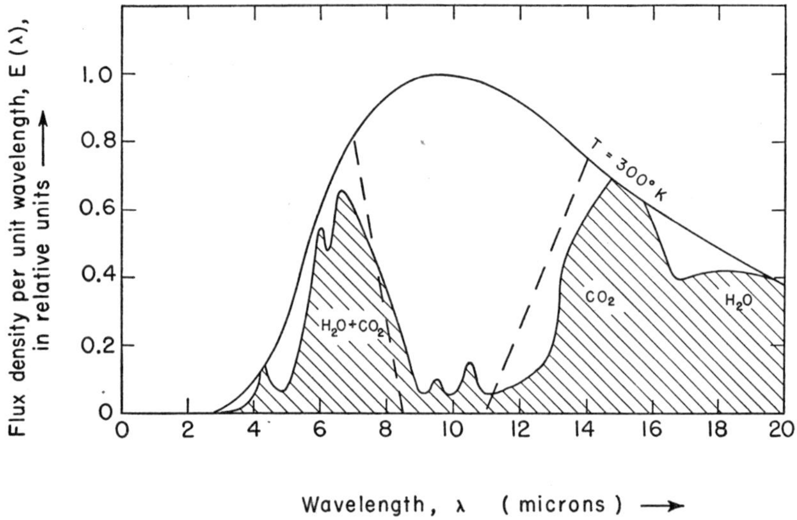

Fig. 4.10 The emission spectrum of the atmospheric gases for a thin layer of atmosphere at sea level. The shaded area is the emission spectrum, whereas the envelope marked $T = 300°$ K is that of a black-body curve at that temperature. Since the area enclosed by the gases is less than that of a black body at the same temperature, it follows that the gases absorb and emit less energy than a black body of the same temperature. The dashed lines are the limits of a simplified emission spectrum for water vapor and carbon dioxide as suggested by Simpson (see text).

optical path of 0.3 millimeter of precipitable water makes for a large number of relatively thin layers in the lower troposphere where the water vapor content is high, and a few thick layers in the upper troposphere where the water vapor content is low.

It is known that a series of shields, each having the same emissivity ϵ, one outer shield of which is illuminated by a black body while the other outer shield radiates to empty space, reaches temperatures such that the equilibrium temperature of the nth shield is

$$T_n = \left[\frac{\epsilon}{(n+1)}\right]^{1/4} T_0 \qquad (4.46)$$

with the last shield (layer in the atmosphere) radiating

$$E = \epsilon\sigma T_n^4 \tag{4.47}$$

units of flux to space for each unit area.

If we apply Eq. 4.47 to the atmosphere with $\epsilon = 0.20$ and T_n equal to the mean temperature of the last layer in the atmosphere that contains 0.3 millimeter of precipitable water, there are serious objections as to the validity of the results. These objections defeat the entire concept of the atmosphere as a gray-body radiator. An analysis(26) based on the concept of the atmosphere's being a gray-body radiator (giving uniform emission independent of wavelength) shows the maximum emission of radiation to take place in the middle troposphere instead of in the lowest levels of the atmosphere wherein there is the greatest concentration of water vapor. This conclusion was so wrong that it was early decided that the wavelength dependence on radiation had to be taken into account. By recognizing the importance of considering the radiation from different parts of the spectra as having widely different values, Simpson(27) was able to present an exceedingly simple but highly effective method for computing the long-wave radiation exchange between the earth, atmosphere, and space. This attempt was a very successful conclusion to his original attempt to consider the atmosphere as a gray body.

Simpson's analysis presents the following argument. An inspection of the infrared spectrum of atmospheric air shows that a black-body radiator at air temperatures (—40 to +40° C) has the majority of its energy concentrated between 3 and 30 microns. In this region, we find absorption due to water vapor and carbon dioxide concentrated at the wavelengths illustrated in Fig. 4.10. An inspection of this figure shows that little appreciable error is made if we consider the regions up to 7 microns and beyond 14 microns as complete absorbers of radiation and the region 8.5 microns $< \lambda <$ 11 microns as completely transparent to radiation of these wavelengths. The intermediate regions 7 to 8.5 microns and 11 to 14 microns are semitransparent, absorbing some fraction near one-half of the radiation. His simplified absorption spectrum is given by the dashed lines of Fig. 4.10. With these ideas in mind, he argues that the earth radiates as a black body at the mean temperature of the locality under consideration, shown as the upper curve of Fig. 4.11. The uppermost layer in the atmosphere over this site will be at a temperature corresponding to the lower black-body curve in the same figure. Now, any radiation that originates at the earth in the region between 8.5 and 11 microns will escape to outer space because of the transparency of the atmosphere to radiation of these wavelengths. The amount of flux is then proportional to the area under the black-body curve at the earth's temperature in Fig. 4.11. In the regions < 7 microns and > 14 microns, radiation from the earth will be in turn absorbed and re-emitted by the successive layers of 0.3 millimeter of precipitable water in the atmosphere until the flux that finally escapes to space will be determined by the temperature of this last layer. In these spectral regions, the uppermost layer radiates flux corresponding to black-body radiation at the

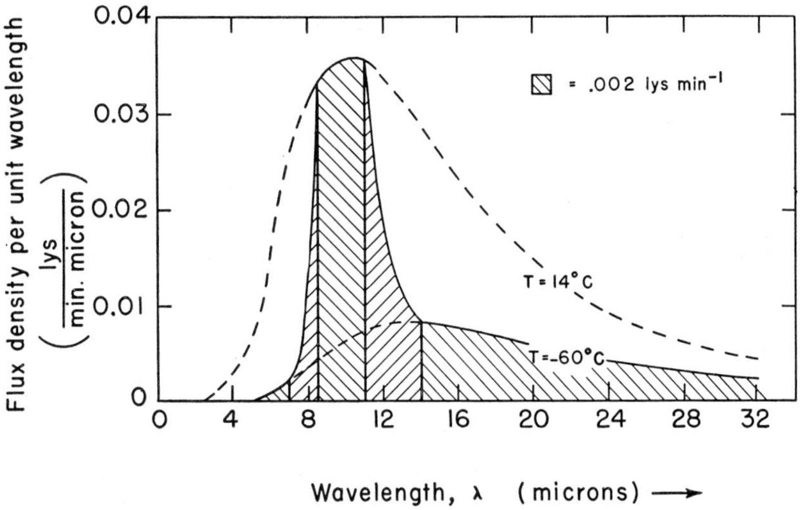

Fig. 4.11 Diagram showing Simpson's method for evaluating the flux of long-wave radiation leaving the earth's atmosphere.

mean temperature of the layer. In the figure, this flux is the area under the lowest curve for $\lambda < 7$ microns and $\lambda > 14$ microns. In the intermediate regions, which are small, the flux emitted is somewhere between the flux that would be emitted by the coldest and warmest temperatures encountered. In the same figure, this last flux is the area shown for 7 microns $< \lambda <$ 8.5 microns and 11 microns $< \lambda <$ 14 microns. The total area under the composite curve (the full heavy line) in flux density units is, in the mean, the long-wave radiation to space at this particular locality. The results obtained from Simpson's method are realistic and easy to apply, and they give a semiquantitative insight into the long-wave radiation process in the atmosphere. Simpson was able to compute the mean radiational heat balance, latitude by latitude, using this method. The results will be discussed more fully in the next chapter when the heat budget of the earth is discussed.

The general theory of radiation transfer by the gases in the atmosphere

Successful as Simpson's method is for treating specific problems in long-wave radiation transfer in the atmosphere, the method lacks generality. The general method makes use of concepts of radiation transfer that we have already studied in both this and the immediately preceding chapter.

Suppose that we set up the equations of long-wave radiation in the atmosphere, conscious of the fact that to a high degree of approximation the distribution of water vapor and carbon dioxide varies only in the vertical. This fact enables us to write the radiation-transfer equations as though the transfer were from a series of parallel layers of air, infinite in extent, and stacked parallel to the surface of the earth. Let the optical path in the vertical be u_z† so that in general any point on a plane u_z units from a selected reference unit area is $u_z \sec \theta$ units of optical path away. θ is the zenith angle of the point on the radiating plane viewed from the reference plane (see Fig. 4.12). The flux from the radiating plane impinging on a unit area of the reference plane is, in accordance with the ideas expressed by Eqs. 4.34 and 4.26,

$$E(0) = \int_0^{2\pi} \int_0^{u_\infty \sec \theta} B^* e^{-u_z \sec \theta} \cos \theta \, d(u_z \sec \theta) \, d\omega. \qquad (4.48)$$

We have assumed that there is an attenuating path (by absorption) between the reference level and the plane located at u_z. Then we sum the contributions of flux from all planes between 0 and u_∞, the greatest extent of the optical path in the vertical direction u_z. B^*, the specific intensity (brightness), of the radiating level at u_z is parallel radiation expressed by Planck's equation, Eq. 4.1. B^* is a function of temperature as well as wavelength. Equation 4.48 is true only for a given wavelength, so that eventually $E(0)$ must be summed over all wavelengths.

Before proceeding further, let us perform a mathematical manipulation that will put Eq. 4.48 into a form which can be evaluated more easily.

† Wavelength subscripts will be omitted because of the profusion of notation required. The discussions of previous chapters should have indicated that equations should always be written for a single wavelength and then the effect of wavelength variation should be taken into account.

As before

$$d\omega = 2\pi \sin \theta \, d\theta. \tag{4.49}$$

The factors in Eq. 4.48 shown below can be written as

$$\cos \theta \, d(u_z \sec \theta) \, d\omega = 2\pi \sin \theta \, d\theta \, du_z. \tag{4.50}$$

Mathematically, as

$$\sin \theta \, d\theta = \frac{d \sec \theta}{\sec^2 \theta}, \tag{4.51}$$

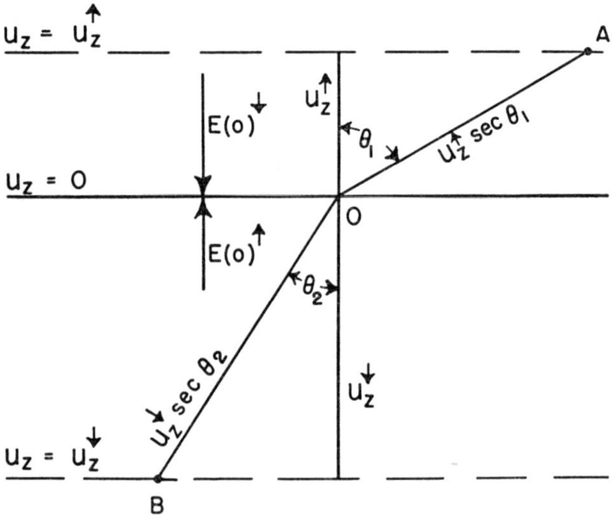

Fig. 4.12 Illustration of the geometry in the transfer of long-wave radiation. $u_z = 0$ is the reference plane. Radiation from point A at level $u_z^{\uparrow}$ and from point B at $u_z^{\downarrow}$ contribute to flux $E(0)^{\downarrow}$ and $E(0)^{\uparrow}$ *measured at* $u_z = 0$. If $E(0)^{\downarrow}$ and $E(0)^{\uparrow}$ refer only to the flux originating in layers $u_z^{\uparrow}$ and $u_z^{\downarrow}$, then $E(0)^{\downarrow} > E(0)^{\uparrow}$ because $u_z^{\uparrow}$ lies nearer to $u_z = 0$ than $u_z^{\downarrow}$. Throughout the discussion it has been assumed that an attenuating medium fills the entire space between $u_z^{\uparrow}$ and $u_z^{\downarrow}$. The arrows show either the direction in which u_z is increasing or the direction in which the flux is flowing.

we find that Eq. 4.48 may be expressed as

$$E(0) = 2 \int_0^{u\,\infty} \pi B^* \int_0^{\frac{\pi}{2}} \frac{e^{-u_z \sec \theta}}{\sec^2 \theta} \, d \sec \theta \, du_z \tag{4.52}$$

through a suitable grouping of terms. If we let $\eta = \sec \theta$, we arrive at a standard form of an integral where

$$H_2(u_z) = \int_0^{\frac{\pi}{2}} \frac{e^{-u_z \sec \theta}}{\sec^2 \theta} \, d \sec \theta = \int_1^{\infty} \frac{e^{-u_z \eta}}{\eta^2} \, d\eta. \tag{4.53}$$

$H_2(u_z)$ can be evaluated for a constant u_z. Then tables of $2H_2(u_z)$ can be made for the various values of u_z. Thus, by use of the notation of Eq. 4.53, Eq. 4.52 can be expressed as

$$E(0) = \int_0^{u_\infty} \pi B^* \, 2H_2(u_z) \, du_z. \tag{4.54}$$

Equation 4.54 is in an interesting form because we have been able by its means to express radiation of magnitude B^* from a diffuse source as diffuse instead of parallel radiation that illuminates our reference area.

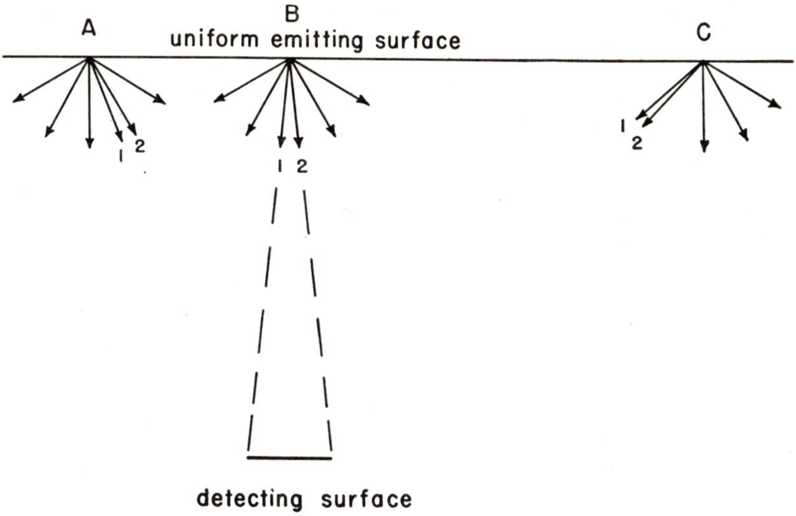

Fig. 4.13 Illustration of the radiation emitted from a diffuse surface and incident on a finite detecting surface. The diffuse radiation from three selected points *A*, *B*, and *C* of the diffuse radiator is shown. Only the bundle of rays between rays 1 and 2 ever reach the detecting surface. These latter rays are essentially parallel radiation. Thus the diffuse radiator may be treated as though it were a source of parallel radiation. The contributions of this parallel radiation to the detecting surface from all locations on the diffuse emitter are summed using Lambert's cosine law.

The magnitude of this diffuse flux per unit area is $E(0)$. Paradoxical as the above statements seem, one must always remember that the components of diffuse flux (magnitude B^*) from a radiant area ultimately reaching a reference level where such flux will be detected and measured are such a small bundle of rays compared to the total bundle of rays of diffuse flux leaving the radiant area, that the small bundle of rays is essentially parallel. Hence, we have parallel radiation received from a unit area of a diffuse emitter, provided that the receiving area (our reference) is small in extent. Figure 4.13 will assist in straightening out this apparent paradox of words. πB^* is sometimes called the diffuse specific intensity, after the ideas expressed by Eq. 4.34*a*.

Now, if we are interested in the steady-state condition of flux transfer at a given reference level in the atmosphere, we have the condition that the flux emitted must by Kirchhoff's laws be equal to the flux absorbed. To express this statement mathematically, we can express the flux absorbed (and hence the flux emitted) as the difference in two streams of flux, one impinging on the reference area from all emitters above the reference plane, and the second directed upward on the reference area from all emitters below. Measuring all distance as positive in a direction away from the reference level, and letting the superscripts$^{\downarrow}$ be from above and $^{\uparrow}$ be from below, we state that the net emitted flux $E(0)^{\uparrow\downarrow}$ at the zero level for any wavelength must be

$$E(0)^{\uparrow\downarrow} = E(0)^{\uparrow} - E(0)^{\downarrow}. \qquad (4.55)$$

$E(0)^{\uparrow\downarrow}$ is generally positive because, as the temperature and moisture conditions are greater in magnitude below than above the reference level, $E(0)^{\uparrow} > E(0)^{\downarrow}$, making $E(0)^{\uparrow\downarrow} > 0$.

Our previous discussion has established the form of $E(0)^{\uparrow}$ and $E(0)^{\downarrow}$, for they are exactly $E(0)$ from Eq. 4.54 taken in the proper vertical direction u_z to make values of $E(0)$ positive. On integrating over all wavelengths, Eq. 4.55 becomes

$$S(0) = \int_0^{\infty} E(0)^{\uparrow\downarrow}\, d\lambda = \int_0^{\infty} E(0)^{\uparrow}\, d\lambda - \int_0^{\infty} E(0)^{\downarrow}\, d\lambda. \qquad (4.56)$$

$S(0)$ is the net flux density when the contributions from all wavelengths have been measured. The evaluation of Eq. 4.56 is difficult and is almost always done by graphical means. Two of the better-known charts for evaluating Eq. 4.56 are the Elsasser radiation chart(8) in this country and the radiation chart by Möller(18,25) used in European countries, especially in Germany. Particular mention of the Elsasser chart will be made later, primarily because it is the chart most likely to be encountered in this country.

In general, as both the optical path and the temperature vary with height, the temperature must be considered a function of u_z. The relationship at any time and place in the atmosphere of the temperature-height-moisture content and hence the derived temperature-optical path function is given by a radiosonde ascent. Therefore, since B^* is an explicit function of temperature, it is an implicit function of u_z, and must so be considered in the evaluation of Eq. 4.56.

It is customary to express ΔS, the net flux absorbed by a reference layer Δz units of height thick, in terms of the cooling rate experienced by this layer through uncompensated radiation processes. Suppose that we calculate from Eq. 4.56 the net flux $S(0)$ transferred through a reference plane at height z and pressure p. Using Eq. 4.56 again, but with new values of optical path consistent with a new reference level, we compute $S(\Delta z)$ at height $z + \Delta z$ and pressure $p - \Delta p$. If $S(\Delta z) - S(0) \neq 0$, which is true in general, then (with the aid of the hydrostatic equation)

$$\frac{1}{\rho}\frac{\Delta S}{\Delta z} = \frac{S(0) - S(\Delta z)}{\Delta p} g. \tag{4.57}$$

$(1/\rho)\,(\Delta S/\Delta z)$ is the flux absorbed in a mass of air Δz units thick whose base has a unit cross-sectional area. In other words, Eq. 4.57 is the flux absorbed per unit mass of air. If we consider that all the absorbed energy must be re-emitted, and that this flux can be considered as heat lost per unit time at the mean pressure of the layer, $p - (\Delta p/2)$, it follows that Eq. 4.57 must be equal to $c_p\ (\Delta T/\Delta t)$. c_p is the specific heat at constant pressure of dry air (a close enough approximation), and $\Delta T/\Delta t$ is the change of temperature per unit time. Thus

$$\frac{\Delta T}{\Delta t} = \frac{g}{c_p}\frac{S(\Delta z) - S(0)}{|\Delta p|} \tag{4.58}$$

where Eq. 4.58 is written in such a fashion that $\Delta T/\Delta t$ is in general negative, the usual case. A negative sign indicates cooling arising from a net upward transferral of flux.

Radiation Charts. Radiation charts now in general use are designed to evaluate Eq. 4.56 and from the evaluation to derive the cooling rate from Eq. 4.58. The Elsasser chart will be taken as an example. For the complete development of the theory of the chart, the student is referred to a monograph(9) by Walter M. Elsasser entitled *Heat Transfer by Infrared Radiation in the Atmosphere*, published in 1942 by Harvard University. The difficulty in constructing a chart of this sort is the difficulty inherent in evaluating Eq. 4.56 realistically. With the aid of Eq. 4.54, each term on the right-hand side of Eq. 4.56 is of the form

$$\int_0^\infty E(0)\,d\lambda = \int_0^\infty \int_0^{u_\infty} \pi B^* 2H_2(u_z)\,du_z\,d\lambda. \tag{4.59}$$

The difficulty in the evaluation becomes apparent if we put the vertical optical path u_z in terms of quantities that are actually measured. If we let k_λ be the *absorption* coefficient per unit of precipitable water w, we find that

$$u_z = k_\lambda w \tag{4.60}$$

and

$$du_z = k_\lambda\, dw. \tag{4.61}$$

Equation 4.59 can then be written as

$$\int_0^\infty E(0)\, d\lambda = \int_0^w \int_0^\infty \pi B^*(\lambda, T) 2H_2(k_\lambda w) k_\lambda\, d\lambda\, dw. \tag{4.62}$$

Since B^* is a function of λ and T, it has been written as $B^*(\lambda, T)$. The order of integration has been interchanged.

Suppose, now, that we wish to integrate Eq. 4.62 over all wavelengths first, and then integrate over the precipitable water. T is a function of the height, as is w, so that at any time $T = f(w)$. Both T and w are independent of wavelength, so that the integrations can be done separately. Thus, we first wish to integrate the part of Eq. 4.62 given by

$$Q(w, T) = \int_0^\infty B^*(\lambda, T) H_2(k_\lambda w) k_\lambda\, d\lambda \tag{4.63}$$

for various values of the temperature and precipitable water content. The values of $Q(w, T)$ that are obtained depend entirely on the form of k_λ as a function of wavelength that is chosen. $B^*(\lambda, T)$ is not an arbitrary function but, as mentioned before, is Planck's radiation law. It is beyond the scope of this book to discuss the various forms that $H_2(k_\lambda w)$ takes as a result of the choice of k_λ. It should be mentioned that the question of the accuracy of the charts hinges on the choice of the absorption function. This was the same problem confronting Simpson in his radiation problem. Water vapor is a band absorber. The bands are composed of a number of lines, many of which are unable to be resolved by optical instruments even by extremely difficult laboratory techniques. However, theory predicts the shape and spacing of these lines as a function of wavelength. Unfortunately, the width of the lines is sensitive to the pressure of the dry air, broadening the half widths† with increasing pressure. The solution of Eq. 4.63 is a problem that must be met and overcome by anyone interested in constructing a radiation diagram. Once this problem has been solved and $Q(w, T)$

† If the absorption coefficient of a gas is plotted as a function of wavelength, the function has the form

$$k_\lambda = \frac{k_0\, \delta^2/4}{\left(\dfrac{1}{\lambda} - \dfrac{1}{\lambda_0}\right)^2 + \delta^2/4}$$

where k_0 is the value of k_λ when $\lambda = \lambda_0$. δ is the half width of the lines. A plot of this function, k_λ versus λ, has a form similar to the Gaussian error curve.

determined for various values of w and T, Eq. 4.62 can be written as

$$\int_0^\infty E(0)\,d\lambda = 2\pi \int_0^w Q(w, T)\,dw. \tag{4.64}$$

Equation 4.64 can be expressed as an area on a chart of $Q(w, T)$ versus w, or, by a suitable transformation, on a chart of w versus T. Such a diagram is called a radiation diagram. Putting Eq. 4.64 into Eq. 4.56 shows that the net radiation $S(0)$ at reference level zero is

$$S(0) = \int_0^w 2\pi Q(w, T)^\uparrow\,dw - \int_0^w 2\pi Q(w, T)^\downarrow\,dw, \tag{4.65}$$

which is the difference of two areas on a radiation diagram.

The Elsasser Chart.[8,9] The Elsasser radiation chart is a diagram, triangular in shape, where ordinates are isotherms of centigrade temperature and abscissas are lines of constant precipitable water. The abscissas converge to a point at $T = -273°$ C. The low-temperature end of the chart is usually not presented, but a table of values is given which, when added to the area of the chart, gives the proper value of $S(0)$. The chart is so constructed that a unit area represents the same amount of energy irrespective of where the area is measured on the chart. A schematic drawing of the radiation diagram is shown in Fig. 4.14. The lines of constant w extend from 0 to ∞, where an infinite amount of precipitable water at any temperature T radiates as and may be used for a black body emitting σT^4 units of radiant flux per unit area. Thus, the entire area to the right of an isotherm is equal to the black-body flux σT^4.

In order for one to use the radiation diagram, the data must be processed and put in suitable form for entry on the chart. From a radiosonde ascent, temperature, pressure, and relative humidity are measured. From these data one finds the temperature, pressure, and specific humidity by standard techniques. These quantities are represented by T, p, and q respectively and are information that is available to us initially. From q, which is defined as the ratio of the density of water vapor to the density of moist air, we can find the precipitable water from the formula

$$w = \frac{\bar{q}\,\Delta p}{g}\ (\text{cm}). \tag{4.66}$$

Δp is always taken as a positive value so as to make w positive. Although w is given as centimeters, a length unit instead of mass, the numerical equivalence of length and mass is true only if the cgs system of units is used, because the density of water is unity in this system.

$\bar{q}$ is the mean value of specific humidity measured at the midpoint of Δp and is a dimensionless quantity in units of grams of water per gram of air. p is measured in dynes per square centimeter. However, if $\bar{q}$ is

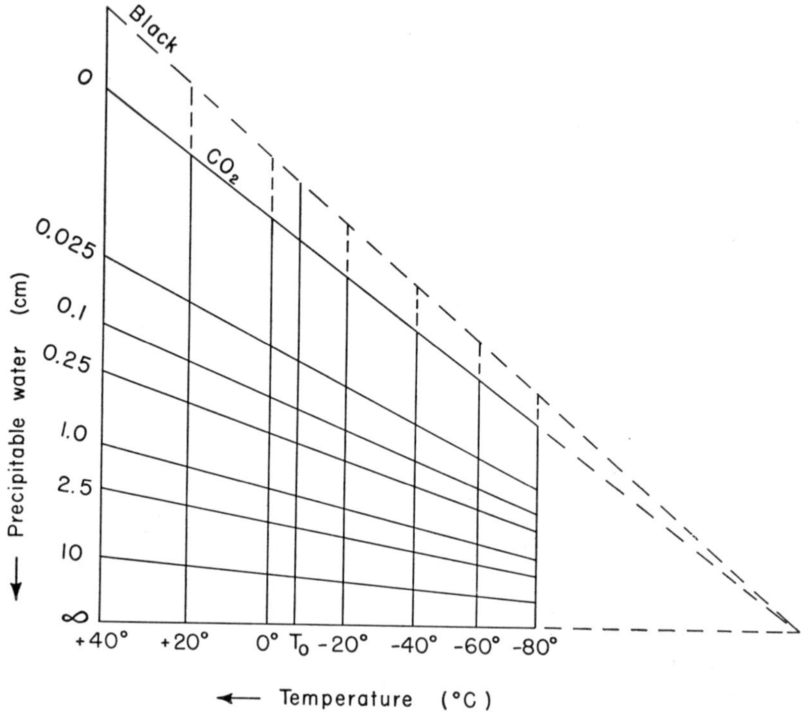

Fig. 4.14 A schematic drawing of the Elsasser radiation diagram.(8,9) The ordinate is in centimeters of precipitable water; the abscissa, temperature in degrees centigrade. The area enclosed by the isotherm T_0, the horizontal line labeled ∞, and the upper dashed line labeled black, all to the right of T_0, is equivalent to $\sigma T_0{}^4$ units of radiative power given by the Stefan-Boltzmann law. The dashed area to the right of $T = -80°$ C is not included. A table of equivalent area is given instead.

measured in grams of water per kilogram of air and p is measured in millibars, the product of the two also gives the correct magnitude in cgs units. Thus, from Eq. 4.66 we can find the amount of precipitable water in the layer Δp units of pressure thick. After w has been found for each layer, w is multiplied by† $\sqrt{p/p_0}$ to compensate for the pressure

† There is some question as to whether $\sqrt{p/p_0}$ or p/p_0 should be used. Latest opinion on the matter indicates p/p_0 is the correction factor. In any case, as w decreases when p/p_0 decreases we find that $w(p/p_0) \approx w\sqrt{p/p_0}$ throughout most of the range of values.

broadening effect on the lines mentioned earlier. p is the pressure at the level where w is measured, and p_0 is standard atmospheric pressure (1000 millibars to a sufficient approximation). This operation gives the effective value of w that is entered on the chart. The operations described are illustrated in Table 4.3 for a few points from an actual atmospheric sounding. Column 7 is the effective precipitable water content for each successive level Δp units of pressure thick. Columns 8, 9, and 10 illustrate data that will be plotted on the radiation diagram. It will be noticed that these three final columns are the summation of the values in column 7, yielding a positive summation when reckoned in a

Table 4.3

Data Sheet for Entry of Effective Moisture versus Temperature on an Elsasser Radiation Diagram

Data are taken from a paper by Walter M. Elsasser in the Supplement to Volume 66 of the *Quarterly Journal of the Royal Meteorological Society,* 1940.

1	2	3	4	5	6	7	8	9	10
Sounding			Basic Reduction of Data				Total Effective w for Reference Level at		
p(mb)	T(°C)	$q\left(\frac{g}{kg}\right)$	$\bar{q}$	Δp	$\sqrt{\frac{p}{p_0}}$	Eff. w	1015 mb	710 mb	505 mb
1015	10	6.7					0	1.36	1.56
			6.3	65	0.99	0.41			
950	7	5.8					0.41	0.95	1.15
			5.5	30	0.97	0.16			
920	7	5.3					0.57	0.79	0.99
			4.6	130	0.93	0.56			
790	−1	3.9					1.13	0.23	0.43
			3.6	52	0.87	0.16			
740	−4	3.3					1.29	0.07	0.27
			2.9	29	0.85	0.07			
710	−2	2.5					1.36	0	0.20
			1.9	45	0.83	0.07			
650	−8	1.3					1.43	0.07	0.13
			1.2	73	0.78	0.07			
580	−14	1.1					1.50	0.14	0.06
			1.0	77	0.74	0.06			
505	−16	0.9					1.56	0.20	0
			0.9	35	0.69	0.02			
470	−18	0.8					1.58	0.22	0.02
			0.5	124	0.64	0.04			
350	−34	0.3					1.62	0.26	0.06
			0.2	49	0.57	0.01			
300	−44	—					1.63	0.27	0.07

direction away from the reference level. This procedure follows from the requirements of Eq. 4.65 for obtaining the net flux $S(0)$ at the chosen reference levels of 1015, 710, and 505 millibars.

When the data of column 8 are entered on a radiation diagram, we obtain a curve that is illustrated schematically by Fig. 4.15. The

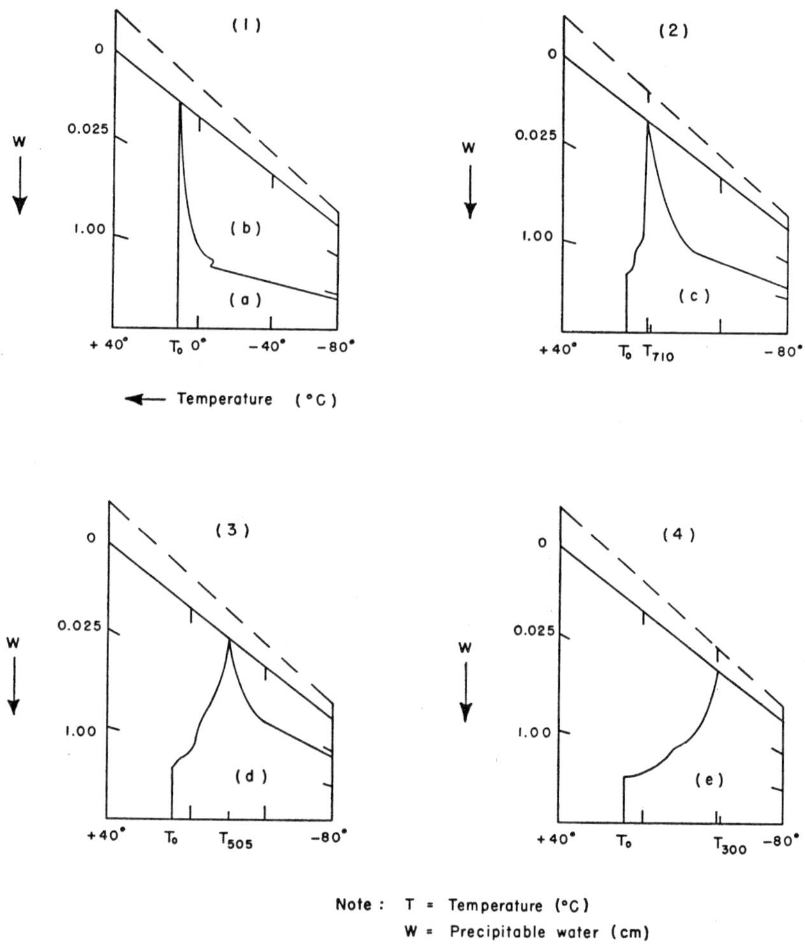

Fig. 4.15 A schematic diagram of the Elsasser chart to accompany Table 4.3. Diagrams 1, 2, and 3 show the net long-wave flux from the earth and atmosphere leaving reference levels 1015, 710, and 505 millibars respectively. Diagram 4 shows the net flux leaving the 300-millibar level to outer space. The ordinate is precipitable water, w, in centimeters, and the abscissa is temperature in degrees centigrade.

relationships of the areas on the diagram to the requirements of Eq. 4.65 are as follows.

The ground is considered as a black body maintained at temperature T_0. The radiation initially leaving the ground is area a plus area b of Fig. 4.15.[9] Of this radiation, an amount of energy represented by area b is returned to the earth from the atmosphere above. The net radiation escaping from the earth's surface, the nocturnal radiation, is just area a. In the construction of the diagram, a mathematical transformation of Eq. 4.65 is made that changes the form of the equation but not the physical sense.

Similarly, the net flux through level 710 millibars and 505 millibars is shown in Fig. 4.15, where now the net flux is area c and d, respectively. The differences in area $c - d$ for the two levels give the difference between $S(0)$ and $S(\Delta z)$ of Eq. 4.58. $\Delta p = 710 - 505 = 205$ millibars. From this information $\Delta T/\Delta t$ can be computed for the mean-pressure level

$$p = 710 - \frac{205}{2} = 608 \text{ millibars.}$$

From the radiation diagram, we are able to make an estimate of long-wave flux leaving a unit area of the earth's surface. The determination of this flux by the Elsasser chart is a new solution to the same problem that Simpson was interested in and it has been discussed earlier in this chapter. If we now choose our reference level at 300 millibars (Table 4.3), there will be little or no water vapor above this pressure level. Hence no radiation will be returned from above, so that all the flux that passes through this level must escape to space. The problem is solved schematically by Fig. 4.15 and gives results that agree in the mean with Simpson's cruder but very straightforward method. Further details and criticisms of the radiation diagram can be obtained from the references given at the end of the chapter.

This chapter has been an attempt to describe the theory and techniques of measurement of the radiant flux exchange in the atmosphere, both short-wave from the sun and long-wave from the earth and atmosphere. In the next chapter, we shall discuss the uses to which such measurements are put, including a discussion of the energy balance at the earth's surface.

References

1. Abbot, C. G., Fowle, F. E., and Aldrich, L. B., "The Distribution of Energy in the Spectra of the Sun and Stars," *Smithsonian Misc. Coll., 74,* No. 7 (1923).

2. Abbot, C. G., Fowle, F. E., Aldrich, L. B., and Hoover, W. H., *Annals of the Astrophysical Observatory of the Smithsonian Institution,* Vols. 1–6 (1900–1942).
3. Anthony, R., "Observation of Non-Rayleigh Scattering in the Spectrum of the Day Sky in the Region 0.56 to 2.2 Microns," *J. Meteorol., 10,* 60–63 (1953).
4. Benedict, W. S., and Plyler, E. K., "Absorption Spectra of Water Vapor and Carbon Dioxide in the Region of 2.7 Microns," *J. Research Natl. Bur. Standards, 46,* 246–265 (1951).
5. Callendar, G. S., "Infra-Red Absorption by Carbon Dioxide with Special Reference to Atmospheric Radiation," *Quart. J. Roy. Meteorol. Soc., 67,* 263–274 (1941).
6. Code, A. D., "Radiative Equilibrium in an Atmosphere in Which Pure Scattering and Pure Absorption Both Play a Role," *Astrophys. J., 112,* 22–47 (1950).
7. Cowling, T. G., "Atmospheric Absorption of Heat Radiation by Water Vapour," *Phil. Mag., 41,* 109–123 (1950).
8. Elsasser, W. M., "An atmospheric radiation chart and its use," *Supplement to Quart. J. Roy. Meteorol. Soc., 66,* 41–56 (1940).
9. Elsasser, W. M., "Heat Transfer by Infrared Radiation in the Atmosphere," *Harvard Meteorological Studies,* No. 6 (1942). Harvard University, Cambridge, Mass.
10. Fowle, F. E., "Water-Vapor Transparency to Low Temperature Radiation," *Smithsonian Misc. Coll., 68,* No. 8, p. 24 (1917).
11. Fritz, S., "Solar Radiant Energy and Its Modification by the Earth and Its Atmosphere," *Compendium of Meteorology,* 13–33 (1951); American Meteorological Society, Boston.
12. Goody, R. M., "A statistical model for water-vapour absorption," *Quart. J. Roy. Meteorol. Soc., 78,* 165–169 (1952).
13. Houghton, H. G., "The solar constant," *J. Meteorol., 8,* 270 (1951).
14. Kaplan, L. D., "On the pressure dependence of radiative heat transfer in the atmosphere," *J. Meteorol., 9,* 1–12 (1952).
15. Kimball, H. H., and Hand, I. F., "Daylight Illumination on Horizontal, Vertical, and Sloping Surfaces," *Monthly Weather Review, 50,* 615–628 (1922).
16. King, J. I., "Line Absorption and Radiative Equilibrium," *J. Meteorol., 9,* 311–321 (1952).
17. King, L. V., "On the Scattering and Absorption of Light in Gaseous Media, with Applications to the Intensity of Sky Radiation," *Phil. Trans. Roy. Soc.* (A), *212,* 375–433 (1913).
18. Möller, F., "Long-Wave Radiation," *Compendium of Meteorology,* 34–49 (1951); American Meteorological Society, Boston.
19. Packer, D. M., and Lock, C., "The Brightness and Polarization of the Daylight Sky at Altitudes of 18,000 to 38,000 Feet above Sea Level," *J. Opt. Soc. Amer., 41,* 473–478 (1951).
20. Planck, M., "Ueber das Gesetz der Energieverteilung in Normalspectrum," *Ann. Physik, 4,* 553–563 (1901).
21. Plass, G. N., "Parallel-beam and diffuse radiation in the atmosphere," *J. Meteorol., 9,* 429–436 (1952).
22. Plass, G. N., "A Method for the Determination of Atmospheric Transmission Functions from Laboratory Absorption Measurements," *J. Opt. Soc. Amer., 42,* 677–683 (1952).

23. Plass, G. N., and Warner, D., "Influence of line shift and asymmetry of spectral lines on atmospheric heat transfer," *J. Meteorol., 9,* 333–339 (1952).
24. Randall, H. M., Dennison, D. M., Ginsberg, N., and Weber, L. R., "The Far Infrared Spectrum of Water Vapor," *Phys. Rev., 52,* second series, 160–174 (1937).
25. Schmidt, K-H., "Prüfung der Strahlungsrechendiagramme von Möller und Elsasser durch Gegenstrahlungsmessungen bei gleichzeitung durchgeführten Radiosondenaufsteigen," *Z. Meteorolog., 5,* 331–340 (1951).
26. Simpson, G. C., "Some studies in terrestrial radiation," *Mem. Roy. Meteorol. Soc., 2,* No. 16 (1928).
27. Simpson, G. C., "Further studies in terrestrial radiation," *Mem. Roy. Meteorol. Soc., 3,* No. 21 (1928).
28. Tousey, R., and Hulburt, E. O., "Brightness and Polarization of the Daylight Sky at Various Altitudes Above Sea Level," *J. Opt. Soc. Amer., 37,* 78–92 (1947).
29. Yamamoto, G., and Onishi, G., "Absorption of solar radiation by water vapor in the atmosphere," *J. Meteorol., 9,* 415–421 (1952).
30. "Emission and Absorption of Radiation in the Atmosphere," *Quart. J. Roy. Meteorol. Soc., 68,* 197–214 (1942). (A collection of papers by various authors.)

Source Books

B1. Forsythe, W. E. (editor), *Measurement of Radiant Energy,* McGraw-Hill Book Company, New York (1937).
B2. Kuiper, G. P. (editor), *The Atmospheres of the Earth and Planets,* University of Chicago Press, Chicago (1949).
B3. List, R. J. (editor), *Smithsonian Meteorological Tables,* sixth edition, Smithsonian Institution, Washington (1951).

Problems

4.1 (*a*) How much energy does a circular cloud 2000 feet in diameter emit toward the earth? Consider the cloud to be an infinitely thin black body whose surface temperature is 7° C. (*b*) How much energy from this cloud is detected on a square centimeter of the earth's surface when the center of the cloud is 2000 feet directly over the receiving surface? Consider a nonabsorbing atmosphere. (*c*) The cloud moves to a position 5000 feet from its position in part *b*. How much flux falls on the square centimeter of part *b*? (*d*) What is the ratio of the fluxes in parts *b* and *c*? How large would the cloud have to be in order for the energy density of the radiation detected at the earth's surface to be the same as the energy density of the radiation emitted from the cloud?

4.2 Let B_H be the diffuse brightness of the horizon sky. The variation of the diffuse sky brightness with zenith angle θ will be taken to be

$$B_D = B_H\,(3 - 2\sin\theta).$$

(*a*) What is the brightness of the zenith sky? (*b*) Find the average brightness of a circular area of the sky contained within a 30-degree angle from the zenith. (*c*) Find the average brightness of the entire sky. (*d*) Repeat the calculations for parts *b* and *c*, but find the average intensity of the radiation on a unit horizontal area. (*e*) Find the average intensity of the radiation incident on a unit horizontal area from a sky of uniform brightness B_H.

4.3 On a certain day, a measurement of the direct solar beam in the 1.5- to 1.6-micron wavelength interval gives the following data.

Zenith angle (degrees)	40°	50°	60°	70°
E(cal cm^{-2} min^{-1})	0.020	0.018	0.015	0.011

E is the flux density of the solar beam measured at the surface of the earth. Find the transmission coefficient a_λ (at normal incidence) and the solar flux density outside the atmosphere in this wavelength interval.

4.4 Consider an atmosphere that is horizontally stratified, infinite in extent, and containing an optical mass u of an absorbing gas, e.g., water vapor. Consider the surface of the earth to be a black-body radiator maintained at temperature T_0. Consider for this problem that all the radiation leaves the earth normal to the surface. We shall consider the radiation in a narrow band of wavelengths $\Delta\lambda$ to have a certain transmissivity $\tau(u)$, a function of the optical mass of the absorbing atmospheric gas. For water vapor $\tau(u)$ is usually given as a function of the precipitable water w. Since w is a function of height z, we may look at τ as a function of either u, w, or z.

a. How much of the energy that initially left the earth's surface is absorbed in a horizontal layer of air Δz units thick whose mean height is z units above the earth's surface? The energy leaving the earth's surface in the narrow interval $\Delta\lambda$ can be written symbolically as

$$B^*(T_0) = B^*(\bar{\lambda}, T_0)\,\Delta\lambda.$$

b. The coefficient of $B^*(T_0)$ obtained in part *a* can be thought of as the absorptivity of the layer Δz. Show that through the use of Kirchhoff's law the energy received at the ground from this layer at height z is of the form

$$\Delta\tau B^*(T) = \Delta\tau B^*(\bar{\lambda}, T)\,\Delta\lambda$$

when T is the mean temperature of the layer. $\Delta\tau$ is the difference in the transmissivities at the top and bottom of the layer at z after the radiation had passed through the intervening atmosphere between the level under consideration and the ground.

c. Deduce that if all the layers above the ground contribute energy to the ground layer, the flux received at the ground is of the form

$$E(0) = \sum_i \Delta\tau_i B^*_i(T).$$

In the calculus notation, this expression may be written as

$$E(0) = \int_{u_0}^{u_\infty} B^*(T)\,\frac{\partial\tau}{\partial u}\,du.$$

$E(0)$ is the total flux density received at the surface at u_0 from all layers from u_0 to u_∞. Compare with Eq. 4.54. If the surface u_0 is at some height z in the atmosphere, there are two streams of flux passing the level u_0, one from above and one from below.

4.5 The results of Prob. 4.4 can be applied to diffuse radiation by averaging the diffuse radiation in such a way that the laws for parallel radiation hold. In a manner analogous to Prob. 4.4*c*, we may write the law for the transfer of diffuse radiation as

$$E(0) = \int_{u_0}^{u_\infty} B_D(T)\,\frac{\partial\tau_D}{\partial u}\,du$$

where $E(0)$ is still the flux density at level u_0. The other quantities are

$$B_D(T) = \pi B^*(\bar{\lambda}, T)\ \Delta\lambda$$

where $B^*(\bar{\lambda}, T)$ can be evaluated from Eq. 4.1, and

$$\tau_D = \tau^b$$

where τ is the transmissivity measured for parallel radiation. Although $1 < b < 2$ so that b in general is not strictly constant, we shall use $b = 1.66$, a compromise value much used in the literature before 1950.

a. How much energy is received at the surface of the earth in the wavelength interval 8–9 microns from an atmosphere having the following sounding?

Pressure, mb	1000	800	750	500	300
Temperature, °C	7°	−5°	−5°	−35°	−67°
Specific humidity, g/kg	1.0	0.8	0.4	0.01	0.002
Height, km	0	1.79	2.29	5.29	8.60

The following table of transmissivity τ as a function of precipitable water is needed. $\Delta\lambda$ is the wavelength interval.

Precipitable Water, cm

$\Delta\lambda$, microns	0.005	0.01	0.03	0.13	0.16	0.25	0.31
5–6	0.22	0.25	0.43	0.55	0.59	0.65	0.68
6–7	0.54	0.69	0.85	0.95	0.95	0.95	0.96
7–8	0.19	0.34	0.42	0.66	0.76	0.83	0.87
8–9	0	0.02	0.02	0.08	0.13	0.35	0.50

b. What is the energy received at the surface from the atmosphere above in the wavelength region 5–9 microns? In this wavelength region, for $\Delta\lambda = 1$ micron, we may write

$$B_D = K_1 e^{K_2\left(1 - \frac{280}{T}\right)}.$$

B_D is in units of watts cm^{-2} when the constants have the following values.

$\Delta\lambda$, microns	5–6	6–7	7–8	8–9
K_1 (watts cm^{-2})	67×10^{-5}	124×10^{-5}	171×10^{-5}	204×10^{-5}
K_2	9.30	7.85	6.82	6.01

c. How much diffuse flux leaves the earth in the wavelength interval 5–9 microns? What is the difference in the incoming and outgoing flux in this interval? Consider the earth to be at a temperature of 7° C.

4.6 Use the data of Prob. 4.5.

(*a*) Find the net flux (the difference between the upward and the downward currents of flux) passing through the 800-millibar surface in the 5–9 micron wavelength interval of the spectrum. (*b*) Repeat part *a* for the 750-millibar surface. (*c*) From the results of parts *a* and *b* find the amount of flux emitted or absorbed in this 50-millibar layer. Express your answer in the number of degrees of cooling or heating of this layer per day that this amount of flux represents for this narrow portion of the spectrum.

4.7 Using the Elsasser chart that accompanies this book, find the net cooling of the atmosphere in the 950- to 920-millibar layer and in the 710- to 650-millibar layer of the sounding given in Table 4.3. Express your answers in degrees centigrade per day.

CHAPTER 5

Radiation Studies and the Heat Budget of the Earth

In the last three chapters, methods have been developed to measure long- and short-wave radiation and to investigate the depletion of this energy by scattering, absorption, and reflection. The topics have necessarily been treated in a piecemeal fashion in order to present particular techniques. In this chapter, we shall make studies of processes in which the interaction and relative importance of the parts to the whole will be considered. For example, suppose that one is walking on a hot afternoon and seeks a place to sit down and rest. The chances are that he will choose a plot of grass shaded by a tree, either near a brook or in a spot exposed to a breeze. One's choice of a favorable microclimate may be instinctive, but the climate itself will be dictated by exposure to the wind, type of soil and vegetation composing the surface, exposure to solar radiation, and evaporation and transpiration of water from the brook and trees, all of which subprocesses affect the balance of heat and moisture which determines the climate. This illustration may be trivial, but the same factors that enter into the picture also enter into the more important scientific and economic problems of the heat budget of the earth.

The local heat budget at the earth's surface

The heat budget at the surface of the earth is probably the region of greatest economic importance and hence of great geophysical importance. By the surface of the earth, we shall mean a layer extending from roughly 2 to 5 meters below the surface of the earth through the surface to the lowest 50 to 100 meters of the atmosphere. Our reference layer will be the earth's surface, which for our purposes will be somewhere in the region between the soil surface and the top of any low vegetation, such as grasses or underbrush. With denuded soil, the soil surface itself will be the reference layer.

Let $E(z)$ = the flux of short-wave radiation; $S(z)$ = the flux of long-wave radiation; $L(z)$ = the sensible flux of heat in the atmosphere, usually by turbulent transfer of a mass of air (eddy diffusion); $B(z)$ = the sensible flux of heat in the soil, due to molecular conduction; and $W(z)$ = the latent heat of phase transformation of water, such as the processes of evaporation, condensation, sublimation, melting, or freezing. These five fluxes refer to a unit horizontal area of a

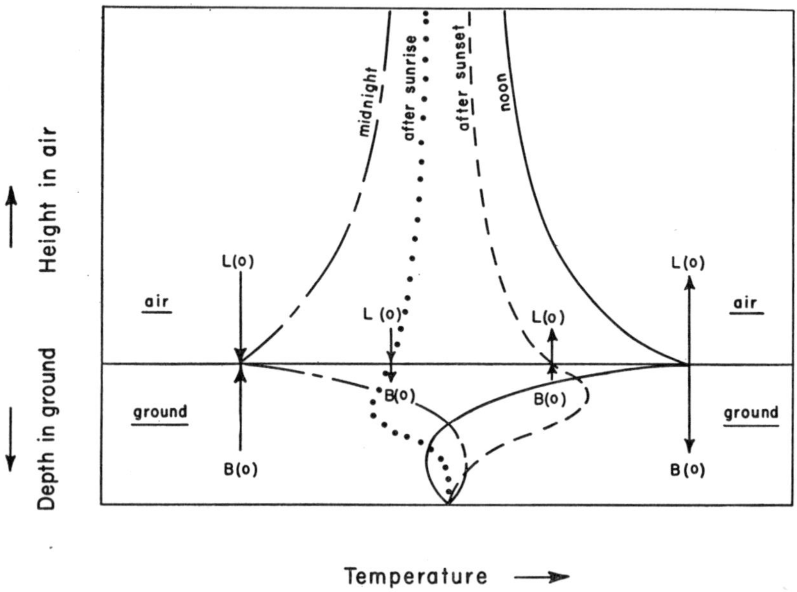

Fig. 5.1 (*Adapted from Lettau.*) Schematic representation of temperature profiles in the ground and in the air at various times of the day. The direction of the fluxes of heat through the surface of the earth from the ground, $B(0)$, and from the air, $L(0)$, is shown by the direction of the arrows. The magnitude is indicated by the length of the arrow. One assumes continuity of the temperature at the virtual surface of the earth.

vertical column extending from the ground at $z = 0$ (the reference level) upward into the air where $z > 0$, and downward into the ground where $z < 0$. Thus, L exists only for $z > 0$, whereas B exists only for $z < 0$. E and S exist only for $z \geqq 0$, whereas W can exist both in the air and in the soil.

At equilibrium,

$$E(0) + S(0) + L(0) + B(0) + W(0) = 0. \tag{5.1}$$

The null argument indicates that equilibrium exists at the surface of the earth where $z = 0$. All fluxes are considered positive when directed toward the reference level. Equation 5.1 is the mathematical formulation of the microclimatic condition expressed in the introductory remarks of this chapter. Figure 5.1 is a schematic drawing of the implications of the local energy balance expressed by Eq. 5.1. The figure shows the balance of energy at different times of the day, emphasizing the necessary reversal of certain fluxes in order to achieve this balance as the day progresses. Let us now examine the theory of measurement behind each of these quantities.

The radiant energies, $E(0)$ and $S(0)$

Chapter 4 was devoted to the measurement of both the short-wave flux $E(z)$ and the long-wave flux $S(z)$. The short-wave flux is composed of direct sunlight plus the scattered and multireflected light from the air, clouds, and earth. The long-wave flux is radiated from the earth and atmosphere. For the latter, a radiation diagram such as the Elsasser chart aids in the evaluation. Because the long-wave radiation term becomes of chief importance at night when short-wave flux is absent, long-wave radiation from the ground level is referred to as nocturnal radiation, irrespective of the time of day.

Simpson, in his usual straightforward manner, gives a direct theoretical approach to estimating the nocturnal radiation. With a shift in emphasis, his argument for finding the nocturnal radiation follows the same pattern as that for finding the outgoing radiation from the top of the atmosphere.

The lower atmosphere contains enough water vapor on the average to absorb completely all long-wave radiation originating at the earth's surface at wavelengths below 7 microns and above 14 microns. The extent of this lowest absorbing layer is not more than 1 kilometer and probably of the order of 100 meters. The mean temperature of the layer on the average is probably not more than 2 or 3° C different from the effective surface temperature of the earth. Little error is made if the absolute temperature of the lowest layer of the atmosphere and that of the earth's surface are considered to be the same. Now, take a black-body curve at the temperature of the earth (Fig. 5.2).† The earth will attempt to radiate this entire black-body energy to space (proportional to the area under the curve). The atmosphere will reradiate back to earth all the energy in the band < 7 microns and > 14 microns. In the transparent region, $8\frac{1}{2}$ to 11 microns, no energy will be absorbed and reradiated back to earth, as this energy will escape directly to space without absorption by the atmosphere. In the partially absorbing regions, $7 < \lambda < 8\frac{1}{2}$ microns and $11 < \lambda < 14$ microns, some of the energy from the earth will be absorbed and

† Figure 5.2 should be used in conjunction with Fig. 4.10, since Fig. 5.2 is an idealization of the properties of Fig. 4.10.

reradiated back to the earth, the remaining energy escaping into space. Simpson estimates the escaped energy in this spectral region to be of the order of one-half the energy leaving the earth. Thus, the energy radiated to space from the surface of the earth will be equal in energy units to the sum of the unmarked plus approximately one-half the single cross-hatched area of Fig. 5.2. In

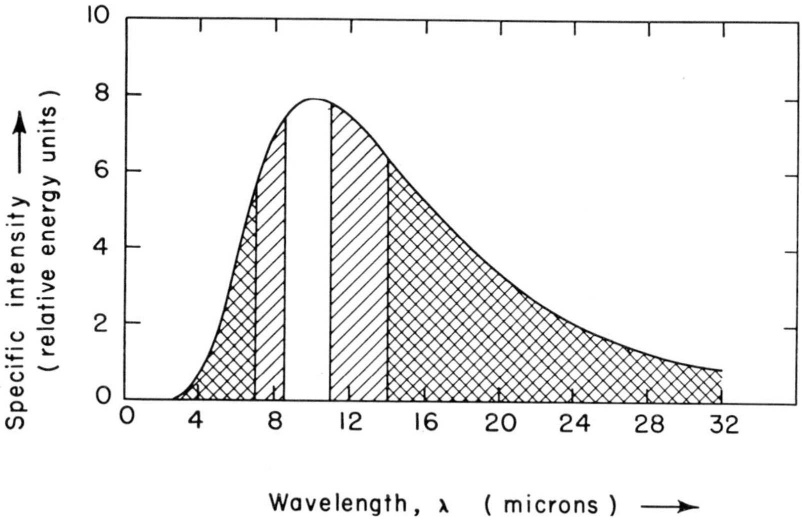

Fig. 5.2 This figure should be used in conjunction with Fig. 4.10. The envelope is the plot of the black-body function for the surface temperature of the earth. The double cross-hatched area represents the region where the atmosphere is opaque to radiation. The single cross-hatched area is the wavelength region where the atmosphere is semitransparent to radiation. The unmarked area is the region where the atmosphere is transparent to radiation.

this method of analysis, the nocturnal radiation will conceivably not be less than the area in the transparent band (moist air case) or more than the area in the transparent plus semitransparent bands (dry air case). This method of analysis suggests that the greatest error in estimating radiational losses in a normal atmosphere is in the estimation of the amount of radiation escaping in the semitransparent regions 7 to $8\frac{1}{2}$ microns and 11 to 14 microns. In order to correct for this deficiency, several empirical formulas and charts have been devised. Most of the techniques for estimating nocturnal radiation are successful because the total water-vapor content of the air varies in the same sense, on the average, as the surface water-vapor pressure or mixing ratio, parameters which can be measured readily at the surface.

Radiation charts based on actual measurements of nocturnal radiation by Brooks,[4] Robinson,[25] and others have been constructed. One chart, prepared under the direction of Robinson, has been in use for a period of at least 5 years at the Kew Observatory, England. Experience gained from direct radiation measurements indicates that the theoretical charts of Elsasser and Möller† may overestimate the amount of flux returned to the earth from the atmosphere, thus underestimating the amount of nocturnal radiation. Measurements indicate that the surface temperature measured by a conventional thermometer gives to within a degree the black-body radiative temperature of the earth's surface as measured by a radiometer. The readings verifying the equivalence of these two temperatures were obtained over a short-grass surface, both for wet and dry grass.[24]

Radiation formulas such as the formula derived by Brunt from theoretical considerations,

$$S(0)^{\downarrow} = \sigma T^4(a + b\sqrt{e}), \tag{5.2}$$

express the fact that the downward radiation is a function of the water-vapor content of the air, which in turn is proportional to the surface vapor pressure e. $S(0)^{\downarrow}$ is the downward radiant flux from the atmosphere, and $\sigma T^4 = S(0)^{\uparrow}$ is the upward black-body flux from the earth at temperature T. a and b are constants that are of the order of 0.5 and 0.06 respectively when e is expressed in millibars. The net long-wave flux leaving the surface is just $S(0)^{\uparrow} - S(0)^{\downarrow}$, making

$$S(0) = \sigma T^4(1 - a - b\sqrt{e}). \tag{5.3}$$

The vertical transfer of heat in the lower atmosphere by eddy diffusion. The evaluation of $L(z)$

The flux of heat in the lower atmosphere through a unit horizontal cross section is given by

$$L(z) = -\rho c_p \kappa_E \left(\frac{\partial T}{\partial z} + \frac{g}{c_p}\right), \tag{5.4}$$

where c_p is the specific heat at constant pressure, ρ the density, and κ_E the eddy diffusivity, all of air. g is the acceleration of gravity, and $\partial T/\partial z$ is the vertical temperature gradient. g/c_p is the dry adiabatic lapse rate, often denoted as γ_D. κ_E measures the rate at which a unit volume of air with a heat content represented by $\rho c_p \, \Delta T$ passes through a unit horizontal area, making $L(z)$ the flux of heat in the atmosphere.

† See Chapter 4.

A common meteorological parameter is the potential temperature Θ, the temperature that a mass of air would attain if it were brought by a dry adiabatic process to the 1000-millibar (mb) pressure surface. If we denote the temperature and pressure of the parcel as T and p respectively, and let the 1000 millibar surface be p_0, the first law of thermodynamics says that for dry air

$$\Theta = T\left(\frac{p_0}{p}\right)^{\frac{R_G}{c_p}}. \tag{5.5}$$

The exponent, the ratio of the gas constant to the specific heat at constant pressure, has a numerical value of 0.286. In general, differentiation of Eq. 5.5 with respect to height gives

$$\frac{\partial\Theta}{\partial z} = \frac{\Theta}{T}\left(\frac{\partial T}{\partial z} + \frac{g}{c_p}\right) \tag{5.6}$$

after the aid of the hydrostatic equation, Eq. 1.9, has been invoked.

Near the surface of the earth where $p \to p_0$, we find that $\Theta/T \to 1$. In this limit Eq. 5.4 may be read as

$$\underset{\text{Limit as } z \to 0}{L(z)} = -\rho c_p \kappa_E \frac{\partial\Theta}{\partial z}, \tag{5.7}$$

the form in which the eddy diffusion equation is most commonly presented. $\partial\Theta/\partial z$ is the lapse rate of potential temperature.

Efforts to evaluate $L(z)$ show that κ_E is not a constant but is a function of elevation, roughness of the ground, wind force, and thermal stratification of the air. However, if we write

$$\kappa_E = \kappa_E'(z + z_0) \tag{5.8}$$

where z_0 is a roughness parameter varying from 0.001 centimeter for smooth surfaces (ice or concrete) to 10 centimeters for ground covered with crops of grasses, we find that κ_E' is a constant for a particular weather situation, i.e., wind speed. Combining Eqs. 5.7 and 5.8 shows that

$$L(z) = -\rho c_p \kappa_E' \frac{\partial\Theta}{\partial \ln (z + z_0)}. \tag{5.9}$$

Equation 5.9 can be evaluated from a graph of potential temperature versus the logarithm of the height. $L(z)$ is proportional to the slope of this curve. For further details on the theory, measurement, and application of the turbulent transfer of heat, more specialized treatments should be consulted.†

† Cf. O. G. Sutton, *Atmospheric Turbulence* (1949), Methuen Co., London. Published in the United States by John Wiley & Sons, New York.

The flux of heat in the soil. The evaluation of $B(z)$

The conduction of heat in the soil under steady-state conditions is given by Newton's law of cooling,

$$B(z) = -\rho_s c_s \kappa_s \frac{\partial T}{\partial z}. \tag{5.10}$$

$B(z)$ is the specific heat flux (i.e., calories min^{-1} cm^{-2} = lys min^{-1}), $\partial T/\partial z$ is the temperature gradient in the soil, and κ_s is the *thermal diffusivity* of the soil. c_s is the specific heat and ρ_s is the density of the soil, respectively. The product $\rho_s c_s \kappa_s$ is a more familiar term, called the *thermal conductivity.* In form, L and B are similar, because in an incompressible fluid or solid the term represented by g/c_p in L is zero in B.

For a particular type of soil, $\rho_s c_s \kappa_s$ is a constant. If, analogous to the evaluation of L, we make a soil sounding of T versus z and measure the slope, we have evaluated $\partial T/\partial z$. However, for $B(0)$, in which we have special interest, the slope would have to be evaluated right at the surface layer. This latter is a difficult thing to do, because the temperature gradient in the soil is changing very rapidly with depth, much more rapidly than the temperature gradient in the overlying air. In practice, therefore, a different mode of attack is used.

Suppose that we consider a column of earth Δz units deep. The mass of earth in a unit cross-sectional area of this column is $\rho_s \, \Delta z$. The internal energy of this mass of earth (strictly the enthalpy) is $\rho_s c_s T \, \Delta z$ + constant. If the temperature of this mass of earth changes by an amount ΔT to a new temperature $T + \Delta T$, the amount of heat that is lost (or gained) in a unit cross section of earth is

$$\Delta Q = -\rho_s c_s \, \Delta T \, \Delta z. \tag{5.11}$$

On a chart of T versus z, $\Delta T \, \Delta z$ would represent an area. Thus, if we take a soil sounding at one hour of the day, and another at a different hour of the day, Δt units of time apart, the two soundings will be different, depending on the amount of heat flowing into the soil from the sun. The area enclosed between the two curves and the surface of the earth represents $\Delta T \, \Delta z$, the summation of all the incremental temperature changes in a time Δt, so that

$$\frac{\Delta Q}{\Delta t} = B(0) = -\rho_s c_s \frac{\Delta T}{\Delta t} \Delta z. \tag{5.12}$$

Equation 5.12 measures $B(0)$, the same heat flux as required by Eq. 5.10. The flux of heat measured by Eq. 5.12 must be essentially vertical, because the unit column of soil used as a reference is surrounded by other unit columns undergoing the same thermal process, thus precluding a horizontal temperature gradient, an indication that no heat is transferred in the horizontal. Figure 5.3 shows a typical soil

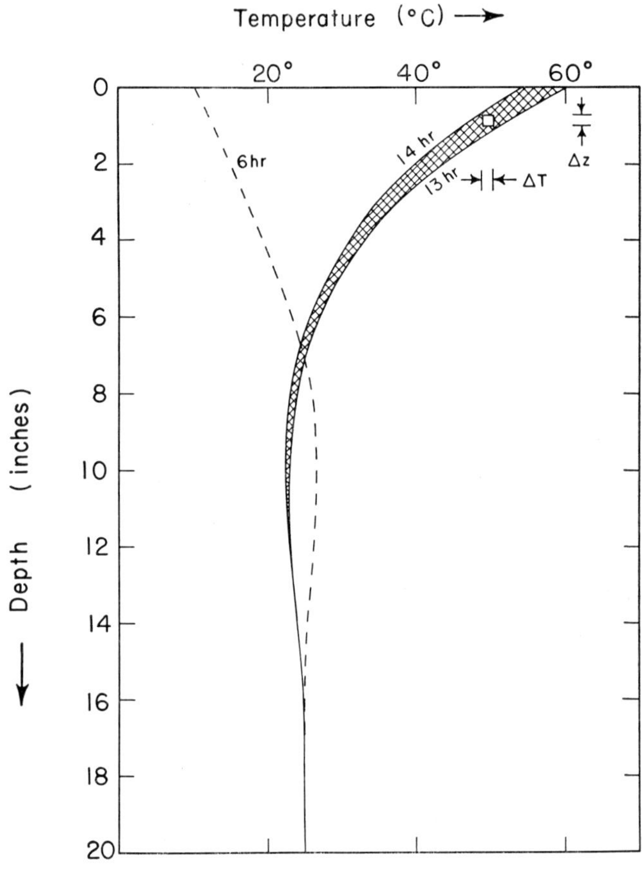

Fig. 5.3 Three soil temperature soundings for 6, 13, and 14 hours local sun time based on a 24-hour clock. The cross-hatched area between soundings for 13 and 14 hours represents the relative energy required to change the temperature of the soil in 1 hour by the amount shown. The tiny rectangle has an area equal to $\Delta T\,\Delta z$ which, by Eq. 5.11, is proportional to ΔQ.

sounding for different times of the day, and illustrates the method of evaluating Eq. 5.12.

In differential form, Eq. 5.12 could be expressed as

$$\frac{\partial B}{\partial z} = -\rho_s c_s \frac{\partial T}{\partial t}. \tag{5.13}$$

By differentiation, Eq. 5.10 can be expressed as (considering $c_s\rho_s\kappa_s$ independent of depth)

$$\frac{\partial B}{\partial z} = -\rho_s c_s \kappa_s \frac{\partial^2 T}{\partial z^2}. \tag{5.14}$$

Eliminating $\partial B/\partial z$ from Eqs. 5.13 and 5.14 makes

$$\frac{\partial T}{\partial t} = \kappa_s \frac{\partial^2 T}{\partial z^2}. \tag{5.15}$$

Equation 5.15 is the fundamental equation that governs transport phenomena in mathematical physics. When T is replaced by c, the concentration of a diffused substance, κ_s is called the diffusivity, and Eq. 5.15 governs the diffusion of one substance through another. The properties of Eq. 5.15 are so well known that in rigorous mathematical treatment it is customary to derive Eqs. 5.4, 5.7, 5.9, and 5.12 as special cases of Eq. 5.15, rather than to deduce Eq. 5.15 from a special case as was done here. κ_s has the dimensions of velocity times distance. $\sqrt{\kappa_s}$ enters into many theoretical developments.†

κ_s is a determining factor in the depth at which a temperature change due to a transport of energy from the surface is noticed in the soil. Two other factors must also influence the temperature change. They are (1) the total mass of earth to be heated in a distance z, and (2) the so-called thermal mass $\rho_s c_s$. Thus, a natural parameter to express the heat flow through a given temperature gradient by different soils, water, snow, or other substances composing the earth's surface is $\rho_s c_s \sqrt{\kappa_s}$, the *thermal property*. Table 5.1 lists the thermal property for different ground types.

An inspection of Table 5.1 will show that still air and snow are good heat insulators, whereas wet soil is a poorer insulator than dry soil. Liquid water is shown to be a very poor thermal insulator. Thus, if we require a certain pattern for the flux of heat from the ground through the surface of the earth, i.e., that $B(0)$ over a period of a day show a sinusoidal variation, Eq. 5.10 says that a good

† In many theoretical treatments, the rate at which the temperature changes in the atmosphere or soil depends on the magnitude of the temperature. If A is a constant, the first statement may be expressed as $\partial T/\partial t = A^2 T$. A solution of Eq. 5.15 under these assumptions is

$$T = T_0 e^{-\frac{A}{\sqrt{\kappa_s}} z},$$

showing the temperature to decrease exponentially with z, a positive number increasing with depth. Here, κ_s is a factor that determines the depth in the soil at which the variation of T with time becomes negligible.

Table 5.1

Specific Heat c_s, Density ρ_s, Thermal Diffusivity κ_s, and Thermal Property $\rho_s c_s \sqrt{\kappa_s}$ for Various Ground Types, Still Water, and Air

After Table 2, H. Lettau in *Trans. A.G.U.*, Vol. 32 (1951), p. 189.

Ground Type	c_s Calories/Degree g	ρ_s g/cm³	κ_s cm²/sec	$\rho_s c_s \sqrt{\kappa_s}$ Calories/Degree cm² sec$^{1/2}$
Turbulent water	1.0	1.0	1 to 100	1 to 10
Concrete	0.22	2.47	0.0107	0.0566
Still water	1.0	1.0	0.0015	0.0388
Sandy clay (15% moisture)	0.33	1.78	0.0037	0.0358
Clayland pasture	$\rho_s c_s = 0.56$		0.0012	0.0194
Quartz sand (medium fine, dry)	0.19	1.65	0.0020	0.0140
Snow	$\rho_s c_s = 0.10$		0.0027	0.0052
Still air	0.24	0.0012	0.16	0.000115
Turbulent air (at 50 meters)	0.24	0.0012	10^4 to 10^5	0.029 to 0.091

insulating ground type requires a large amplitude in the variation of the surface temperature during the period in which the variations of $B(0)$ are considered. Conversely, a good heat conductor will show small variations of the surface temperature over the same period of time. We have assumed the temperature at some depth z to have remained constant while these surface temperature variations were occurring. Measurements[13] in sandy soil at Gila Bend, Arizona, show the temperature to be constant at approximately 18 inches below the surface when diurnal (daily) surface temperatures are considered. Other localities and soil will show different numerical values of the depth at which a constant temperature is reached.

The transfer of heat through phase transformations. The evaluation of W(z)

The least work of a theoretical nature and perhaps of a practical nature has been accomplished on this phase of the heat budget of the earth, particularly when one is considering the heat budget at a specific locality. Experiments such as measuring the rate of evaporation from pans of water have little bearing on the amount of water transpired by trees and plants to be evaporated into the air. However, measurements have been made of the amount of water evaporated by the ground and grasses under natural conditions. The experiment consists of supplying known amounts of water to large tubs of earth in which plants are growing and then measuring the amount of water that has to be supplied to keep the earth at a certain wetness. The water that cannot be accounted for in the growing process must be the water transpired by the plants and subsequently evaporated.

Measurements can be and have been made of the amount of dew and frost deposited on surfaces from the air. Other estimates, which necessarily must be on a grand scale, measure the amount of rainfall and the surface-water runoff. The difference between these two quantities is taken to be the amount of water supplied to the soil. If soil reaches on the average some maximum state of dryness, the water supplied to the soil will be a measure of the moisture that will be evaporated from this same soil on a sequence of dry days. Because of the difficulty of theoretical treatment, most local studies of the heat balance at the surface of the earth are carried out in deserts where $W(0) \approx 0$. The same term is neglected in studies that involve the prediction of minimum temperatures during the night.

Some studies involving the heat budget at the surface of the earth

The Nocturnal Radiation. For much of the remainder of the chapter, we shall be concerned with the evaluation of Eq. 5.1 in specific circumstances. The simplest condition to treat is that of nocturnal radiation. At night, over relatively dry soil, $E(0) = 0$ (no sun) and $W(0) = 0$ (no evaporation or condensation: we will assume no freezing of the ground). Equation 5.1 reduces to

$$S(0) + B(0) + L(0) = 0. \tag{5.16}$$

The long-wave radiation leaving the ground surface is balanced by heat conducted to the surface from deep in the ground plus the heat transferred from the air to the ground. The magnitudes of $B(0)$ and $L(0)$ usually add, because an inversion of temperature rapidly forms in the air under low wind speeds. (Remember that by our convention flux is positive when directed toward the surface layer.) Under low wind speeds, where κ_E, the coefficient of eddy diffusion is so small that $B(0) \gg L(0)$, we may write Eq. 5.16 as

$$|S(0)| = |B(0)|. \tag{5.17}$$

Most treatments of nocturnal radiation use Eq. 5.17 as the heat balance equation. For strong winds, $L(0)$ is not negligible as compared to $B(0)$, so that Eq. 5.16 must apply. Experience verifies theory to show that extremely low temperatures due to radiational cooling are not attained under conditions given by Eq. 5.16, whereas very low temperatures can be obtained when Eq. 5.17 prevails.

Brunt[5,B1] used the condition imposed by Eq. 5.17 to predict minimum temperatures at night due to radiation from the surface of the earth. He proceeded as follows.

If Eq. 5.17 is substituted in Eq. 5.10 and the resulting equation solved for the temperature gradient in the soil, the result is

$$\left(\frac{\partial T}{\partial z}\right)_0 = -\frac{S(0)}{\rho_s c_s \kappa_s}. \tag{5.18}$$

$(\partial T/\partial z)_0$ is the temperature gradient in the surface layer of the ground. Beyond some depth z_∞, the temperature is constant, so that the temperature gradient at $z = \infty$ must be

$$\left(\frac{\partial T}{\partial z}\right)_\infty = 0. \tag{5.19}$$

Equations 5.18 and 5.19 are boundary equations that a solution of the differential equation of heat flow, Eq. 5.15, must satisfy. Through differentiating Eq. 5.15 partially in respect to z and rearranging, the resulting equation may be expressed as a function of temperature gradient:

$$\frac{\partial}{\partial t}\left(\frac{\partial T}{\partial z}\right) = \kappa_s \frac{\partial^2}{\partial z^2}\left(\frac{\partial T}{\partial z}\right). \tag{5.20}$$

Brunt's solution of Eq. 5.20,† subject to the boundary conditions given for $z = 0$ and $z = \infty$, is

$$T = T_0 - \frac{2}{\sqrt{\pi}} \frac{S(0)}{\rho_s c_s \sqrt{\kappa_s}} \sqrt{t} \tag{5.21}$$

where T_0 is the surface temperature at time $t = 0$, and T is the surface temperature at time t. When his expression for the nocturnal radiation, Eq. 5.3, is substituted into Eq. 5.21,

$$T - T_0 = -\frac{2\sigma T_0^4}{\sqrt{\pi}} \frac{[1 - a - b\sqrt{e}]}{\rho_s c_s \sqrt{\kappa_s}} \sqrt{t}. \tag{5.22}$$

$T - T_0$ is the fall of temperature during a night of t seconds duration when the vapor pressure in the surface layer of the atmosphere is e millibars. All other symbols are constants for a given locality.

Frost[10] attacks the problem of finding the minimum temperature at night by considering that all the heat radiated from the layer of air touching the earth's surface is supplied by turbulent transfer of heat downward from layers of air above the earth's surface. His

† See D. Brunt, *Physical and Dynamical Meteorology,* Cambridge Press, 1944.

condition follows directly from Eq. 5.16 with $B(0) = 0$, making

$$|\,S(0)\,| = |\,L(0)\,|. \tag{5.23}$$

The temperature fall during the night as derived by Frost under the assumption of Eq. 5.23 is

$$T - T_0 = -\frac{S(0)\,[(m+1)^2 at]^{m/(m+1)}}{am\rho c_p \Gamma[1/(m+1)]}. \tag{5.24}$$

$m = \frac{1}{3}$ and $a = 40$ are roughness parameters, ρ and c_p are the density and specific heat at constant pressure of air. $\Gamma\left(\frac{1}{m+1}\right)$ is the mathematical Gamma function whose argument is $1/(m+1)$. The other symbols are the standard ones of this chapter.

In an effort to improve the predictive value of either Eq. 5.22 or Eq. 5.24, both of which give reasonable values for minimum temperatures, Knighting[(16)] gives a formula for the minimum temperature due to radiational cooling based on the three terms of Eq. 5.16. The reader is referred to the original article for the formula and details of the derivation.† Measurements at Gila Bend have shown that $L(0)$ cannot be neglected in radiational cooling formulas. The $L(0)$ term amounts to 20 to 40 per cent of $S(0)$. From the foregoing discussion, the factors that aid in maximum nocturnal cooling of the ground surface are clear skies, low water-vapor content of the atmosphere, low wind velocity, and a low value of the thermal property, $\rho_s c_s \sqrt{\kappa_s}$.

Work has also been done at Gila Bend on maximum temperatures attained during the day because of insolation from the sun. The governing equation in this work is Eq. 5.1, with $W(0) = 0$, leaving

$$E(0) + S(0) + B(0) + L(0) = 0. \tag{5.25}$$

During high sun $E(0) > S(0)$, so that many times the $S(0)$ term is neglected.

Lettau[(17)] has been able to explain theoretically the phase lag of temperature in respect to both the diurnal and monthly mean flux of both heat conducted through the soil to the surface and the eddy flux of heat to the surface from the air. Curves of $B(0)$, $L(0)$, and $T(0)$, shown as Fig. 5.4, are from measurements made by Albrecht in Potsdam in 1903. The curves show that the maximum temperature in summer at Potsdam occurs 3 hours after the maximum flux of heat is conducted from the surface through the soil. Similarly, the maximum monthly mean temperature occurs in July, $1\frac{1}{2}$ months after the maximum flux of

† E. Knighting, "A Note on Nocturnal Cooling," *Quant. J. Roy. Meteorol. Soc.*, *76* (1950), p. 173.

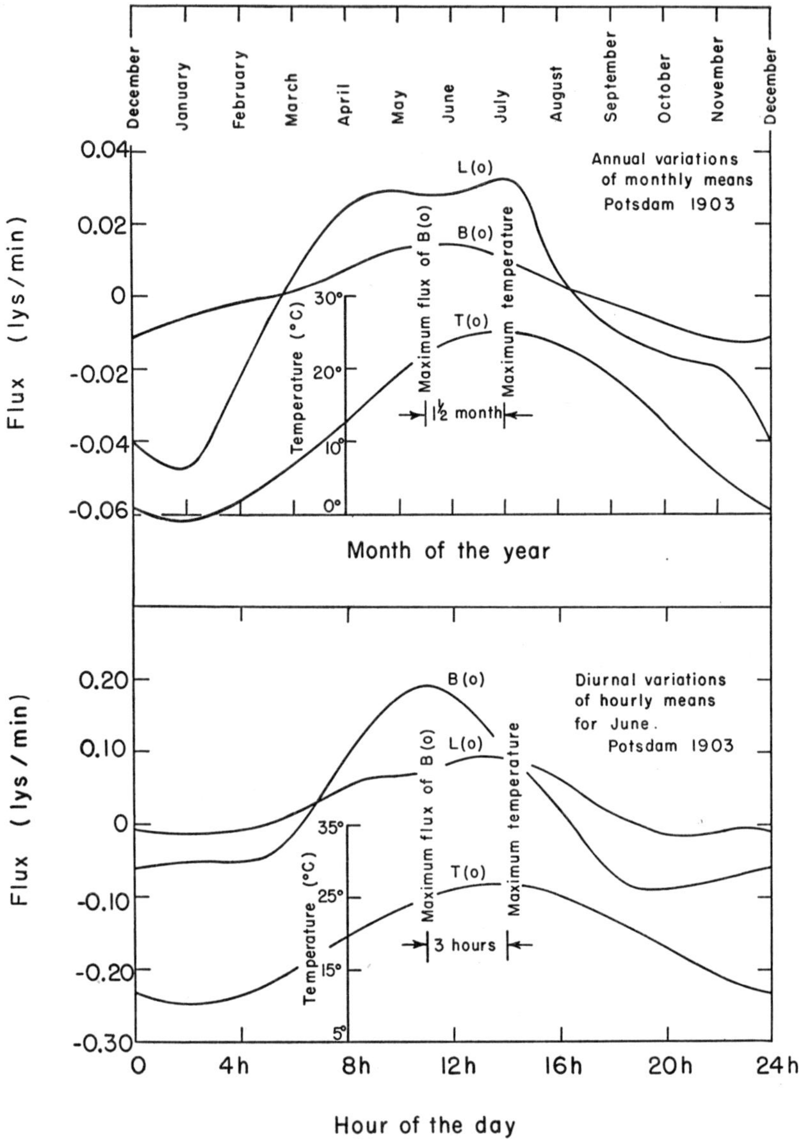

Fig. 5.4 (*After Lettau.*) A typical trace of the flux of heat through the surface of the earth to the soil, $B(0)$, and the air by eddy diffusion, $L(0)$. The course of the surface temperature, $T(0)$, is also shown. Note the lag in time of the maximum temperature behind the maximum flux of heat in the surface soil.

$B(0)$ which occurs in the middle of May. These results are very consistent with examples of heat conduction in other engineering fields. It is a well-known fact that thick walls on a building not only minimize the flow of heat from inside to outside, keeping the inside temperature relatively constant, but that the maximum heat flow through the walls from the outside occurs after the time of maximum heating of the outside wall by the sun. The thicker the wall, the later the maximum heat load inside the house occurs.

The temperature of the lower stratosphere

It has been stated by Gold (1909) and Emden (1913)† and has subsequently become a generally accepted fact that the troposphere is a region where heat exchange is predominately by convection and the turbulent mixing of air with different thermal histories. On the other hand, it is stated that the stratosphere is a region where the heat exchange is mainly through the process of radiative transfer. These two predominating processes lead to a mean lapse rate in the middle troposphere (500–700 millibars) of 6.5° C per kilometer, a lapse rate that is remarkably independent of latitude or longitude. Near the upper troposphere the lapse rate approaches the dry adiabatic lapse of 9.8° C per kilometer. Observations show the lower stratosphere (< 20 kilometers) to be a region where the lapse rate is nearly zero, leaving the stratosphere to be considered in the mean as isothermal. Because of the presence of ozone centered above 20 kilometers, the temperature at higher levels increases again. A discussion of the problem will be left to a later chapter on the upper atmosphere. At this point in the discussion, it is only necessary to realize that a heat source exists at altitudes greater than 20 kilometers. This heat source is a result of absorption of solar radiation by ozone in the wavelength region of the spectrum between 0.22 and 0.32 micron and of terrestrial radiation in a narrow region centered at 9.7 microns. For the discussion that follows, the absorption by ozone in the 9.7 microns band is important in determining a radiation balance at the tropopause.

The observational facts quoted have led to the classical problem of attempting to explain by theory the temperature of the stratosphere and the height at which the tropopause occurs. The tropopause by definition is the boundary surface between the troposphere and stratosphere, and is identified as the surface where the decrease of temperature with height changes to a nearly isothermal lapse rate. Mathematically, the statement implies that the temperature gradient

† See reference 11.

at the tropopause is discontinuous. The temperature itself is single valued at the tropopause, implying continuity of temperature at this point. The foregoing remarks on conditions at the troposphere are true only under average conditions. Inspection of daily weather charts shows that more than one tropopause may exist at a given locality at a given time. At other times, no tropopause may be evident. In the mean, no tropopause occurs in the vicinity of 30° latitude because of the presence of a strong current of air of high velocity (the jet stream) at the elevation (200 millibars) at which the tropopause would normally occur. The tropopause is higher than the jet-stream center at latitudes less than 30°, and lower than the jet stream at latitudes greater than 30°.

As a tropopause at a fixed height is a permanent feature of the mean atmosphere, it is reasonable to assume that the mean tropopause is maintained as a result of thermal equilibrium; i.e., as much energy leaves the region of the tropopause as enters it. A word of caution should be inserted at this point concerning mean values. If the mean heat balance is required at a point (say at the tropopause) the instantaneous heat budget should be computed, and from averaging several instantaneous budgets a mean heat budget should be computed. It is not correct in general to obtain the mean heat budget by using mean values of the components (such as temperature, lapse rate, or wind shear) entering the heat budget. This concept of means has been discussed before (see footnote, p. 76) and care should always be exercised in their computation. In other words, if we say that at the tropopause located at height z, Eq. 5.1 must hold, it is not true in general that

$$\overline{E(z) + S(z) + L(z) + W(z)} = \overline{E(z)} + \overline{S(z)} + \overline{L(z)} + \overline{W(z)} \quad (5.26)$$

where the bar indicates the quantities over which the average value is taken. As no ground surface exists at the tropopause, it is evident that $B(z) = 0$. The tedium attending the evaluation of the left side of Eq. 5.26 is so great that in general the left side is not evaluated. Instead, the left-hand side is approximated by the evaluation of the right-hand side of Eq. 5.26, where the averages of the individual terms are used as an approximation to the true average.

In the lower stratosphere, it has been determined that in the mean the wind shear (the change of the velocity of the horizontal wind with height) is of one order of magnitude less than the wind shear in the upper troposphere. This fact indicates that κ_E of Eq. 5.4 is small, so that L, the eddy diffusion of heat, is small. Therefore, we assume $L(z) = 0$. Measurements of humidity in the lower stratosphere indicate small values of water vapor. These measurements, together with the absence of clouds in the stratosphere, indicate that to a good approximation the latent heat of phase transformation, $W(z) = 0$. Finally, estimates will show that the absorption of solar radiation by the gases of the lower stratosphere, carbon dioxide, water vapor, and to a small degree ozone, is small compared to the infrared absorption of these same gases, so that we may say that $E(z) = 0$. Equation 5.26 is then written as

$$S(z) = 0. \tag{5.27}$$

Equation 5.27 is taken as the condition of equilibrium at the base of the lower stratosphere and hence is coincident with the height of the tropopause. Equation 5.27 implies a balance between the long-wave radiation absorbed on entering the region of the tropopause from the stratosphere above and the troposphere below, with the amount of energy reradiated away from the same level. At the tropopause, Eq. 5.27 may be written in the expanded form of

$$(S^{\downarrow} + S^{\uparrow})_{H_2O} + (S^{\downarrow} + S^{\uparrow})_{CO_2} + (S^{\downarrow} + S^{\uparrow})_{O_3} = 2S_{\text{emitted}} \tag{5.28}$$

where $S = S(z)$ is to be understood for each term. For each gas (subscript), $S^{\downarrow}$ means the contribution of energy to a unit area at the tropopause by the entire stratosphere above the level, $S^{\uparrow}$ the contribution to the same unit area by the entire troposphere below.

In general, the contribution $S^{\uparrow}$ from the troposphere is mainly from the water vapor in the troposphere plus some energy from the earth itself. The latter energy will be absorbed by the O_3 in the 9.7-micron absorption band. An Elsasser chart or Simpson's method (subject to the errors of each method) will give an estimation† of $S^{\uparrow}$, because $S^{\uparrow}$ is equal to the flux emitted from the ambient distribution of temperature and water vapor in the troposphere. Hence, the evaluation of $S^{\uparrow}$ implies a difficult calculation of the energy flux that arrives

† The Elsasser chart, which gives a more accurate evaluation of long-wave flux than Simpson's approximate method, is strictly accurate only at pressures near atmospheric (1000 mb). The exact amount of emission by CO_2 and H_2O at lower pressures is not known exactly and is a matter of great concern at the present. One can say, however, that since the Elsasser chart considers carbon dioxide to radiate as a black body only in the wavelength region, 13.3–17.1 microns, true only in respect to the radiation by water vapor at 1000-mb pressure, for average conditions in the troposphere the relative importance of radiation from carbon dioxide to that from water vapor does not stay constant with increasing height. Since the decrease of the mass of water vapor takes place at a rate the order of four times that of the corresponding decrease in carbon dioxide, the Elsasser chart must underestimate the importance of the contribution of carbon dioxide in respect to water vapor at tropopause elevation.

at the tropopause from the radiating gases of the troposphere. The term $2S_{emitted}$ is the flux emitted by the gases at tropopause temperature for equilibrium conditions.

Figure 5.5 illustrates a graphical solution to the energy balance in the stratosphere.(8) The area under the heavy solid curve of the figure represents the energy arriving at the stratosphere from the troposphere and the surface of the earth. The method of arriving at this curve is discussed in Chapter 4 and illustrated by Fig. 4.11.

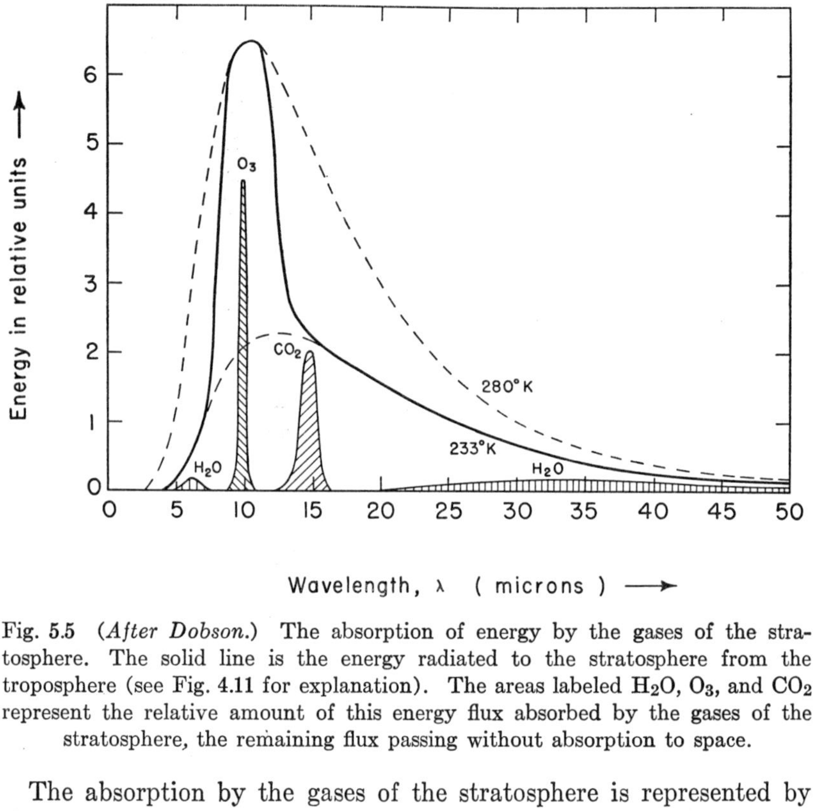

Fig. 5.5 (*After Dobson.*) The absorption of energy by the gases of the stratosphere. The solid line is the energy radiated to the stratosphere from the troposphere (see Fig. 4.11 for explanation). The areas labeled H_2O, O_3, and CO_2 represent the relative amount of this energy flux absorbed by the gases of the stratosphere, the remaining flux passing without absorption to space.

The absorption by the gases of the stratosphere is represented by the cross-hatched areas of the same figure. The sum of energy represented by the cross-hatched areas is just the left-hand side of Eq. 5.28. Setting this value equal to $2S_{emitted}$, the right-hand side of the same equation, and through Eq. 4.8 setting $S_{emitted}$ equal to σT^4, we can arrive at a value of the equilibrium temperature in the stratosphere.

The principal difficulty with this method is the approximate nature of the Simpson method, so that any temperature deduced from the procedure outlined is probably an average temperature for the lower stratosphere because of the assumption that all the energy comes from below and none from above the level in question. However, the method is sensitive enough to show the proper latitudinal and seasonal variation of mean stratospheric temperature. The variation occurs principally through the variation in the amounts of water vapor and ozone. Little is known about the variation of the CO_2 in the stratosphere, but it is assumed to be a constant amount.

In a study of radiative equilibrium at the tropopause, Goody[11] was able to show good agreement with mean tropopause temperatures and heights when using reasonable distributions of temperature, pressure, moisture, carbon dioxide, and ozone in the atmosphere. His results are given in Table 5.2.

Table 5.2

The Height and Temperature of the Tropopause from the Condition of Radiative Equilibrium

(After Goody)

Ground Temperature, °K		Tropopause Temperature, °K		Tropopause Height, km	
		Calculated	Observed	Calculated	Observed
Equatorial region	300	205	193	14.6	16.5
↓	290	208	208	12.6	12.6
	280	210	214	10.8	10.2
	270	213	217	8.8	8.2
Polar region	260	214	219	7.1	6.3

The Heat Budget of the Atmosphere. It has long been recognized that the sun is the ultimate source of the energy that drives our atmospheric heat engine, shown as the large-scale circulation patterns of the earth's atmosphere. With the increase in knowledge of the properties of thermal radiation, it was recognized that the radiant energy that drives the atmosphere and ocean in reality is the difference between the short-wave radiation absorbed by the earth and atmosphere and the long-wave radiation reradiated to space by the same earth and atmosphere. In Chapter 4, it was assumed that when averaged over the entire earth through several years of time, the long- and short-wave heat fluxes were equal. Over shorter periods of time and at different sites on the earth, there is observed a definite unbalance in the two fluxes of radiation. This radiative unbalance, if allowed to persist, would eventually lead to a condition where equatorial regions increased in mean temperature while polar regions decreased in mean temperature, a conclusion contrary to observation. A transport[27] of the excess energy from equator to pole must arise in order for this radiation unbalance to be redistributed over the earth. Thus, the conditions of equilibrium make it necessary to postulate a circulation in the atmosphere and oceans that will transport just enough heat in all of its forms to balance the radiation excess or deficit in the various regions of the earth. The heat may be transported in the atmosphere as the latent heat stored in water vapor before condensing to water, as the enthalpy of the air, a quantity that depends on the temperature and specific heat of the air, as potential and kinetic energy, or by the ocean currents. At any latitude we must be able to write to a high degree of precision an equation analogous to Eq. 5.1 but applying to horizontal rather than local vertical motions. This equation is

$$E + S = W + H + P.E. + K.E. + M \tag{5.29}$$

where E = the flux of short-wave radiation absorbed by earth and atmosphere; S = the flux of long-wave radiation emitted by earth and atmosphere; W = the net flux of latent heat of phase transformation of water; H = the flux of sensible heat (enthalpy flow); $P.E.$ = the flux of potential energy; $K.E.$ = the flux of kinetic energy; and M = the flux of heat transported by ocean currents (at least 10 per cent of the total flux).

When Eq. 5.29 is averaged over the entire earth through a period of years, observation shows that, to measurable accuracy, both sides of the equation are identically zero.

The quantitative evaluation of the radiation balance and an estimation of the total transport of heat across parallels of latitude (meridional transport) was first calculated by G. C. Simpson[26, Chapter 4, pp. 26, 27] in 1928. Later, in Germany, F. Bauer and H. Phillips[3] conducted a joint investigation of the same problem, publishing their results in 1935. Subsequently, a number of investigators attacked the problem. The results quoted will be after H. G. Houghton[14] (1952) and P. Raethjen[23] (1950), which show substantial agreement with other estimates published since 1949 by Albrecht[2] and by London.[18]

The estimate of the amount of absorbed solar radiation is one of the principal differences between the 1930 and the 1950 groups. Another difference, which is a corollary problem of the first, lies in the estimation of the amount of radiation reflected back to space by the clouds. The revised estimate for 1950 has come about not by a change in the estimate of mean cloudiness (50 to 52 per cent by all groups for a hemisphere) but through better measurements of the reflectivities and absorptivities of different generic cloud types. Some values of reflectivities and absorptivities for different type clouds are given in Table

2.3 of Chapter 2. The early workers had only a single set of data for estimation purposes. These data came from measurements of the reflectivity of stratocumulus clouds taken in 1919 by Aldrich.† His results gave a mean reflectivity for an overcast sky of 78 per cent.

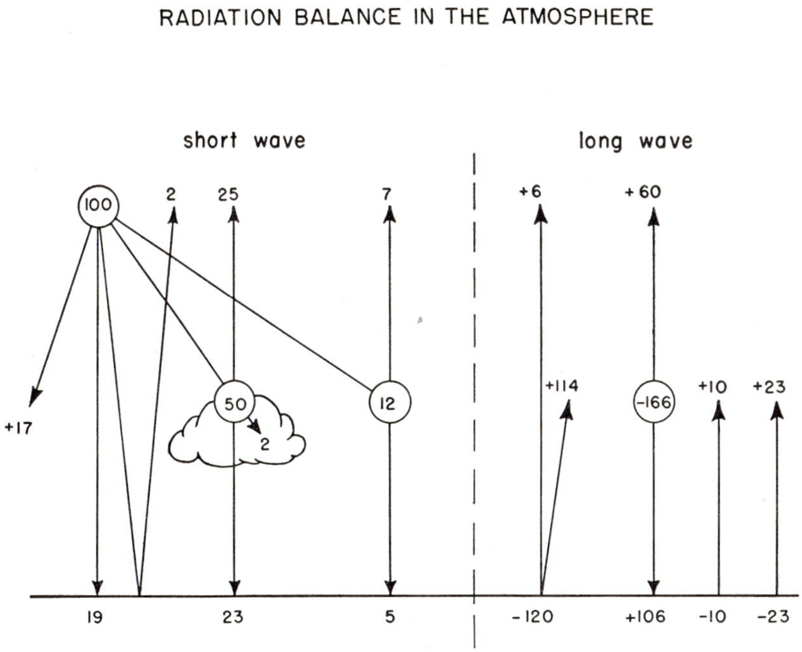

Fig. 5.6 An estimate of the mean annual heat budget of the atmosphere of the northern hemisphere. Values given are in per cent of the average insolation at the top of the atmosphere over the entire earth. A value of 100 units represents 0.485 langley minute^{-1}. We have the following sums in the heat balance:

Albedo of earth = 2 + 25 + 7 = 34%.
Absorption by atmosphere = 17 + 2 = 19%.
Absorption by ground = 19 + 23 + 5 = 47%.
Long-wave radiation to space = 6 + 60 = 66%.
Nocturnal radiation = 120 − 106 = 14%.
Radiation emitted by atmosphere = 17 + 2 + 114 + 10 + 23 = 166%.

The resulting estimates for the albedo of the earth (scattered plus reflected solar radiation) are much lower (34 to 36 per cent) by the 1950 group than the estimates of the 1930 group, which were taken to be 42 to 43 per cent.

The annual mean radiation heat balance for the northern hemi-

† Location cited. Chapter 4.

sphere is shown in Fig. 5.6. The estimates of Houghton and London were used in constructing the diagram, which follows closely the form of a diagram published by London.† The results are given in percentages through letting 100 units equal 0.485 langley per minute, the value of the mean flux of energy when the incoming solar radiation is distributed equally over the entire surface of the earth instead of localized to the hemisphere whose cross section intercepts the insolation. For the short-wave radiation balance, we see that 34 units of energy are reflected or scattered back to space, whereas 47 units are absorbed by the earth and 19 units are absorbed by the atmosphere. Of the 19 units, only 2‡ are absorbed by clouds, the other 17 by the gases of the atmosphere. The $47 + 19 = 66$ units of absorbed energy eventually are reradiated back to space as long-wave radiation. These 66 units of solar energy equal to 0.320 langley minute^{-1}, in the interim between being absorbed and reradiated to space, cause a meridional component of the winds to be impressed on the general west-east circulation of the atmosphere. The latter circulation arises from the rotation of the earth. The meridional component of the circulation is required to transport the equatorial excess of radiation to the polar regions. The mean annual excess of radiation can be inferred from Fig. 5.7 as being directly related to the difference between the curves labeled incoming solar radiation (short-wave) and outgoing terrestrial radiation (long-wave). The abscissa is plotted as the sine of the latitude, so that the spacing of latitude lines is proportional to the area on the earth contained between any two parallels of latitude φ_1 and φ_2. With a plot of this type, the area underneath the radiation curve gives the correct relative amounts of radiation being received by a latitude zone $\Delta\varphi$ degrees wide. As the long- and short-wave radiation must be equal over the entire hemisphere, the areas under the two curves between 0° and 90° must be the same. Mathematically, there is also some mean ordinate that gives the same area as under each of the two radiation curves. The mean ordinate (460 langleys per day) must be identified with the 66 units of flux shown as radiated to space in Fig. 5.6.

Monthly or daily estimates of the excess of short- over long-wave

† J. London, "Study of the Atmospheric Heat Balance" (1951), New York University, College of Engineering Report 131.06.

‡ Individual measurements of absorption by clouds show values of 5 to 9 per cent of the incident solar radiation absorbed by stratus and up to 20 per cent absorbed by deep clouds. However, in the heat-budget calculations only 50 per cent of the sky is covered and the sun shines for only one-half the time on any spot. These facts can cut the average absorption from 8 per cent to, for example, $\frac{1}{2} \times \frac{1}{2} \times 8 = 2$ per cent.

radiation as a function of latitude show displacements of the pattern of Fig. 5.7, with a corresponding difference in magnitude of flux. The inference is that in anything less than annual averages the southern hemispheric heat balance must also be considered in these studies. In

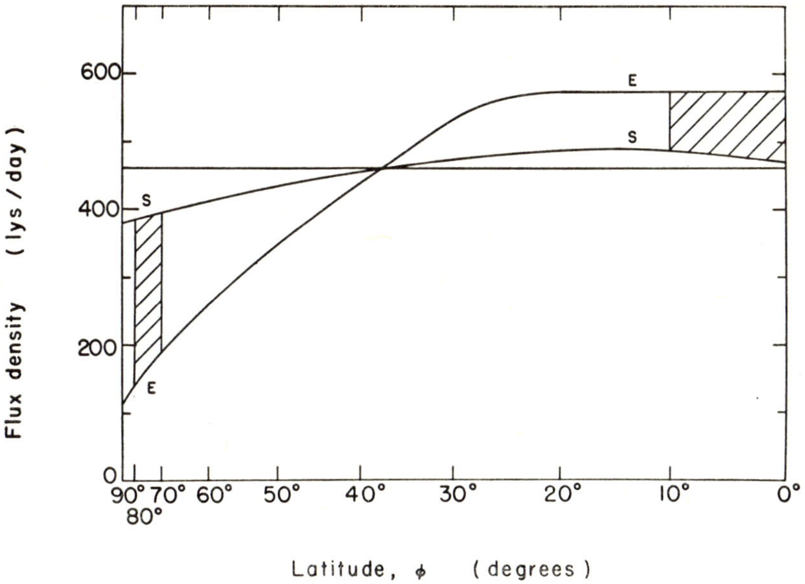

Fig. 5.7 (*After Houghton.*) The annual mean radiation as a function of latitude. E represents the short-wave solar energy and S represents the long-wave terrestrial radiation. As the ordinate is spaced as the sine of the latitude, area under the curve represents energy received per unit time by a latitude zone $\Delta\varphi$ units wide. The straight line at 460 langleys per day is the average value of both E and S. The difference $E - S$ over any incremental latitude difference $\Delta\varphi$ is the amount of energy that must be transported per day in the mean over a latitude circle $_\varphi$. The hatched areas show the relative energies that must be transported across entire latitude circles located at 5° and 75°. The latitude zones are taken as 10° wide.

the few studies that have been made, a transport of heat across the equator is called for from radiation considerations, the sign of the transport changing from winter to summer.

The 66 units of long-wave radiation to space contributed by the earth and atmosphere arise from geographical sites much different from those in which the short-wave radiation was absorbed. The principal difference between the short- and long-wave radiation balances lies in the lack of transparency of clouds and the atmosphere to long-wave radiation. This feature was discussed in Chapter 4. The complicated interchange of long-wave radiation among earth, atmosphere, and space leads to the long-wave balance pictured in Fig. 5.6. From this diagram, we see that while 66 units of radiation ultimately are returned to space, at any average instant 120 units leave the earth's atmosphere and are absorbed by the atmosphere. At the same time, the atmosphere is returning 106 units of this radiation to the earth, leaving a net loss by radiation (the so-called nocturnal radiation) of $120 - 106 = 14$ units.

On recalling the band structure of water vapor and Simpson's simple picture of nocturnal radiation earlier in the chapter, one is led to the conclusion that an appreciable part of the 14 units is radiated directly to space in the $8\frac{1}{2}$- to 11-micron region of the spectrum. In addition to direct radiational losses from the earth to the atmosphere, the atmosphere shows further heat gains resulting from the condensation of water vapor to clouds and rain. This gain amounts to 23 units. The turbulent flux of heat from the ground to the air adds an additional 10 units of flux. The latter heat fluxes plus the 19 units of short-wave radiation absorbed directly by the atmosphere from the sun complete the assumed heat balance of the northern hemisphere. The balance sheet is noted in the text under Fig. 5.6 for space, atmosphere, and earth.

It will be noticed that the direction of the eddy flux of heat is reversed from the 1930 to the 1950 results. The eddy flux of heat is the last component of the heat balance that is estimated. Independent estimates of the eddy flux have been made by workers interested in atmospheric turbulence. Because of the stronger winds and consequently greater vertical mixing during the daylight hours, the net transfer of heat by eddy diffusion is inferred by them to be away from the earth's surface. The earlier radiation studies of Bauer and Phillips required the incorrect sign of the eddy flux of heat, a transport to rather than away from the earth's surface.

Table 5.3 is included to show the difference in the radiation balance averaged over 6-month periods instead of over the entire year. The table is derived from Raethjen. His values have been converted to langleys day^{-1} (cal cm^{-2} day^{-1}) for comparison with Houghton's results in Fig. 5.7. The table is self-explanatory. The transport terms are derived from considering the total energy excess or deficit (column 8 multiplied by the area of the zone which equals column 10). The energy of column 10 must be transported across a latitude circle in order to keep the zone from changing its mean temperature. Thus, dividing column 10 by the length of the latitude circle across which the transport is to occur gives the value of the transport per unit length

Table 5.3

A Table of the Radiation Balance of the Northern Hemisphere (Earth Plus Atmosphere) According to P. Raethjen[23]

Values in columns 3 and 4 can be compared with values obtained by Houghton, which are plotted as Fig. 5.7, if one averages the summer and winter values for each zone to obtain a mean yearly average. Column 6 can be compared directly with the differences in flux in Fig. 5.7.

Column

1. *Zone*
2. f Fractional area of the earth's surface by zones
3. S Long-wave emission over a zone
4. E Absorbed solar radiation
5. $F = E - S$ Radiation balance (half-year average)
6. F_y Radiation balance (yearly average)
7. $F_s = \frac{1}{2}(F - F_y)$ Storage of radiant energy
8. $F_T = F - F_s$ Transport of energy
9. φ Latitude circle through which energy is transported

10, 11. Transport. Energy-transport current (direction shown by arrows).

The values for the hemisphere (lowest line) are the zonal values of radiation weighted by the relative area. The mean albedo of the earth is calculated to be 40 per cent (Raethjen), whereas Houghton gives 34 per cent.

	1	2	3	4	5	6	7	8	9	10	11
	Zone	f	S	E	F	F_y	F_s	F_T	φ	Trans. 10^{18} $\frac{\text{cal}}{\text{day}}$	Trans. 10^{8} $\frac{\text{cal}}{\text{cm day}}$
			ly/day		(Averaged over half year)						
Winter	90° to 60°	0.07	421	33	−388	−240	−74	−314			
									60°	↑ 98	↑ 492
Winter	60° to 40°	0.11	421	169	−252	−93	−80	−172			
									40°	↑ 180	↑ 587
Winter	40° to 20°	0.15	450	372	−78	+38	−58	−20			
									20°	↑ 191	↑ 510
Winter	20° to 0°	0.17	432	510	+78	+126	−24	+102			
									0°	↑ 115	↑ 285
Summer	0° to 20°	0.17	426	600	+174	+126	+24	+150			
									20°	↑ 11	↑ 26
Summer	20° to 40°	0.15	456	610	+154	+38	+58	+96			
									40°	↓ 55	↓ 181
Summer	40° to 60°	0.11	437	503	+66	−93	+80	−14			
									60°	↓ 49	↓ 250
Summer	60° to 90°	0.07	442	350	−92	−240	+74	−166			
	Hemisphere	0.50	437	437	0	0	0	0			

	Zone	Albedo after Raethjen
Winter	90° to 60°	0.66
Winter	60° to 40°	0.51
Winter	40° to 20°	0.38
Winter	20° to 0°	0.37
Summer	0° to 20°	0.36
Summer	20° to 40°	0.36
Summer	40° to 60°	0.44
Summer	60° to 90°	0.55
	Hemisphere	0.40

	Zone	Albedo after Raethjen	Albedo after Houghton
Yearly average	90° to 60°	0.60	0.52
Yearly average	60° to 40°	0.47	0.39
Yearly average	40° to 20°	0.36	0.30
Yearly average	20° to 0°	0.36	0.31
Hemisphere		0.40	0.34

of latitude that is indicated in column 11. The values of column 11 correspond to the energy in the left-hand side of Eq. 5.29, which must be transported across a latitude circle by means of the various transport processes shown as the right-hand side of the same equation.

References

1. Albrecht, F., "Untersuchungen ueber den Waermehaushalt der Erdoberflaeche in verschiedenen Klimagebieten," *Reichsamt f. Wetter dienst,* Berlin, *8* (1940).
2. Albrecht, F., "Über die Wärme- und Wasserbilanz der Erde," *Ann. Meteorol., 2,* 129–143 (1949).
3. Bauer, F., and Phillips, H., "Der Wärmehaushalt der Lufthülle der Nordhalbkugel," *Gerlands Beitr. Geophys., 42,* 160–207 (1934); *45,* 82–132 (1935); *47,* 218–223 (1936).
4. Brooks, F. A., "Observations of atmospheric radiation," *Mass. Inst. of Tech., Papers* in *Physical Oceanography* and *Meteorology, 8,* No. 2 (1941).
5. Brunt, D., "Radiation in the Atmosphere," *Quart. J. Roy. Meteorol. Soc.,* Supplement to Vol. *66,* 34–40 (1940).
6. Czepa, O., "Über die nächtliche Abkühlung der Erdoberfläche und der bodennahen Luftschicht," *Z. Meteorol., 4,* 359–362 (1950).
7. Czepa, O., "Über die Energieleitung durch langwellige Strahlung in der bodennahen Luftschicht," *Z. Meteorol., 5,* 292–300 (1951).
8. Dobson, G. M. B., Brewer, A. W., and Cwilong, B. M., "Bakerian Lecture: Meteorology of the lower stratosphere," *Proc. Roy. Soc.* (A), *185,* 144–175 (1946).
9. Fritz, S., and MacDonald, T. H., "Measurements of Absorption of Solar Radiation by Clouds," *Bull. Am. Meteorol. Soc., 32,* 205–209 (1951).
10. Frost, R., *Prof. Notes,* Meteorological Office, London, No. 95 (1948) (see Knighting, reference 15).
11. Goody, R. M., "The thermal equilibrium at the tropopause and the temperature of the lower stratosphere," *Proc. Roy. Soc.* (A), *197,* 487–505 (1949).
12. Goody, R. M., and Robinson, G. D., "Radiation in the troposphere and lower stratosphere," *Quart. J. Roy. Meteorol. Soc., 77,* 151–187 (1951).
13. Guild, W. R., "Note on heat transfer at the soil surface," *J. Meteorol., 7,* 140–144 (1950).
14. Houghton, H. G. Unpublished studies of atmospheric radiation, Dept. of Meteorology, Mass. Inst. of Technology, Cambridge, Mass. (1952).
15. Jehn, K. H., and Gerhardt, J. R., "A preliminary study of the eddy transfer of heat near the earth's surface," *J. Meteorol., 7,* 343–346 (1950).
16. Knighting, E., "A note on nocturnal cooling," *Quart. J. Roy. Meteorol. Soc., 76,* 173–181 (1950).
17. Lettau, H., "Theory of Surface-Temperature and Heat-Transfer Oscillations Near a Level Ground Surface," *Trans. Am. Geophys. Union, 32,* 189–200 (1951).
18. London, J., "Study of the Atmospheric Heat Balance," Progress Report 131.06, Research Division, College of Engineering, New York University (1951) (see also *J. Meteorol., 9,* 145–151 (1952)).
19. Middleton, W. E. K., "Note on the reflection of diffuse daylight from still water," *Quart. J. Roy. Meteorol. Soc., 78,* 627–628 (1952).
20. Möller, F., "Der Wärmehaushalt der Atmosphäre," *Experienta, 6,* 361–367 (1950).

21. Neiburger, M., "Reflection, Absorption, and Transmission of Insolation by Stratus Cloud," *J. Meteorol., 6,* 98–104 (1949).
22. Priestly, C. H. B., and Sheppard, P. A., "Turbulence and transfer processes in the atmosphere," *Quart. J. Roy. Meteorol. Soc., 78,* 488–529 (1952).
23. Raethjen, P., "Wärmehaushalt der Atmosphäre," *Kurzer Abriss der Meteorologie dynamische geshen,* Teil 2, Hamburg (Universitat Hamburg).
24. Rider, N. E., and Robinson, G. D., "A study of the transfer of heat and water vapour above a surface of short grass," *Quart. J. Roy. Meteorol. Soc., 77,* 375–401 (1951).
25. Robinson, G. D., "Notes on the measurement and estimation of atmospheric radiation—2," *Quart. J. Roy. Meteorol. Soc., 76,* 37–51 (1950).
26. Simpson, G. C., "Distribution of terrestrial radiation," *Mem. Roy. Meteorol. Soc., 3,* No. 23 (1929) (see also *3,* No. 21 (1928) and *2,* No. 16 (1928) referred to in Chapter 4).
27. White, R. M., "On the Energy Balance of the Atmosphere," *Trans. Am. Geophys. Union, 32,* 391–396 (1951).

Source Books

B1. Brunt, D., *Physical and Dynamical Meteorology,* Cambridge University Press (1944) (Chapters V through VII).
B2. Sutton, O. G., *Atmospheric Turbulence,* Methuen's Monographs, John Wiley & Sons, New York (1949).

Problems

5.1 During the course of a day a unit area of ground surface receives $E(0)$ units of flux from the sun and reaches a temperature T. The air immediately overlying the surface has a vapor pressure e. For certain selected hours of the day, these values are as follows.

Hour of the Day

	6 hr	9 hr	12 hr	15 hr	18 hr	24 hr
$E(0)\left(\frac{\text{cal}}{\text{cm}^2\text{ min}}\right)$	0.11	0.84	1.21	0.87	0.13	0
T, °C	16°	23°	26°	27°	24°	18°
e, mb	12	13	15	16	15	14

Find the difference between the long and short wave radiation throughout the day. Assume $a = 0.5$ and $b = 0.06$ in the Brunt radiation formula.

5.2 Find the eddy flux of heat in the atmosphere at both the 6-foot and the 35-foot levels from the following data.

Height above surface (feet)	3	6	12	20	35	55	80
Temperature, °C	27.7°	28.2°	28.0°	26.9°	26.2°	26.2°	26.1°

Take $z_0 = 0.1$ ft and $\kappa_E' = 1.2 \times 10^3$ cm^2 sec^{-1} per ft. The surface pressure is 1018 millibars. The air is assumed to be dry.

5.3 Consider the following soil soundings taken 3 hours apart.

Depth, inches	0	1	2	4	6	9	12	18
Temperature, °C	34.0°	38.5°	41.0°	36.0°	30.0°	25.5°	25.0°	25.0°
Temperature, °C	18.5°	24.0°	28.0°	32.0°	31.0°	27.5°	26.0°	25.0°

What is the average flux of heat per unit area through the soil surface during this period? The surface is cooling throughout the period. The soil constants are those applicable to quartz sand (Table 5.1).

5.4 The steady-state flux of heat by eddy diffusion can be written as

$$L(z) = L(0) - \int_0^z \rho c_p \left(\frac{\partial \Theta}{\partial t} \right) dz.$$

This equation is related to Eq. 5.7 through the same formal system of equations that relates Eq. 5.12 to Eq. 5.10. $L(0)$ is the heat added to or taken from the air at ground level. A reasonable value for $L(0)$ is 4×10^{-3} cal cm^{-2} sec^{-1}.

a. Find $L(z)$ at 6 feet and at 35 feet from the following soundings spaced 1 hour apart. The surface temperature is increasing with time.

Height above surface (feet)	3	6	12	20	35	55	80
Temperature, °C	26.8°	26.9°	26.0°	25.7°	25.5°	25.6°	25.1°
Temperature, °C	28.4°	28.1°	28.0°	27.3°	27.2°	26.9°	26.7°

b. With Eq. 5.7 (or Eq. 5.4) as the defining equation, find κ_E, the eddy diffusivity of air, at the 6-foot and 35-foot levels. Assume dry air at a surface pressure of 1018 mb.

5.5 During a good radiation night, the net radiation leaving the earth is 0.3 of black body. At the beginning of the radiative cooling period, the air next to the earth (and thus presumably the earth's surface) has a temperature of 12° C and a pressure of 1010 mb. What is the temperature of the surface air 8 hours later (*a*) using Brunt's formula over a quartz sand surface (Table 5.1); and (*b*) using Frost's formula?

CHAPTER 6

Refraction and Diffraction by Atmospheric Suspensoids. Atmospheric Optics

The physical explanation of rainbows, haloes, coronas about the sun and moon, and other diffraction phenomena due to suspensoids (usually rain or ice crystals) in the atmosphere was known before 1900. The classic book† in this field is Pernter-Exner's[B5] *Meteorologische Optik,* which should be consulted for details of derivations and other fine points. In this chapter, an attempt will be made to present the classical explanation of several optical effects, and to indicate the limits beyond which the simple explanations must be modified to be in agreement with a more complete and rigorous treatment of the investigated phenomena. The discussion will lead us from the simple geometric optics of large (compared to wavelength) spheres and prisms to more complicated diffraction problems. The latter arise as the suspensoids become smaller. In the limit of very small particles, the scattering problem, already discussed in Chapter 2, occurs. The integrating theory for all of these size effects is the Mie[Chapter 2, 5] theory mentioned before.

For the range of particles usually found in the atmosphere, i.e., of the order of 1 millimeter (1000 microns) for a raindrop, a simple calculation shows that the ratio of diameter to wavelength is about 2000 to 1 for a raindrop illuminated by visible light and is 1 to 30 when the same drop is illuminated by a 3-centimeter wavelength radar. The latter condition has been discussed in Chapter 2 and is clearly a case where scattering (for radar, it is back scattering) theory applies. When the drop is illuminated by visible light, the raindrop acts like a small spherical lens with internal reflection, and the methods of ray tracing from geometric optics give very accurate results. Subsequent sections will illustrate more fully these statements.

† W. E. Humphreys' *Physics of the Air* contains excellent sections on classical meteorological optics and has the obvious advantage for American students of being written in English.

Rainbows, fog bows, glories

Figure 6.1 shows the geometry attending the refraction and internal reflection of one typical ray from the many ray pencils composing the parallel light from the sun in their passage through a spherical rain-

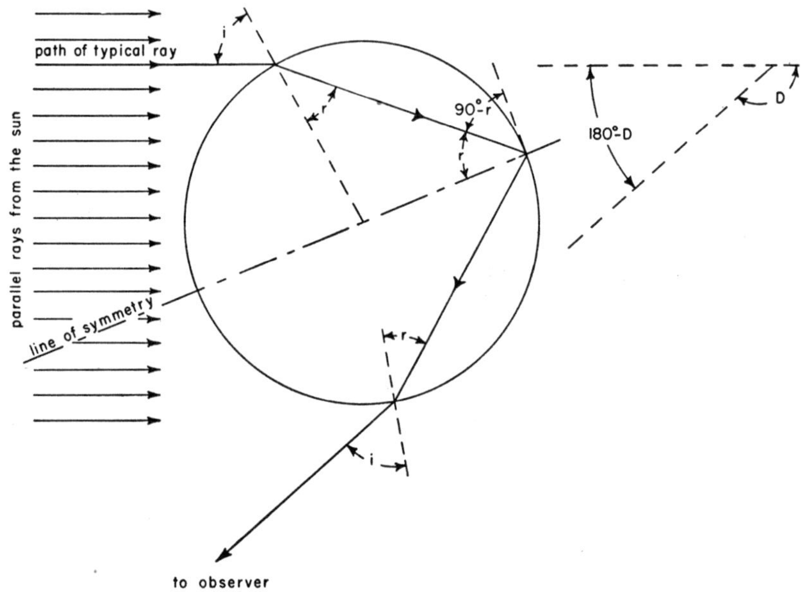

Fig. 6.1 The geometrical path of a typical ray entering into the rainbow problem. At each interface between the air and the water droplet, there are three rays, an incident, reflected, and refracted ray. At each interface, only the reflected or refracted ray that is ultimately detected by the observer is shown. The refracted rays, although not shown, account for most of the energy. The internally reflected rays are rather weak.

drop. For a single ray we can write the total angle through which the ray is turned from the point at which it enters the raindrop to the point at which it emerges. This, called the angle of deviation, is angle D in Fig. 6.1. D will be seen to be equal to twice the difference between the angle of incidence and the angle of refraction plus twice the angle formed with the tangent on undergoing one internal reflection. For n internal reflections, there are n such internal angles. Thus, if i is the angle of incidence and r is the angle of refraction, we can write

$$D = 2(i - r) + 2n\left(\frac{\pi}{2} - r\right). \tag{6.1}$$

If one traces the path of several of the sun's rays through a spherical drop, using the ray-tracing methods of geometric optics, it immediately becomes apparent that there is a concentration of rays denoting high intensity of light about the ray whose angle of deviation in passing through the drop is measured to be least. This measurement bears out the prediction from diffraction theory that the greatest intensity of light should be concentrated about the path of minimum deviation. Positions other than those for minimum deviation show much less intensity, the intensity falling off sharply when the path of minimum deviation is left. Little error is made then in describing the properties of the rainbow, through a knowledge of the properties of the minimum deviation.

The minimum deviation of a bundle of rays is found by differentiating Eq. 6.1 and setting the result equal to zero. This leads to

$$\frac{dD}{di} = 2\left[1 - (n+1)\frac{dr}{di}\right] = 0. \tag{6.2}$$

Snell's law, derived from Eq. 1.1, relating the index of refraction m to the angles of incidence and refraction, is

$$\sin i = m \sin r. \tag{6.3}$$

The derivative of Eq. 6.3 shows that

$$\cos i = m \cos r \frac{dr}{di}, \tag{6.4}$$

so that Eq. 6.2 may be written as

$$(n+1)\cos i = m \cos r. \tag{6.5}$$

We have eliminated dr/di from Eq. 6.2 by using Eq. 6.4. Equation 6.5 signifies the condition that must exist for the path of the most intense pencil of light being refracted by a raindrop.

We can eliminate one of the angles by separately squaring Eqs. 6.3 and 6.5 and adding the results. We obtain

$$\sin^2 i = m^2 \sin^2 r \qquad \text{(on squaring 6.3)}$$

$$(n^2 + 2n + 1)\cos^2 i = m^2 \cos^2 r \qquad \text{(on squaring 6.5)}$$

whose sum is

$$(n^2 + 2n)\cos^2 i + (\sin^2 i + \cos^2 i) = m^2(\sin^2 r + \cos^2 r).$$

As $\sin^2 i + \cos^2 i = \sin^2 r + \cos^2 r = 1$, the latter expression simplifies to

$$\cos i = \sqrt{\frac{m^2 - 1}{n^2 + 2n}}. \tag{6.6a}$$

Substituting Eq. 6.6*a* into 6.5 gives the corresponding equation for *r*, namely that

$$\cos r = \frac{n+1}{m}\sqrt{\frac{m^2-1}{n^2+2n}}. \tag{6.6b}$$

The second derivative† of Eq. 6.1 indicates that the values of *i* and *r* derived from Eqs. 6.6*a* and 6.6*b* give minimum rather than maximum deviation.

Once the index of refraction of the drop is known, plus the number of internal reflections *n*, the minimum deviation can be found from Eqs. 6.6*a*, 6.6*b*, and 6.1. For the formation of a primary rainbow, $n = 1$. A water droplet illuminated by the red $H\alpha$ line ($\lambda = 656.3$ mμ) has an index of refraction, $m = 1.3318$. From the three equations just

Table 6.1

Angles of Incidence, Refraction, and Angular Radius of the Rainbow, 180° − *D*, for Primary, Secondary, and Tertiary Rainbows

The wavelengths used are identifiable lines in the mercury emission spectrum.

Color	Index of Refraction	Angle	Primary ($n = 1$)	Secondary ($n = 2$)	Tertiary ($n = 3$)
Violet	1.3435	i	58° 48′	71° 30′	76° 36′
$\lambda = 404.7$ mμ		r	39° 33′	44° 54′	46° 23′
		$180° - D$	40° 36′	53° 36′	142° 8′
Green	1.3352	i	59° 17′	71° 46′	76° 47′
$\lambda = 546.1$ mμ		r	40° 5′	45° 19′	46° 48′
		$180° - D$	41° 46′	51° 38′	139° 10′
Yellow	1.3341	i	59° 21′	71° 48′	76° 49′
$\lambda = 577.0$ mμ		r	40° 10′	45° 23′	46° 52′
		$180° - D$	41° 58′	51° 18′	138° 42′
Red	1.3318	i	59° 29′	71° 53′	76° 52′
$\lambda = 656.3$ mμ		r	40° 19′	45° 31′	46° 58′
		$180° - D$	42° 18′	50° 40′	138° 0′

named, we find $\cos i = 0.5078$ and $\cos r = 0.7626$, making $i = 59°29'$, $r = 40°19'$ and $D = 137°42'$.

A primary rainbow may be observed when one faces a distant rain shower with the sun at one's back. Some of the sunlight striking the

† A positive value of the second derivative is the calculus condition for a minimum point.

rain shower is deviated to the observer's eye by the raindrops, meeting the conditions for minimum deviation. The locus of all the pencils of sunlight undergoing minimum deviation to the eye is a circle whose angular radius is $180° - D$. This latter angle is the angle formed between the line (or ray) joining a point on the rainbow and the observer's eye, with the line (or ray) containing the apparent center of the sun, the observer's eye, and the center of the rainbow. The angle, $180° - D$, is 42°22′ for the red bow of a primary rainbow. As

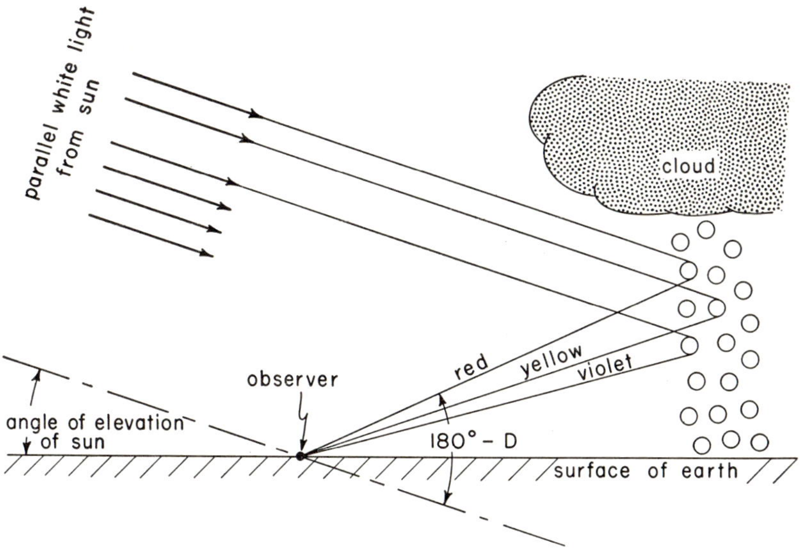

Fig. 6.2 Schematic drawing of a cross section of the ray geometry in a primary rainbow.

$180° - D$ varies but little from 42° for all colors, the primary rainbow is known as a 42° bow. Because the index of refraction is a function of wavelength, the minimum deviation is also a function of wavelength. Table 6.1 illustrates the variation of $180° - D$ as a function of index of refraction and number of internal reflections.

The variation of the minimum deviation with wavelength indicates that the geometry is favorable for drops located at a particular point in the sky to deviate only one color to an observer on the ground. The other colors deviated by this array of drops will be undetected by him. An inspection of Table 6.1 and Fig. 6.2 shows that drops low in the rain sheet should give the blue contribution of the sun's light, whereas drops higher up should give the red contribution. This separation of colors from white incident sunlight causes the primary rainbow (the most intense bow that is observed) to have a sequence of colors ranging from red on the outside through yellow to violet on the inside of the bow.

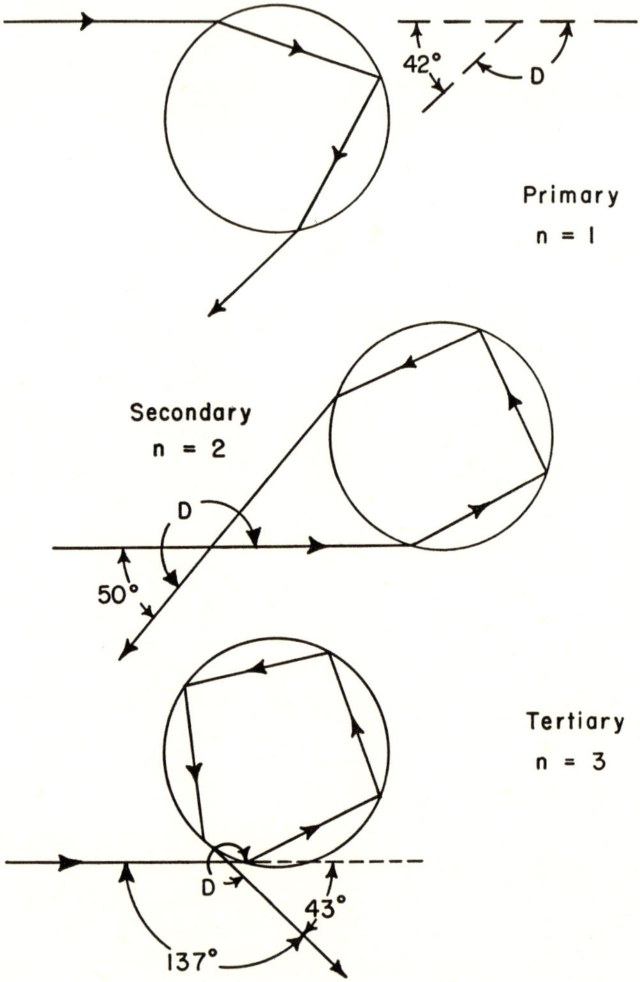

Fig. 6.3 The ray geometry required for the formation of primary, secondary, and tertiary rainbows.

Rainbows are classified by the number of internal reflections that a ray must undergo before emergence and eventual detection by an observer. Physical optics tell us that reflections from an air-water interface are not complete, since most of the energy passes through the interface instead of being reflected. From this it follows that the more internal reflections in the drop the weaker the emerging beam. Thus, if n is the number of internal reflections, the primary bow, characterized by $n = 1$, is the most intense rainbow observed. The secondary bow has $n = 2$, whereas the tertiary bow has $n = 3$, etc. The ray paths for these higher-order bows are shown in Fig. 6.3.

From the geometry and physical conditions just discussed, we deduce the following conditions necessary for observing rainbows.

1. Showery conditions necessitated by the simultaneous occurrence of rain and sunshine.
2. Primary bows: sun at back; rain in front of observer.
3. Secondary bows: sun at back; rain in front; bow observed above the primary bow with primary bow colors reversed.
4. Tertiary bows: faint; in direction of sun; same sequence of colors as in primary bow.
5. Higher orders: rarely observed; too weak.

Since rain is needed to produce a rainbow, only fragments of rainbows may be observed when local showers are occurring. When no bow is observed, it may be either because no rain is occurring at the particular angle required for seeing a rainbow or because of some obstruction in the ray path. An intervening cloud or even a part of the earth may constitute the obstruction. The latter, particularly the earth's horizon, precludes observing completely circular rainbows. By confining the discussion to the primary bow and utilizing Fig. 6.2, we deduce that, for the sun on the horizon, the incident rays are parallel to the earth's surface, and only the upper half of the bow can be seen. As the sun rises the bow appears to sink so that when the elevation angle of the sun is equal to $180° - D$, the angular radius of the rainbow, no rainbow will be observed. Thus, rainbows are not observed in general from the ground around noon. Occasionally, complete "rainbows" are reported by airplane passengers. The bows are observed to be encircling the shadow of the airplane cast on the top of clouds. This phenomenon is not a true rainbow but is a "glory," described later in this chapter.

The complete rainbow phenomenon is not describable by simple geometric optics. A complete description[13,B2] of the rainbow requires the aid of diffraction theory to explain the intensities of the colors plus the presence of the various supernumerary bows that are occasionally observed inside the primary bow. Let us consider the wave fronts of the sunlight on emerging from a drop, shown graphically in Fig. 6.4. We notice that the cusplike wave fronts in the figure show alternate regions of reinforcement and interference of the wave-front segments, leading to alternating regions of light and darkness. Further inspec-

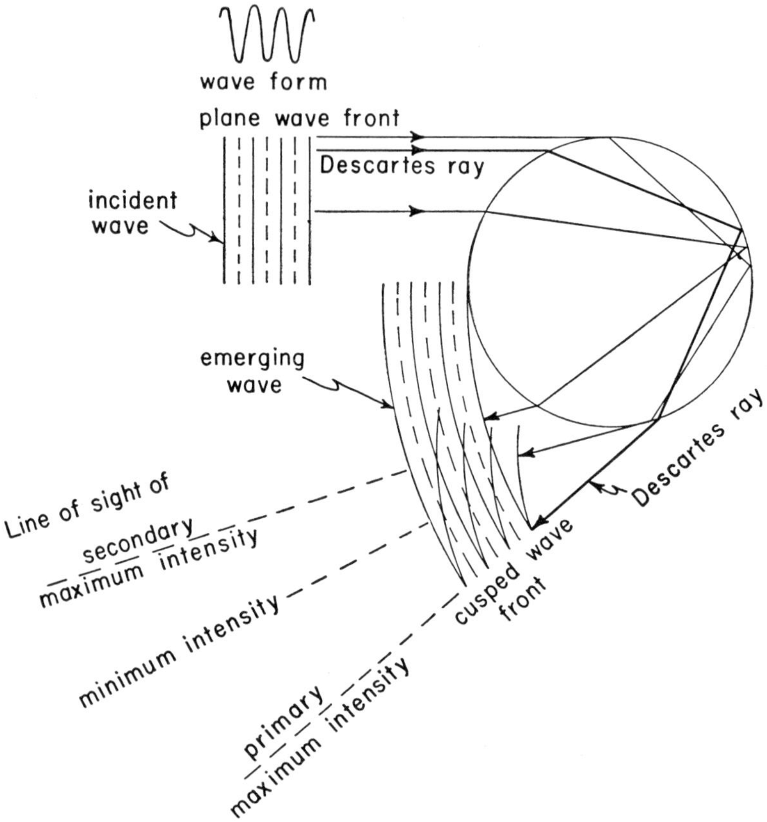

Fig. 6.4 The wave-front patterns for a primary rainbow. The initial plane wave, shown as a series of solid lines for the position of maximum amplitude and dashed lines as the position of minimum amplitude, becomes a cusped spherical wave after passage through a raindrop. The cusped portion of one wave interferes with the spherical portion of the wave next to it to form alternating regions of light and dark, the supernumerary rainbows. The ray that undergoes minimum deviation is called the Descartes ray and is the locus of the apex of the cusp. The three rays illustrate the path that the three rays travel *in the same period* of time. This means that the locus of the end points of the rays is a line (or in three dimensions a surface) of constant phase.

tion shows that this diffraction pattern should fall within the area encircled by the primary bow. The elements of the diffraction pattern (see Fig. 6.5), a series of colored rings because of the variation in wavelength of visible light, are the supernumerary rainbows.

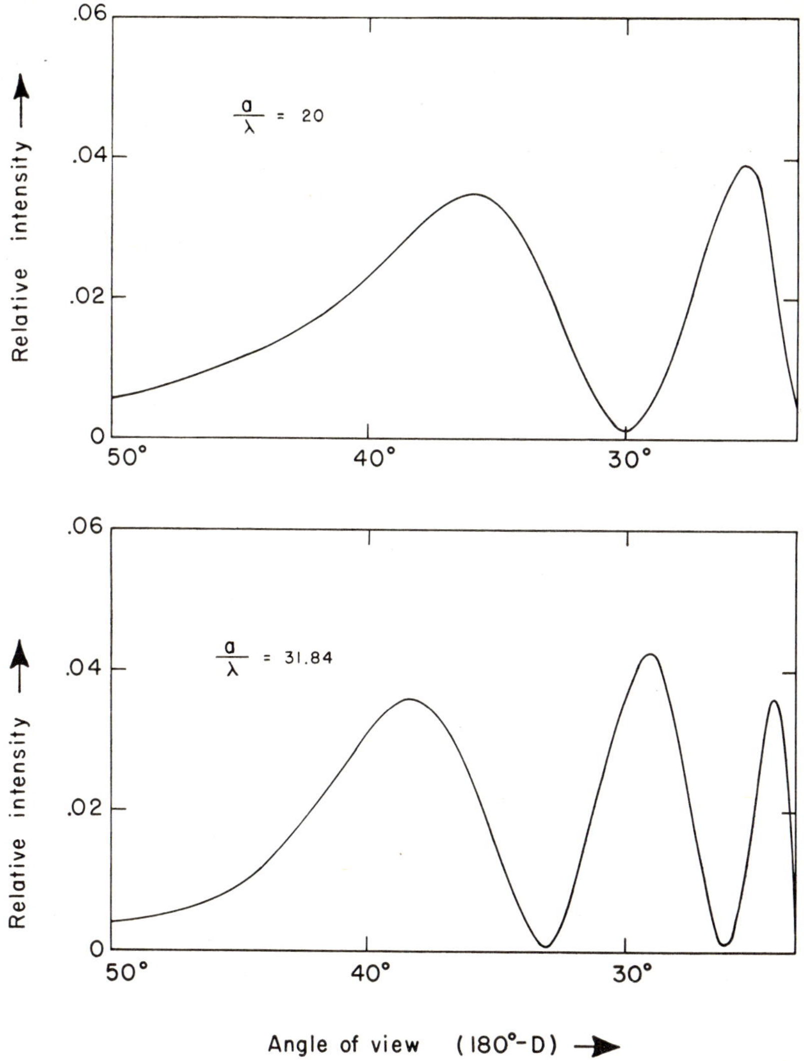

Fig. 6.5 (*After Malkus et al.*) The variation of the relative light intensity of a rainbow with the angle of view. a/λ is the ratio of drop diameter to wavelength of the incident parallel radiation. The size of a droplet giving the lower diffraction pattern is of the order of 0.022 millimeter, which is a cloud drop rather than a raindrop. From drops as small as this, mist bows rather than rainbows are formed. This figure illustrates the several diffraction maxima possible when the rainbow problem is solved completely, leading in the case of small raindrops to the supernumerary bows.

As the size of the raindrops becomes so small as to approach large cloud drops, the ratio of size to wavelength is no longer great. Simple geometric optics must be replaced by the more exact diffraction theory to predict effects. The effect of size on the number and intensity of rainbows has been summarized by Minnaert,† who is quoted below. For raindrops of the mean diameters given at the left, the appearance of the rainbow is

"1–2 mm	Very bright violet and vivid green; the bow contains pure red, scarcely any blue. Supernumerary bows are numerous (five, for example) violet-pink alternating with green, merging without interruption into the primary bow.
"0.50 mm	The red is considerably weaker. Fewer supernumerary, violet-pink and green again alternating.
"0.20–0.30 mm	No more red; for the rest, the bow is broad and well developed. The supernumerary bows become more and more yellow. If a gap occurs between the supernumerary bows, the diameter of the drops is 0.20 mm. If a gap is formed between the primary bow and the first supernumerary bow, the diameter of the drop is less than 0.20 mm.
"0.08–0.10 mm	The bow is broader and paler, and only the violet is vivid. The first supernumerary bow is well separated from the primary bow by a fairly wide gap and clearly shows white tints.
"0.06 mm	The primary bow contains a distinct white stripe.
"$<$0.05 mm	Mist-bow. White."

When the size gets below 0.10 millimeter (100 microns), the drops begin to be of the order of cloud droplets. The rainbow is then in reality identical with the glory, previously mentioned. This is often observed by aviators, when the airplane is in the center of a colored circle positioned on the top surface of an underlying cloud deck.

Theory and observation also show that the portion of the sky between the primary and the secondary rainbow is darker than the rest of the sky. An inspection of Fig. 6.4 shows that light originally equally distributed on both sides of the Descartes Ray is brought to one side of this ray, the cusped wave front side, leaving the other side depleted of light. This effect leads to supernumerary bows on the inside of the primary bow, a net decrease in light to the outside of the primary bow.

Since unpolarized light is polarized by reflection, one would cor-

† M. Minnaert, *Light and Colour in the Open Air,* G. Bell and Sons, London (1940).

rectly expect the internally reflected light from the rainbow to be polarized. As discussed in Chapter 2, the plane of polarization is the plane containing the sun, the raindrop, and the observer. The intensity of light from the rainbow polarized normal to this plane divided by the total intensity of light from the rainbow is the polarization Γ. For a primary bow, calculation shows the polarization to be of the order of 0.95 and about 0.89 for the secondary bow.

The optical effects of large ice particles in the atmosphere. Haloes

At temperatures below —20° C in the atmosphere, clouds are composed predominantly or entirely of ice crystals rather than water droplets. Often, these clouds have low concentrations of ice particles, presenting a tenuous appearance in the sky. The solar or lunar disks may be clearly visible through this cloud cover. Under favorable conditions at night a colored ring many lunar diameters in width encircles and is concentric with the moon. A similar ring may encircle the sun by day. Such a phenomenon is called a halo. As with the rainbow, the halo has a simple explanation from geometric optics. For the explanation, we accept the basic assumption that the cloud is composed of ice needles of a hexagonal cross section, every linear dimension of which is large compared to a wavelength of light. Observation shows that such ice needles exist in abundance in ice clouds. The smallest dimension is of the order of microns. On falling through the air, the crystals orient themselves with the longest axis parallel to the ground (like pencils on being dropped) and are continuously spinning on this horizontal axis. The optical properties of these hexagonal crystals are in general the same as a prism with a 60° prism angle. At any instant of time, enough of the spinning crystals have the correct position in space to make the following analysis valid.

In a prism whose prism angle is A, of Fig. 6.6, the deviation D for any ray through the prism is

$$D = (i - r) + (i' - r') = i + i' - A. \tag{6.7}$$

As $A + \left(\frac{\pi}{2} - r\right) + \left(\frac{\pi}{2} - r'\right) = \pi$ from the properties of a triangle, we have deduced that

$$r + r' = A. \tag{6.8}$$

The energy is concentrated about the ray, which undergoes minimum deviation. By differentiating Eq. 6.7 and setting $dD/di = 0$, we find that minimum deviation occurs when

$$\frac{dD}{di} = 0 = 1 + \frac{di'}{di}. \tag{6.9}$$

Differentiation of Eq. 6.8 gives

$$\frac{dr'}{dr} + 1 = 0. \tag{6.10}$$

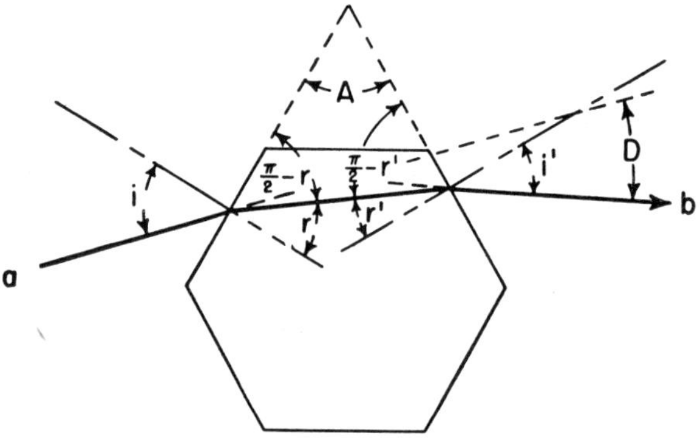

Fig. 6.6 The geometry attending the deviation, angle D, of a narrow pencil of light rays ab through a prism whose angle is A. The dashed lines for an ice crystal are extensions of the sides.

Equating Eq. 6.9 and 6.10 indicates that at minimum deviation, the ray must satisfy the identity

$$\frac{di'}{di} = \frac{dr'}{dr}. \tag{6.11}$$

An obvious solution of Eq. 6.11, identical with the one desired, is

$$i = i' \tag{6.12}$$

and

$$r = r'. \tag{6.13}$$

From Eqs. 6.12 and 6.7, the condition for minimum deviation becomes

$$i = \frac{D + A}{2}. \tag{6.14}$$

Through substituting (6.13) in (6.8), the angle of refraction at minimum deviation is

$$r = \frac{A}{2}. \tag{6.15}$$

When Eqs. 6.14 and 6.15 are substituted in Eq. 6.3, Snell's law becomes

$$\sin\frac{D+A}{2} = m\sin\frac{A}{2}\,. \qquad (6.16)$$

Equation 6.16 is the basic relationship for finding minimum deviation from knowledge of the index of refraction and the prism angle. One additional restriction implicitly imposed on the derivation is that the light ray must strike the crystal in the same plane as the plane in which A is measured; namely, the cross section. This will be the only case considered in detail.

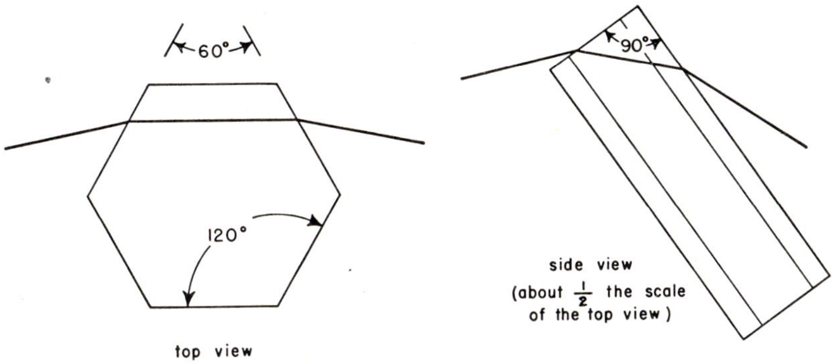

Fig. 6.7 An illustration of the two possible ways in which light undergoes minimum deviation in a hexagonal ice crystal. There is no way in which a ray can undergo minimum deviation in an ice crystal when the prism angle is 120°.

There are three *a priori* ways in which light can be visualized as refracted through an ice needle. The two possible ways are illustrated in Fig. 6.7. A, the prism angle, is respectively 60°, 90°, and 120°. Substitution of these values for A in Eq. 6.16 leads to the values of D reported in Table 6.2.

Calculation shows that no real minimum deviation exists for a prism angle of 120°. The maximum value that Eq. 6.16 can have is

$$\sin\frac{D+A}{2} = 1 = m\sin\frac{A}{2}\,. \qquad (6.17)$$

For $m = 1.31$, the condition imposed by Eq. 6.17 shows that

$$A \leqq 2\,\text{arc}\sin\frac{1}{1.31} = 99°28'.$$

Table 6.2

A Table of Minimum Deviations *D* for Visible Light Refracted through Ice Crystals

The prism angle is A. m is the index of refraction for visible light.

Color	m	$A = 60°$	$A = 90°$	$A = 120°$
Violet	1.317	22°22′	47°16′	No Descartes ray
Yellow	1.310	21°50′	45°44′	No Descartes ray
Red	1.307	21°34′	45°6′	No Descartes ray

For prism angles greater than 99°28′, no Descartes ray† exists. Therefore, we commonly observe only two haloes, the 22° halo arising from the 60° prism angle and the much less common 46° halo arising from the 90° prism angle. The latter halo results from the crystals falling with their long axes vertical, an unstable and hence less often observed mode of fall.

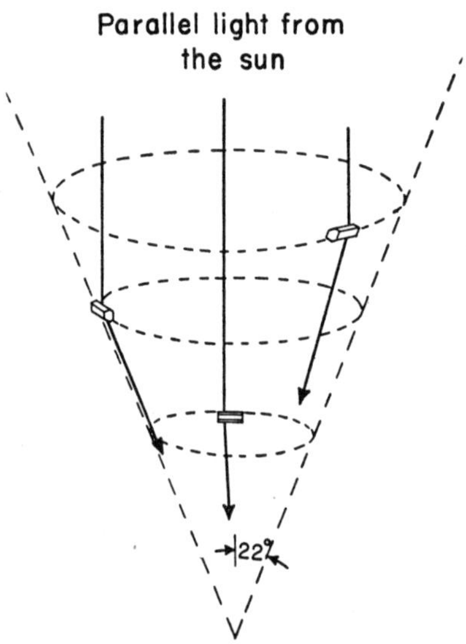

Fig. 6.8 Schematic drawing of the deviation of sunlight by ice crystals to produce the 22° halo. The halo will be observed along the dashed circles at the height of the ice-crystal cloud.

Figure 6.8 is a schematic drawing of the geometry producing the common 22° halo. The solid line shows the path of the principal ray for a single index of refraction. The dashed line is an extension of the line along which the color seems to originate. Table 6.2 shows that the red ray is deviated less than the blue. The difference in deviation leads to the red color's being observed nearest the sun (inner edge of the halo), with blue to the outside. The other colors are in their normal sequence from red to blue. As is common with most optical phenomena in the atmosphere, the red color is strong and readily observed, whereas the blue end of the spectrum is usually weak or absent. In haloes, the blue end is so weak because the tipping of the crystals from the horizontal causes the rays to pass through a crystal whose prism angle is close to but

† The Descartes ray is that ray which would be expected to undergo minimum deviation.

larger than 60°. The tipping effect plus diffraction effects by small crystals lead to a halo lacking in purity, with the red being the purest color observed.

Parhelia, or Sun Dogs. When the sun is close to the horizon and thin cirrus clouds are present, denoting ice crystals in the air, colored streaks 22° from and at the same elevation as the sun may sometimes be observed. This optical phenomenon is called parhelia of 22° or, commonly, sun dogs. They are explained as follows.

Plates and needles are common forms of ice crystals. The only difference between a plate and a needle is that the greatest length of the plate is measured across the hexagonal face of the crystal rather than as in the needle normal to this face. This fact of geometry causes the plates to fall face downward on the average, much in the manner of a falling leaf. Thus, when the sun is on the horizon, we have the same geometry for the crystals in the clouds close to our horizon as was described for the 22° halo. Since we are unable to see the other rays from the sun deviated by clouds higher above the horizon, we observe only a concentration of light on either side of the sun on the horizon, the sun dogs. Because of nonuniform crystal orientations, the sun dogs are not spots but may have some vertical extent. Sun dogs may show a reddish color on the inner edge, but are more commonly white. As the sun rises, the sun fails to strike the crystals edge on. The geometry discussed above no longer holds; hence, sun dogs are not oberved when the sun is high in the sky.

Sun Pillars. Another phenomenon is observed under the same circumstances as sun dogs, but arising from a different optical principle. Occasionally, one notices a streak of white light in the same vertical plane as the sun. This streak of light, which may be above, below, or both above and below the sun, is called a pillar. It is a result of the reflection of the sun's rays from the flat horizontal faces of ice plates in clouds above and below the apparent position of the solar disk.

Diffraction in the Atmosphere. So far, we have dealt with and illustrated some of the common scattering processes in the atmosphere, in addition to other phenomena arising from refraction and reflection. In the course of the development of these topics, it was necessary to mention the modifying effects of diffraction, although diffraction in itself was never expressly defined. Certain phenomena are primarily diffraction effects, the most common of which is the corona about the sun or moon. We would correctly expect that microwaves will also undergo diffraction, but that the diffracting obstacles will be much larger than those objects diffracting light.

Diffraction occurs through the disturbance of the flux of radiant energy immediately adjacent but exterior to the edge of a foreign object in the energy field. A simple principle due to Huygens is helpful in explaining diffraction effects. Huygens' principle states that every point on a wave front may be considered the source of spherical waves. As a corollary to the principle, we consider that when two infinitesimal segments of a wave front reach the same point in space with the same phase, the amplitudes of the two waves add. Conversely, when two wave fronts are 180° out of phase, the amplitudes subtract. For two waves of equal intensity 180° out of phase, complete cancellation of phase amplitude occurs. Diffraction, which is the study of phase effects, occurs only when the amplitudes are referred to the same initial wave front. The radiation must arise from the same source.

One immediate corollary of Huygens' principle becomes evident. It is this. In an optically homogeneous medium† such as air, application of the principle shows that the wave front is propagated in the same manner as its previous history would indicate; i.e., a plane wave continues to be propagated as a plane wave, a spherical wave as a spherical wave, etc. No new information is obtained.

However, let us examine the situation occurring when a diffracting edge such as that shown in Fig. 6.9 interrupts a plane wave. Application of Huygens' principle shows contributions to the light at P or Q from all points on the wave front Oy. The wave front Oy' contributes nothing to our sample points P and Q, being reflected or absorbed by the wall at $x = 0$. The paths from points on the wave front Oy to P, i.e., the lines OP, aP, and bP, in traveling different distances, arrive at P with varying phases. The intensity of light at P is the sum of the contributions of energy from all points of wave front Oy. The phase of each wavelet is considered in calculating the sum as the wavelets arrive at P. This method of summing intensities gives a variation of intensity along the y co-ordinate at x, the familiar diffraction pattern. The diffracted intensity as a function of y at x is shown in Fig. 6.9b. The dashed lines show the abrupt transition from light to shadow when diffraction is not considered.

Let us now examine the diffraction pattern resulting from replacing the semi-infinite wall Oy' by a wall whose y' dimension extends only to O'. We consider the distance OO' to be only a few wavelengths of the incident radiation in extent. There are now intensity contributions at P from all points of wave front $O'y'$ as well as from points on Oy. The diffraction pattern along the yy' direction at x becomes symmetrical about an axis normal to and at the midpoint of the opaque wall

† An optically homogeneous medium is one in which, for a stated wavelength, the index of refraction is a constant.

OO'. The diffracted intensity is shown in Fig. 6.9*c*. The intensity pattern may be visualized as being the *phase* sum of the diffracted intensities from the edges O and O'.

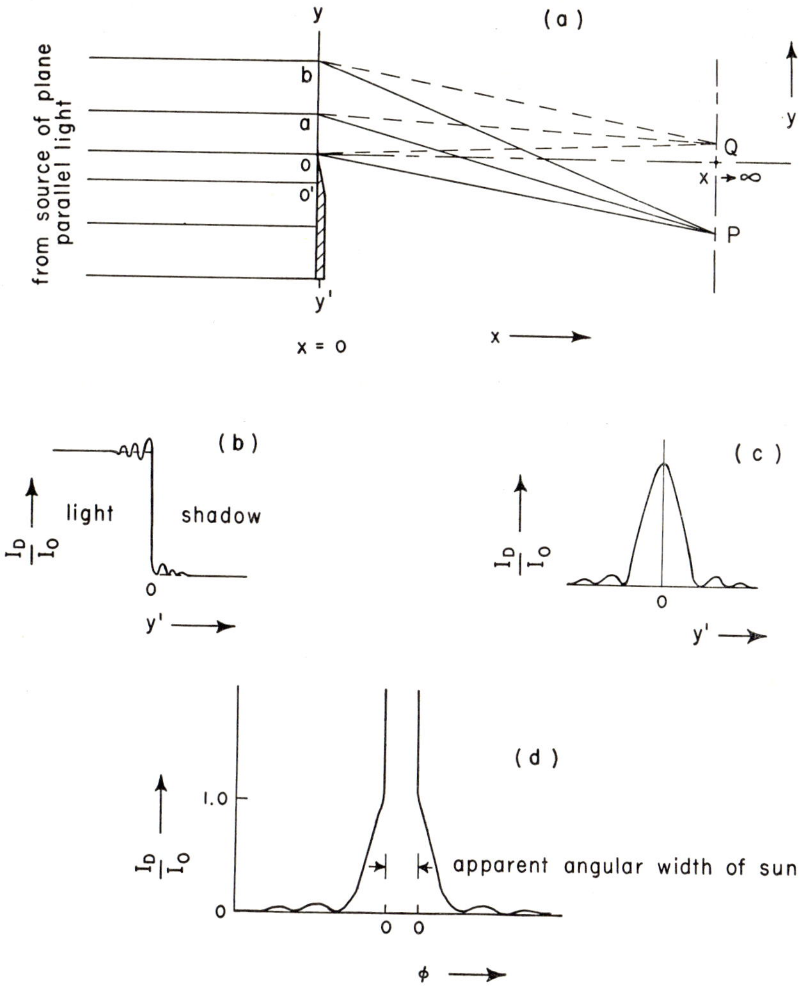

Fig. 6.9 Schematic representation of geometry and intensity patterns of (*a*) geometry of diffraction and (*b*) intensity pattern for a straight edge. (*c*) intensity of diffraction pattern from two edges (needle) and (*d*) intensity pattern of corona about the sun.

In these analyses, we have considered the distance Ox to be so large in respect to the wavelength of the incident radiation as to approach infinity. By so doing, we have confined the discussion to Fraunhofer diffraction, the case most commonly observed. We have assumed that the barriers Oy' and OO' extend to infinite distance perpendicular to the plane of the paper. Otherwise, there will be diffraction effects in that plane also. Qualitatively, the latter is just what happens when OO' is the diameter of a cloud drop illuminated by sun or moonlight. The pattern of Fig. 6.9*c* is circularly symmetrical, with the central maximum split because of the finite angular diameter (about $\frac{1}{2}°$) of the luminary. A pattern such as is shown in Fig. 6.9*d* is observed. Furthermore, for white incident light the diffraction ring is colored. Dispersive effects on white light by a drop of a given size are to be expected because the ratio a/λ assumes a range of values as λ varies from $0.4 < \lambda < 0.7$ micron. The maximum intensity for the various colors will not appear at the same point along y; a dispersion of the spectral colors is indicated. Let us examine the mathematics attending the diffraction of light by a circular object of diameter a. In practice, a is the diameter of a cloud drop and λ, the wavelength of the incident radiation, is for visible light from the sun or moon.

Mathematical theory of the corona

The mathematical development of the diffraction pattern from a circular aperture illuminated by a plane wave follows. Babinet's principle, which states that Fraunhofer diffraction from a circular aperture shows the same diffraction pattern as that from an opaque disk or sphere of the same diameter, justifies the discussion of the mathematically simpler case. The results will be applied to the opaque disk or sphere.

We shall consider the contribution to the intensity at P, in Fig. 6.10, of the flux from every incremental element of area $r\,dr\,d\psi$ in the circular aperture of diameter a. As the aperture is initially a surface of constant phase (coherent) source, there will be gradual changes in the phase of the light arriving at P from the successive incremental areas across the face of the disk. Let us choose the ray OP as our reference ray and relate all other rays to it by a phase difference δ. δ is a function of the position (r, ψ) of the aperture from which it originates. As the final intensity is independent of time, we shall start with the time averaged values, the root mean square values of the electric vector described in Chapter 2.

The contribution to the magnitude of the electric vector at P from any element $r\,dr\,d\psi$ having a phase difference δ with the central ray is

$$dy = \cos \delta \; r\,dr\,d\psi. \tag{6.18}$$

When the contribution to the intensity at P from the entire aperture is considered, the amplitude of vibration is y, the integral of Eq. 6.18. The integral,

$$y = \int_0^{2\pi} \int_0^{a/2} \cos \delta \, r \, dr \, d\psi, \tag{6.19}$$

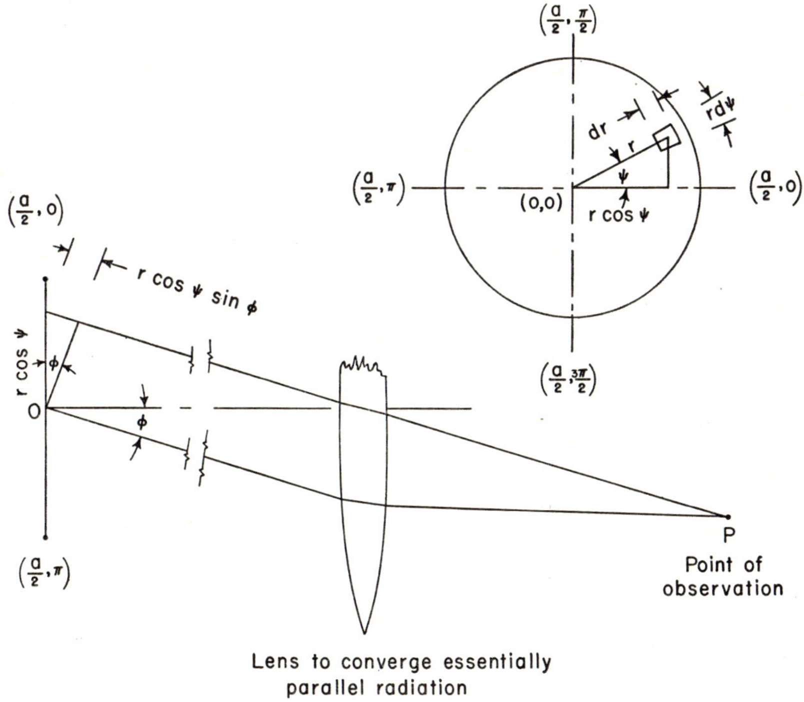

Fig. 6.10 Ray and wave-front geometry for setting up the diffraction pattern from a circular aperture.

is directly proportional to the electric vector at point P, whereas the square of Eq. 6.19 is the relative intensity of the diffracted light at P. We have considered for convenience a unit amplitude of vibration of the light, a constant factor for all rays.

In order to evaluate Eq. 6.19 we must find δ as a function of r and ψ. This can be done by considering Fig. 6.10. From the geometry of Fig. 6.10 we see that the difference in length between the reference ray OP and the variable ray is $r \cos \psi \sin \phi$. The ratio of this length to a wavelength is some fraction of a complete wavelength of 2π radians angular measure. By definition, the angle so stated is the phase angle δ. We have therefore

$$\delta = \frac{2\pi}{\lambda} \sin \phi \, r \cos \psi, \tag{6.20}$$

where, for purposes of the following integration, $(2\pi/\lambda) \sin \phi$ is a constant. Substitution of Eq. 6.20 in Eq. 6.19 shows

$$y = \int_0^{2\pi} \int_0^{a/2} \cos\left(\frac{2\pi}{\lambda} \sin \phi \, r \cos \psi\right) r \, dr \, d\psi. \tag{6.21}$$

Equation 6.21 can be integrated through expanding the cosine as a power function and integrating the resulting series in r, term by term, substituting the appropriate limits. The resulting equation in $\cos \psi$ can also be integrated term by term and limits can be substituted. The result of this tedious integration is a power series in $(\pi a/\lambda) \sin \phi$, called a Bessel function of the first order, the notation for which is $J_1(x)$. Tables of Bessel functions are tabulated and published, just as in the case of the more elementary trigonometric functions $\sin x$ and $\cos x$. x is called the argument of the function and is the variable term x^n in the power series expansion. In this case, $x = (\pi a/\lambda) \sin \phi$.

$\pi a/\lambda$ is commonly denoted as α in the literature. The solution of Eq. 6.21 is

$$y = \frac{\pi a^2}{4}\left[\frac{2}{\alpha \sin \phi} J_1(\alpha \sin \phi)\right]. \tag{6.22}$$

The intensity I_D has been defined as the square of y. If I_D is the intensity of the function given by Eq. 6.22 when $\phi = 0$, the relative intensity i_D is

$$i_D = \frac{I_D}{I_0} = \left[\frac{2J_1(\alpha \sin \phi)}{\alpha \sin \phi}\right]^2. \tag{6.23}$$

An investigation† of I_0 shows it to be equal to $(\pi^2 a^4/16\lambda^2)E_0$. The factor E_0 is the flux density of the parallel radiation incident on the plane of the drop. For n drops of the same size

$$\frac{I_D}{E_0} = \frac{n\pi^2 a^4}{16\lambda^2}\left[\frac{2J_1(\alpha \sin \phi)}{\alpha \sin \phi}\right]^2 \tag{6.24}$$

where I_D now refers to the coronal intensity from n drops. If several sizes of drops are present, we must write I_D as the sum over i classes

† Bricard, J., "Lumière diffusée en avant par une goutte d'eau sphérique," *J. phys. radium,* Ser. 8, 4, 57–66 (April 1943).

of drop size, so that the relative intensity i_D becomes,

$$i_D = \frac{I_D}{I_0} = \frac{\sum_{i=1}^{\infty} n_i a_i^4 \left[\frac{2J_1(\alpha_i \sin \phi)}{\alpha_i \sin \phi}\right]^2}{\sum_{i=1}^{\infty} n_i a_i^4}. \tag{6.25}$$

Equation 6.25 refers only to monochromatic light. For completeness we must have an integration over all wavelengths when we consider the drops to be illuminated by white light.

Let us consider Eq. 6.25, the coronal pattern when a cloud or fog of uniform drops is illuminated by a single wavelength. Under this assumption, a is constant so that Eq. 6.25 becomes identical with Eq. 6.23. A plot of Eq. 6.23 is shown in Fig. 6.11. The maximum and minimum values of I_D/I_0 occur when

$$\alpha \sin \phi = 0, 5.15, 8.47, \cdots \text{ (maximum value)} \tag{6.26}$$

and

$$\alpha \sin \phi = 3.84, 7.02, 10.18, \cdots \text{ (minimum value)} \tag{6.27}$$

or approximately

$$\alpha \sin \phi = (n + 0.22)\pi \qquad \text{(minimum value)}. \tag{6.28}$$

The first maximum at $\alpha \sin \phi = 0$ is usually obscured by the finite size of the source. The first maximum observed is usually when

$$\alpha \sin \phi = 5.15, \tag{6.29}$$

or, writing $\alpha = \pi a/\lambda$, when

$$\sin \phi = \frac{5.15}{\pi} \frac{\lambda}{a}. \tag{6.30}$$

From Eq. 6.30, we have our explanation for the corona about the sun. For clouds whose drops are of a uniform size (never completely realized) we see that the position of the maximum intensity varies as the wavelength, i.e., $\sin \phi_{\max} \sim \lambda$, indicating that $\phi_{\max}$ increases as λ. Because of this relationship, the maximum intensity of red light is found at a greater angle from the directly transmitted beam than the maximum intensity for blue light. The corona has circular symmetry about the sun or moon, just as in the halo. The edge of the luminary is essentially the origin of angle ϕ. Consistent with the above arguments, the corona about the sun or moon shows a white inner ring with a reddish-brown outer ring. This part of the corona, the most commonly observed, is called the aureole.

The above equation correctly describes the relative angular positions of the colors of the corona. It fails to emphasize that these colors are seldom of high purity and they may often be so pale as to cause the corona to appear as a bright white ring close to the sun or moon. If one refers to Fig. 6.11 and plots the diffraction pattern

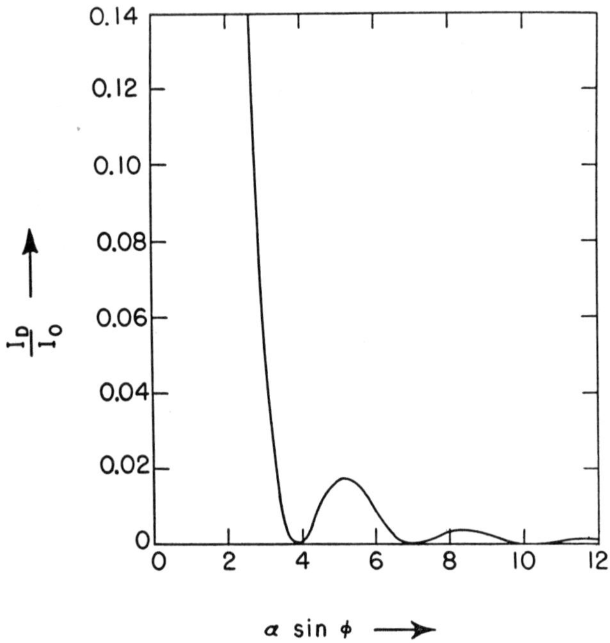

Fig. 6.11 A plot of the function $\frac{I_D}{I_0} = \left[\frac{2J_1(\alpha \sin \phi)}{\alpha \sin \phi}\right]^2$ where $\alpha = \frac{\pi a}{\lambda}$. At $\phi = 0$, the relative intensity of the diffracted light $\frac{I_D}{I_0} = 1$. a is the diameter of a drop, λ is the wavelength of the incident and diffracted light and ϕ is the angle at which the diffracted light is viewed off the transmitted beam.

for several colors, even with the same drop diameter, he will find that any angular position ϕ will contain not one but all colors. The intensities of the colors will not all be the same, however. For this condition, colorimetry principles indicate a dominant color of low saturation or purity. The purity of the coronal colors is further decreased by increasing the spread of the distribution of cloud-drop sizes. As a result, one expects white coronas about the disk of the sun or moon when shining through a cloud of nonuniform droplets.

One cannot rule out entirely the possibility that an array of ice needles, uniform in size, may cause coronas. The ice needles can diffract light in the same manner as a number of slits in a transmission diffraction grating. In support of this, coronas having the finest coloring and best development have been observed occasionally in thin cirrus clouds. The point is debatable, since water droplets may exist to temperatures considerably below 0° C.

Typical Corona Manifestations. Most coronas are observed about the moon. The most frequent corona manifestation is the *aureole,* a brownish-red light in the outer portion of the corona and white over the inner area next to the luminary. On occasion, in well-developed coronas, four distinct series of colors have been observed. These groups or orders are as follows, I being closest to the luminary, IV farthest away:

I Aureole (bluish): white: (yellowish): brown-red.
II Blue: green: (yellow): red.
III Blue: green: red.
IV Blue: green: red.

The colors in parentheses may or may not be present.

In general, then, most coronas consist only of the aureole. The latter is formed by the biggest drops present, with the corona having a radius that varies between 1° and 5°. The extreme values which have been recorded for the best-developed coronas are in the neighborhood of 10°.

Other diffraction manifestations are the glory or Brocken-bow, Bishop's ring, and iridescent clouds. The *glory* is a ring of colored light that is occasionally observed about the shadow of a person's head cast upon the upper surface of a cloud bank or fog. It is caused by white light entering the drops, undergoing one internal reflection, and then emerging toward the observer. The drop acts as a prism, separating the colors into the spectrum observed. Although this is a diffraction effect, and the intensities must be computed on such a basis, the appearance is akin to that of a primary rainbow. *Bishop's ring* is a corona produced by particles other than water drops. In particular, a brilliant corona may be produced by suspended volcanic ash. This phenomenon was first described by the man whose name is associated with it. *Iridescent clouds* are very high cirrus clouds with very brilliant colors. It is thought that the colors are due to fragments of coronas formed by the cloud particles.

Coronal measurements have been used to estimate the size of cloud drops by using the relationship

$$\sin \phi = (n + 0.22) \frac{\lambda}{a} \tag{6.31}$$

where ϕ is the position of the nth minimum of light of wavelength λ. a is the diameter that is to be determined. This relationship supposes monochromatic light and uniform drops, neither condition being completely fulfilled in practice. In general, then, such values of a as are found from the sin ϕ relationship are an average value of an approximate nature and give no specific information as to the distribution of drop sizes. Values of 10 to 70 microns found by this method are in the generally accepted range of values for cloud and fog drops.

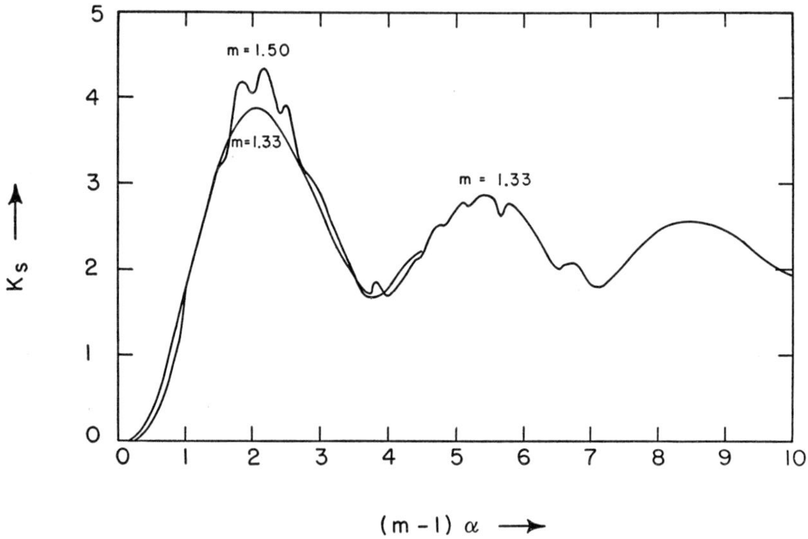

Fig. 6.12 A plot of the transmission cross section for nonabsorbing spherical particles with indices of refraction, $m = 1.33$ and $m = 1.50$. For the nonabsorbing case, the transmission cross section is identical with the scattering cross section, K_s. $(m - 1)\alpha = (m - 1)\pi a/\lambda$ where a/λ is the ratio of diameter to wavelength. For a given m and a, the figure is a plot of K_s versus the frequency $(\sim 1/\lambda)$ of the incident light. Approximate limits for the following optical regions are as follows:

Rayleigh region	$(m - 1)\alpha < 0.1$
Diffraction region	$3.0 < (m - 1)\alpha < 20$
Geometric optics	$(m - 1)\alpha > 20$

The Mie Theory. (1,2,3,14,17) In the discussion that has preceded this section, we have considered the consequences of physical optics when light is intercepted by droplets having large α. The result is a corona the sine of whose angular half width is inversely proportional to the drop diameter. In the limit of very large drops, the corona width becomes undetectably small, and the transmission cross section becomes equal to the geometric cross section. In Chapter 2, we discussed the case of very small α, wherein the transmission cross section (or scattering cross section) becomes proportional to the inverse fourth power of the wavelength, i.e., Rayleigh scattering.

The solution for the intermediate region is shown in Fig. 6.12. The regions of Rayleigh scattering and of diffraction optics are shown. The region in between these two limits is a region of rapidly varying transmission cross section, where many interesting and anomalous optical properties can be predicted. For example, if we could conceive of the moon shining through a dust cloud of approximately spherical particles whose index of refraction was 1.50, we might under a favorable size of particles see a horizon moon whose color was other than a deep yellow. For example if the particle diameter was grouped about $a = 0.75$ micron, red light would be scattered much more strongly than blue, giving a "blue moon." Such phenomena have been reported.[6,16,19]

References

1. Aden, A. L., "Electromagnetic Scattering from Spheres with Sizes Comparable to the Wavelength," *J. Appl. Phys., 22,* 601–605 (1951).
2. Aden, A. L., and Kerker, M., "Scattering of Electromagnetic Waves from Two Concentric Spheres," *J. Appl. Phys., 22,* 1242–1246 (1951).
3. Brillouin, L., "The Scattering Cross-section of Spheres for Electromagnetic Waves," *J. Appl. Phys., 20,* 1110–1125 (1949).
4. Buchwald, E., "Regenbogenfarben," *Ann. Phys.,* 5 Folge, *43,* 488–493 (1943).
5. Dietze, G., "Die anormale Trübung der Atmosphäre September/Oktober 1950," *Z. Meteorol., 5,* 86–88 (1951).
6. Gelbke, W., "Bemerkungen zum Phänomen der blauen Sonne," *Z. Meteorol., 5,* 82–84 (1951).
7. Gumprecht, R. O., and Sliepcevich, C. M., "Tables of Light-Scattering Functions for Spherical Particles," Engineering Research Institute, University of Michigan, Ann Arbor (1951).
8. Gumprecht, R. O., Sung, N., Chin, J. H., and Sliepcevich, C. M., "Angular Distribution of Intensity of Light Scattered by Large Droplets of Water," *J. Opt. Soc. Amer., 42,* 226–231 (1952).
9. Günther, S., "Die Himmelshelligkeit in der Nähe der Sonne," *Optik, 5,* 240–257 (1949).
10. Houghton, H. G., and Chalker, W. R., "The Scattering Cross Section of Water Drops in Air for Visible Light," *J. Opt. Soc. Amer., 39,* 955–957 (1949).
11. Maier, W., "Kristallhalo," *Z. Meteorol., 4,* 111–119 (1950).
12. Maier, W., "Das Phänomen der Sonnendoppelschatten," *Z. Meteorol., 4,* 16–21 (1950).
13. Malkus, W. V. R., Bishop, R. H., and Briggs, R. O., "Analysis and Preliminary Design of an Optical Instrument for the Measurement of Drop Size and Free-Water Content of Clouds," *Natl. Advisory Comm. Aeronaut.,* Tech. Note 1622, Washington, D.C. (1948).
14. Montroll, E. W., and Hart, R. W., "Scattering of Plane Waves by Soft Ob-

stacles. II. Scattering by Cylinders, Spheroids, and Disks," *J. Appl. Phys., 22,* 1278–1289 (1951).
15. Neuberger, H., "General Meteorological Optics," *Compendium of Meteorology,* American Meteorological Society, Boston, 61–78 (1951).
16. Runge, H., "Blaue Sonne—blauer Mond," *Z. Meteorol., 5,* 60–62 (1951).
17. Van der Hulst, H. C., "Scattering in the Atmospheres of the Earth and the Planets," Chapter III of *The Atmospheres of the Earth and Planets,* G. P. Kuiper, editor, University of Chicago Press, Chicago (1952).
18. Wexler, H., "The Great Smoke Pall—September 24–30, 1950," *Weatherwise, 3,* 129–134 (1950).
19. Wilson, R., "The Blue Sun of 1950 September," *Monthly Notices Roy. Astron. Soc., 3,* No. 5, 478–489 (1951).

Source Books

B1. Dorsey, N. E., *Properties of Ordinary Water-Substance,* Am. Chem. Soc. Monograph Series, Reinhold Publishing Corp., New York (1940).
B2. Humphreys, W. J., *Physics of the Air,* McGraw-Hill Book Co., New York (1940). (Chapters III, IV, V, and VI of Part IV.)
B3. Minnaert, M., *Light and Color in the Open Air,* G. Bell and Sons, London (1940).
B4. Neuberger, H., *Introduction to Physical Meteorology,* Chapter V, the Pennsylvania State College School of Mineral Industries, State College (1951).
B5. Pernter, J. M., and Exner, F. M., *Meteorologische Optik,* 2. Aufl., W. Braumüller, Wien u. Leipzig (1922).

Problems

The following quantities are defined for the problems in this chapter.

F_0 = the flux of plane parallel light intercepted by an area equal to the geometric cross section of a spherical drop.
F_s = the total scattered flux from a single drop.
F_D = the total diffracted flux from a single drop.
F_R = the total reflected plus refracted flux from a single drop.
E_0 = the flux density of the incident radiation in the plane of the flux-scattering drop.
I_s = the intensity of the scattered energy computed from the Mie theory.
I_D = the intensity of the diffracted energy found from diffraction theory.
I_R = the intensity of the reflected plus the refracted energy from a spherical drop.

The other symbols are those from the text, unless otherwise stated.

6.1 For a large droplet, i.e., $\alpha > 40$, the Mie theory predicts

$$F_s \approx F_D + F_R \approx 2F_D \approx 2F_R \approx 2F_0.$$

In the limit as $\alpha \to \infty$, the relationship is exact. The scattering area coefficient, sometimes called the scattering cross section or transmission cross section, is a dimensionless number defined as

$$K_s = \frac{F_s}{F_0},$$

the ratio of the total scattered flux to the flux incident on the geometric cross section of a spherical droplet.

From diffraction theory, the diffracted intensity from a sphere is found to be

$$I_D = \frac{F_0}{4\pi}\alpha^2\left[\frac{2J_1(\alpha \sin\phi)}{\alpha \sin\phi}\right]^2.$$

From the Mie theory, the scattered intensity is

$$I_s = \left(\frac{i_1 + i_2}{2\pi\alpha^2}\right)F_0.$$

Show that

$$K_s = \frac{1}{\alpha^2}\int_0^\pi (i_1 + i_2)\sin\phi\, d\phi$$

and

$$\frac{F_D}{F_0} = \frac{\alpha^2}{2}\int_0^\pi \left[\frac{2J_1(\alpha \sin\phi)}{\alpha \sin\phi}\right]^2 \sin\phi\, d\phi.$$

6.2 Gumprecht,(8) using National Bureau of Standards (NBS) notation, gives from the Mie theory the following values of i_1 and i_2 for $m = 1.33$ and $\alpha = 40$. i_1 and i_2 are proportional to the polarized components of the intensity of the scattered energy. (Note: $\phi = 180° - \gamma$, where γ is NBS notation.)

ϕ	0°	2°	4°	6°	8°
i_1	641,100	346,500	22,700	24,470	26,850
i_2	641,100	339,900	19,280	25,400	24,840

ϕ	10°	20°	30°	40°	60°
i_1	1,848	6,539	3,587	1,627	270.8
i_2	1,962	5,773	2,734	1,353	365.7

ϕ	90°	120°	150°	180°
i_1	33.40	45.53	155.4	517.2
i_2	43.60	10.32	117.3	517.2

Plot and find K_s by graphical integration. Hint: Note that $\sin\phi\, d\phi = -d(\cos\phi)$.

6.3 If we call i_D the relative intensity of the energy diffracted by a spherical drop or circular disk, then, by Eq. 6.23,

$$i_D = \left[\frac{2J_1(\alpha \sin\phi)}{\alpha \sin\phi}\right]^2.$$

Evaluation of i_D gives the following table.

$\alpha \sin\phi$	0	0.4	0.8	1.2	1.6	2.0
i_D	1.000	0.961	0.850	0.690	0.508	0.333
$\alpha \sin\phi$	2.4	2.8	3.2	3.6	3.8	4.4
i_D	0.188	0.086	0.027	0.003	0.000	0.009
$\alpha \sin\phi$	5.2	6.0	7.0			
i_D	0.017	0.009	0.000			

Find F_D/F_0 by graphical integration for $\alpha = 40$. $i_D \to 0$ for all values of $\alpha \sin\phi > 7$ (see Prob. 6.1).

6.4 Wiener,† by the use of geometric optics combined with the Fresnel laws for

† Wiener, C., "Die Helligkeit des klaren Himmels und die Beluchtung durch Sonne, Himmel und Rückstrahlung," *Abhandl. Naturforsch.*, *73*, 1–240 (1907). (Sometimes given as 1900.)

the intensity of the reflected and refracted light, was able to find the ratio I_R/F_0 for a large water droplet. Selected values from his work are given in the following table.

ϕ	0°	2°	4°	6°	8°	10°	20°
$\frac{I_R}{F_0}$	1.295	1.276	1.244	1.199	1.140	1.069	0.667
ϕ	30°	40°	60°	90°	120°	150°	180°
$\frac{I_R}{F_0}$	0.370	0.189	0.036	0.002	0.006	0.027	0.032

Find F_R/F_0 by graphical integration. Note that this pattern is independent of size, provided that $\alpha > 40$.

6.5 Show that

$$K_s = \frac{F_D}{F_0} + \frac{F_R}{F_0} \approx 2$$

from the results of Probs. 6.2, 6.3, and 6.4. Explain the origin in respect to the sphere of the diffracted and reflected flux. Explain the difficulties of separating F_D and F_R as α becomes small. $K_s = 2$ is a statement of the conservation of energy, because then $F_D = F_R = F_0$.

6.6 The scattering cross section at any angle ϕ is defined as

$$K_s\ (\phi) = 4\pi \frac{I_s}{E_0},$$

where I_s is a function of the angle ϕ. $K_s\ (\phi)$ has the dimensions of area.

a. Show that alternate forms for $K_s\ (\phi)$ are

$$K_s\ (\phi) = \frac{4\pi}{E_0}\frac{dF_s}{d\omega} = 4\pi r^2 \frac{E_s}{E_0}.$$

r is the distance from scatterer to detector. ω is the solid angle subtended by the detector when viewed from the scatterer. F_s, E_s, and I_s are measured at angle ϕ to E_0.

b. When $\phi = 180°$, $K_s(180°) = K_s'$, the back scattering or the radar cross section. Find K_s' from the following data for $m = 1.33$ and $\alpha = 1$ if the wavelength of the incident radiation is 300 millimicrons. The incident ultraviolet light is unpolarized.

ϕ	90°	120°	150°	180°
i_1	0.03395	0.02693	0.02259	0.02115
i_2	0.00004	0.00597	0.01654	0.02115

See Prob. 6.1 for additional definitions.

c. Find K_s' from the Rayleigh theory and compare with part *b*. Is the Rayleigh theory valid for this value of α?

6.7 How does the diffraction pattern from an array of drops whose relative numbers are

Diameter (microns)	4	5	6	7
Per cent by number	30	40	20	10

compare with the diffraction pattern from an array of 5-micron drops? Compare the results by plotting the intensities after the central intensity at $\phi = 0°$ has been normalized to unity. The incident energy is light whose wavelength is 628 millimicrons. Use the table for i_D given in Prob. 6.3.

CHAPTER 7

The Physical Conditions Attending the Formation of Cloud Particles

The study of the formation of cloud and rain is perhaps the section of physical meteorology that is most useful to its sister subject of synoptic meteorology. Synoptic meteorology faces the problem inherent in all engineering—that of using currently available knowledge to arrive at an immediately useful result, i.e., to forecast the weather. Physical meteorology attempts to provide synoptic meteorology with many of the necessary concepts and much of the information to whet the prognostic tools of the weather forecaster.

The study of the mechanism of the formation of rain has been accelerated by the discovery that under favorable conditions man has some small but perhaps economically important control over the production of rain. Although the final judgment on the degree of success that can be expected to attend these studies of precipitation mechanisms cannot be made at this time, many of the basic physical factors entering into the formation of cloud and rain are now known in quantitative as well as qualitative fashion.

As interesting and exciting as the contemplation of the "triggering off" of rainfall by man himself may be, the study of cloud physics is not primarily concerned with this phase of the subject. Rather, since the study and understanding of the process of natural rainfall will be reflected ultimately in better weather forecasts, therein lies much of the economic value of the study of cloud physics.

There are other more specialized applications of cloud physics. These include the dissipation of cloud and fog in the vicinity of airfields, the minimizing of ice formation on aircraft, and the study of atmospherically induced electrical effects on communications. It is the aim of the subsequent sections to provide an introduction to many of the underlying physical principles that must be considered when one is faced with a problem involving clouds or precipitation in the atmosphere.

The Classical Condensation Theory.[22,B1] Although not usually explicitly stated as such, the theory of cloud and rain formation based on classical thermodynamics is still the large-scale guidepost that governs our thinking when discussing cloud physics. The details of the theory have been shown to need modification, and the theory as originally stated is no longer taken seriously. One fact is assuredly true, that although the classical thermodynamic theory has little predictive effect in cloud-physics work, the cloud with the moisture and air composing it is a thermodynamic system subject to the laws of thermodynamics. These laws express how the energy is distributed among the various phases of the cloud and in which directions changes in the system cannot proceed.

Let us examine the classical condensation theory and see what details are incorrect or inconclusive and need modification. The classical condensation theory states that (1) all processes are adiabatic† and reversible; (2) a parcel of air having a constant mass of water held therein as water vapor is cooled on rising under adiabatic conditions, until a temperature is reached such that the air is saturated by its water vapor; (3) further cooling of the air proceeds at a slower rate, although adiabatic, because of the condensation of the water vapor to liquid water. The amount of water vapor condensed is determined by the condition that the air be just saturated during the entire condensation process; (4) when the temperature of the ascending air reaches 0° C, the melting point of pure water, all liquid water held in suspension in the air freezes; and (5) further rising of the air produces direct sublimation of the water vapor to the ice phase as the air parcel tends to maintain saturation. We may raise the following objections to the classical condensation theory, taking them in order.

1. The reversible adiabatic assumption, although leading to a useful approximation in the free atmosphere, by its very nature cannot predict the effect of heat sources and sinks on the behavior of a cloud system. These thermal sources and sinks provide the driving mechanism of starting and eventually stopping a cloud system. Starting and stopping imply acceleration and accelerations imply external forces and energies. The adiabatic assumption says that the thermal energies that might be used to accelerate the system are identically zero.

2. By the second assumption of the theory, the modification of the reference air parcel by its being mixed with other air masses of differ-

† An adiabatic process is one in which a system changes state without the flow of heat across the boundaries of the system itself. It is assumed that the student is familiar with the basic thermodynamic concepts of system, process, state, and reversibility.

ent thermal and moisture states is neglected. It is inherent in the classical condensation theory that there can be no addition of modifying air masses to the system under consideration. Too, the theory ignores the principle of continuity of mass implicit in all flow problems of gases or liquids.

3. Supersaturation in the atmosphere is ruled out classically. In present theory, initial supersaturation is needed. The amount of supersaturation is small but significant.

4. Water does not, in the absence of the ice phase, freeze at 0° C. Undercooling† of liquid water to many degrees below the melting point at 0° C is a common phenomenon. The undercooling phenomenon is important in condensation theory, since it allows the coexistence in air of ice and water particles at temperatures below 0° C. This fact has important theoretical implications in explaining the onset of precipitation from clouds.

5. Observations show that even at temperatures several degrees below the melting point of ice, water vapor may condense to undercooled liquid water rather than sublime directly to ice. The ultimate low-temperature limit of this phenomenon has not been unequivocally proved, but is at least as low as —40° C. Of course, in many cases, of which there are an increasing number as lower temperatures are reached, freezing of the water may occur almost immediately after condensation. Therefore, in many practical applications, the presence or absence of the intermediate liquid state may have indistinguishable thermal effects on a process.

In summary, we may conclude that the mechanism of growth of an individual cloud is not described by the classical condensation theory. This must be supplemented by a more rigorous theory that will include the details of the condensation process. Let us examine the physical laws that any successful theory of cloud growth and precipitation must recognize.

Nuclei of Condensation.[16,23,29] The maximum amount of water vapor that the atmosphere can hold in the presence of liquid water is dependent only on the temperature of the air and the chemical composition of the particles suspended in it. The air pressure itself has a negligible effect on the saturation temperature of the air. From thermodynamics, we find that in the absence of suspensoids in the air, air is deemed saturated when in the presence of liquid water the partial

† Undercooling rather than supercooling is the word preferred to describe the existence of the liquid state of a substance at temperatures below the melting point of the solid phase.

pressure of water vapor e reaches a maximum value of e_s, the saturation vapor pressure of water. e_s is a unique function of temperature. The ratio $(e \times 100)/e_s$ is called the relative humidity. Saturation in respect to a pure water surface is attained when $e = e_s$. The relative humidity is then 100 per cent.

It is possible to conduct experiments in which the relative humidity reaches values up to 500 per cent with no condensation occurring. Relative humidities of this order of magnitude can be attained by the rapid adiabatic expansion and ensuing cooling of a mass of saturated air. The production of high relative humidities by this means governs the operation of the Wilson cloud chamber used for detecting atomic radiations. As soon as condensation occurs, two phases, liquid water and water vapor, are present in the chamber. The relative humidity drops toward 100 per cent with the attendant condensation of the vapor, reaching 100 per cent at the conclusion of the condensation process. *Condensation does not begin until the water vapor has a suitable surface on which to condense.* The surface of condensation is called a *nucleus of condensation,* and the process of introducing these surfaces into a vapor (also liquid) phase is called *nucleation.* If the nucleus of condensation is other than a water surface, *heterogeneous nucleation* occurs. The ions on which the moisture condenses in a cloud chamber are an example of heterogeneous nuclei. On the other hand, if water vapor condenses on a liquid water surface, no matter if the surface is submicroscopic in size, the process is called *homogeneous nucleation.* In the free atmosphere, heterogeneous nucleation is the only important nucleation process. Homogeneous nucleation is ruled out because of the prohibitively high (>500 per cent) relative humidities required. In the freezing of undercooled water, another important problem, homogeneous nucleation may play an important role.

In natural atmospheric processes, indirect evidence based on vertical velocities indicates that relative humidities of about 0.1 per cent greater than saturation over a pure water surface are not to be expected. These data therefore rule out ions and all nonhygroscopic dusts and salts as nuclei because of the high relative humidities required before they will act as condensation nuclei.† These dusts, or kerns, as they are called, which make up most of the suspensoids in the atmosphere and are important for their light scattering effects on visibility, contribute little or nothing to the condensation process.

† This statement applies to particles less than 10^{-4} centimeter (1 micron) because of the increase in equilibrium vapor pressure of water with decrease in radius of curvature below this critical size.

It is for this reason that dust counts made by instruments such as the Aitken kern counter have little correlation with the number of active condensation nuclei in the atmosphere.

All evidence[27,28,29] to date points to sea salt as being the principal nucleus of condensation, with sulfurous and nitrous acids playing a secondary role. To support this statement is the fact that rain has a chloride ion content such that when a mass of rainwater is divided by the number of raindrops making up the mass, the chloride ion content per raindrop leads to a reasonable size for a condensation nucleus. Direct counts of chloride ions in cloud-free air give large numbers of salt nuclei, even though these counts are not restricted to the immediate vicinity of the seashore. The sea-salt theory is weakened through an estimation by Simpson[26] of the number of nuclei put into the air by the evaporation of water from sea spray, the principal mechanism envisaged. His values seem to indicate too few nuclei produced for the number required in the atmosphere, postulating at least one nucleus per droplet of condensate.

Table 7.1

Ranges of Sizes of Condensation Nuclei and Kerns Occurring in the Atmosphere

Origin of Nucleus	Number per Cubic Centimeter of Air	Mass, gram	Diameter, micron
Sea salt	10 to 10^3†	3.5×10^{-16} to 2.8×10^{-13}	6×10^{-2} to 8×10^{-1}
Combustion	10 to 10^3	2×10^{-16} to 5×10^{-15}	2×10^{-2} to 2×10^{-1}
Location of kern count:			
Oceans	10^3 to 10^4	Sizes greater than microscopic ($>\frac{1}{2}$ micron diameter)	
Country	5×10^3 to 40×10^4		
Urban	150×10^3 to 400×10^4		

† Estimate based on the number of cloud droplets per cubic centimeter of air.

Most of the sulfurous and nitrous acid determinations, being made as they were near industrial centers, cannot be acceptable as indicative of the average amount in nonurban areas. The estimate is undoubtedly much too high. The presence of the acid nuclei in uninhabited areas cannot be ruled out entirely, because acid production in the atmosphere is accomplished, but on an unknown scale, by forest fires, volcanic eruptions, and lightning. Table 7.1 shows some measurements of natural nuclei made by various investigators. The size of the nucleus is given both in terms of mass and as the equivalent spherical diameter of a particle of the same mass.

There appears to be some discrepancy, undoubtedly based on the method of observation, in the number of effective condensation nuclei per unit volume of air. The chloride ion content of clouds determined by the titration of Cl^- gives numbers of nuclei equal to the number of cloud droplets, i.e., 50 to 500 per cubic centimeter of air. Direct catches of nuclei on glass slides give 10 to 20 per liter of air but of mass of the order of 10^{-10} gram. The direct evidence does not rule out submicroscopic nuclei in sufficient numbers to have each cloud drop initially condense on one nucleus. Investigations(1,18) in Japan using an electron microscope have verified the existence of large numbers of particles at least as small as 5×10^{-2} micron in equivalent spherical diameter existing in fog droplets and snowflakes. Only solid particles can be measured, since volatile products are evaporated in the high vacuum of the electron microscope. Much of the residue appears to be nonhygroscopic, leaving the sea-salt nuclei question still unanswered.

The Initial Condensation Stage. Admitting the existence of a sufficient number of condensation nuclei of finite size ($a > 10^{-6}$ centimeter) in the atmosphere on which condensation can proceed, we must reconcile two facts of chemistry in determining the initial stage of droplet growth. Both concern the magnitude of the equilibrium vapor pressure, the quantity in a cloud system which determines the mass of liquid remaining in equilibrium with a given mass of its vapor.

From physical chemistry we know that the equilibrium vapor pressure of water is reduced when a salt is dissolved in liquid water, the reduction being expressed as

$$\frac{e}{e_s} = 1 - CM \tag{7.1}$$

where e_s is the saturation vapor pressure of the pure water, e is the equilibrium vapor pressure of the molar aqueous salt solution containing M moles of solute, and C is a constant determined by the particular electrolyte used and the temperature of the solution.

By definition, the molar concentration M of a salt solution is defined as the number of moles of solute per liter of solution. At the equilibrium condition expressed by Eq. 7.1, the solution is so dilute that the density of the solution is not appreciably different from D, the density of water. Thus, the mass of solution in one drop is very closely $\frac{4}{3}\pi D r^3$, so that the number of drops in a liter of solution is $10^3/\frac{4}{3}\pi r^3$, provided that r is measured in centimeters. The number of drops multiplied by the mass of solute per drop gives the mass of solute per liter of solution. The number of moles of solute is given by the mass of solute in the liter of solution divided by the molecular weight of the solute. By combining these definitions, we see that the equilibrium vapor pressure over a droplet of a salt (the solute) such as NaCl or $MgCl_2$ dissolved in water (the solvent) is given approximately by

$$\frac{e}{e_s} = 1 - \frac{\text{Constant}}{r^3} \tag{7.2}$$

where r is the radius of the water drop and the constant is a function of the temperature and of the chemical nature of the solute. Equation 7.1 is in a form analogous to but not identical with *Raoult's law* for determining the reduction of equilibrium vapor pressure over solutions containing nonelectrolytical solutes. The vapor over the salt solution is pure water vapor, because the dissolved salt, a solid when in undissolved form, has a negligibly small vapor pressure. Equation 7.2 was developed from Eq. 7.1 on the implicit assumption that the magnitude of the reduction in vapor pressure is not a function of the radius of curvature of the salt solution, a direct contradiction to the experimental and theoretical work of Sir William Thomson (Lord Kelvin). His famous formula relating the radius of curvature of a surface to its equilibrium vapor pressure was developed for a pure substance in equilibrium with its vapor, i.e., a water drop (not an aqueous salt solution) in equilibrium with water vapor. There are many ways of deriving the Thomson formula, but the one based on classical thermodynamics will be illustrated.

Let σ denote the bulk surface tension of water. Surface tension is defined as the force applied normal to a unit length of a surface in order to keep the surface from collapsing. As an example of the definition, we note that a droplet of water in air has a spherical shape. Mechanical attempts to break up or permanently distort a water drop are resisted much in the same way that the skin of a balloon filled with air resists distortion. This skin effect of the boundary of a substance is related to the surface tension. For a sphere of radius r exerting a pressure p because of its surface structure, the definition of surface tension states that

$$\sigma = \frac{\text{Force}}{\text{Circumference}} = \frac{\pi r^2}{2\pi r} p = \frac{r}{2} p. \tag{7.3}$$

The product of pressure and area is force, where in this case the area must be the area enclosed by a circumference.

In order to obtain the Thomson formula, let us examine the energy required to expand a spherical drop of water against a surrounding atmosphere of water vapor. The atmosphere has a vapor pressure of such magnitude that an equilibrium state is maintained between the drop and the atmosphere; i.e., the drop neither evaporates nor grows. From Fig. 7.1 we see that the volume through which the liquid expands

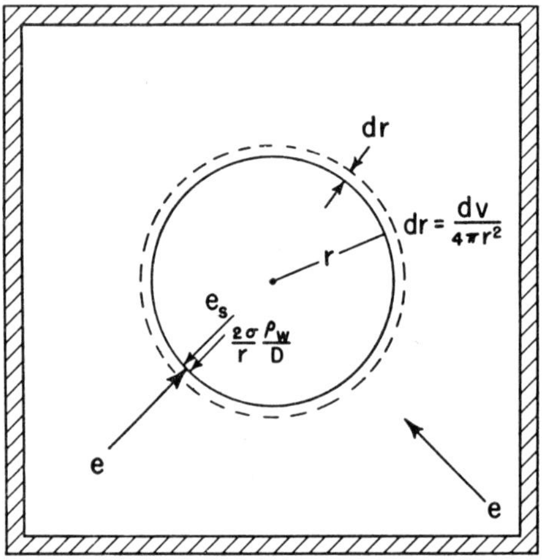

Fig. 7.1 The representation of the change in volume of a drop by the expansion of the drop against a surface film exerting a pressure $p = 2\sigma/r$. e_s is the internal and e the ambient vapor pressure. p represents the difference in pressure that must be exerted by the surface film to maintain equilibrium. At equilibrium and in the limit where $e - e_s \to 0$, thermodynamics requires that $e - e_s = (\rho_w/D)p$.

is the same volume through which the volume of vapor contracts. If there is no heat exchanged between the drop and its vapor, and if the temperature of the drop and vapor are the same and do not change during the process described, the first law of thermodynamics states that the work by the drop on the water vapor atmosphere is equal to the work expended by the water vapor on the liquid drop. Under these conditions, we find that the work of contracting a unit mass of the vapor from a specific volume v to a very small volume that approaches zero is

$$\int_0^v e\,dv = \int_{e_s}^e \frac{1}{\rho_w}\,de = R_W T \int_{e_s}^e \frac{de}{e} = R_W T \ln \frac{e}{e_s}\,. \tag{7.4}$$

It is inherent in the assumption that as the liquid drop expands, the volume that the vapor occupies becomes smaller because the sum of the volumes of the drop and vapor remains constant. Thus, it is assumed that as the specific volume of the vapor approaches zero, the liquid drop has grown so large that its surface is a plane surface. The equilibrium vapor pressure over a plane surface is of course the saturation vapor pressure e_s. The water-vapor density is ρ_w.

The liquid water drop can expand only by doing work against the surface film of water as this film is stretched. The film exerts a constant pressure p against any forces tending to displace the film. There is no transferal of mass between the liquid and its vapor by diffusion. A change in volume of a unit mass of liquid at constant pressure represents the work in stretching the surface film, being

$$\int_0^v p\,dv = pv = \frac{p}{D}\cdot \tag{7.5}$$

The volume change of the liquid was, by Fig. 7.1, to be through the same volume as the vapor. v is the volume of a unit mass of liquid and is equal therefore to the reciprocal of the density. Equating Eqs. 7.4 and 7.5 satisfies the thermodynamics of the vapor-liquid system. We find therefore that

$$R_W T \ln \frac{e}{e_s} = \frac{p}{D}\cdot \tag{7.6}$$

Substituting p in Eq. 7.3 for its equal in Eq. 7.6, and simplifying shows that

$$\ln \frac{e}{e_s} = \frac{2\sigma}{DR_W T}\frac{1}{r}\cdot \tag{7.7}$$

Equation 7.7 is the Thomson formula, showing the dependence of vapor pressure on the radius of curvature.

For values of $e/e_s \to 1$, $\ln e/e_s$ may be written as $\ln \left[\left(\frac{e}{e_s} - 1\right) + 1\right]\cdot$ The latter expression is in the form of $\ln (x + 1)$ which for small values of x is just equal to x itself. Thus, in this limit Eq. 7.7 may be written

$$e - e_s = 2\sigma \frac{\rho_w}{D}\frac{1}{r} \tag{7.8}$$

after substituting from the ideal gas law for the density of the saturated vapor.

At a fixed temperature, σ and D are constants, and the logarithm of the relative humidity varies as the reciprocal of the droplet radius. Equation 7.7 does not hold for very small values of r, because the concept of bulk surface tension has no meaning. Thus, the results of Eq. 7.7 are invalid for a droplet so small that it contains only a few molecules in its structure. At the other limit, where $r \to \infty$, the condition for a plane surface, we see that $e \to e_s$. This latter is the correct limit because we have defined saturation vapor pressure as the equilibrium vapor pressure over a plane liquid surface.

Calculations show that the effect of radius of curvature becomes important for drops having diameters less than 1 or 2 microns. For drops of greater diameter the equilibrium vapor pressure is the saturation vapor pressure given in tables. This fact is an important simplification that enters into the law of droplet growth by diffusion of water vapor to its surface. This law will be discussed in the next section.

The Thomson equation must be modified when other forces are present. An important modification occurs when the droplet carries an electrical charge of Q electrostatic units on its surface. The equilibrium vapor pressure becomes

$$DR_W T \ln \frac{e}{e_s} = \frac{2\sigma}{r} - \frac{Q^2}{8\pi r^4}. \tag{7.9}$$

Calculations show that for reasonable values of Q and r the term $Q^2/8\pi r^4$ has a negligible effect on the vapor pressure of a drop. Thus, it appears that even in thunderstorms where drops carry a relatively high electric charge density on their surfaces, the equilibrium vapor pressure is the same as the saturation vapor pressure over a plane surface for drops greater than 2 microns in diameter.

In the atmosphere, a condensation nucleus of the order of $r = 10^{-6}$ centimeter condenses water on its surface, dissolves, and forms a concentrated solution in droplet form whose radius may lie between $10^{-6} < r < 10^{-4}$ centimeter. By the time the droplet has grown to 10^{-4} centimeter (1 micron) in radius, the concentration of solute is so small that the drop may be considered as pure water. In the range of sizes $10^{-6} < r < 10^{-4}$ centimeter the true equilibrium pressure over a droplet is a combination of the lowering of vapor pressure by an aqueous solution, Eq. 7.2, and the radius of curvature effect of pure water, Eq. 7.7. These two equations are plotted in Fig. 7.2, with the resulting equilibrium curve, a combination of the two effects, drawn as the solid line. We interpret Fig. 7.2 as follows.

A condensation nucleus of size given by point A is placed in an atmosphere whose relative humidity e/e_s is the value given at B. In this atmosphere, the nucleus will condense moisture on its surface, grow from size A to a concentrated aqueous solution whose size is given at B, and then stop growing. Equilibrium with the ambient vapor pressure e has been reached. If the relative humidity then be changed to the value at C, supersaturation in respect to a plane water surface, the drop will grow to the size at C and come to equilibrium. Now, if the ambient† relative humidity is brought to the value at D, the drop will grow successively to positions E, F, G, and so on, never attaining an equilibrium state. By the time the drop has grown to

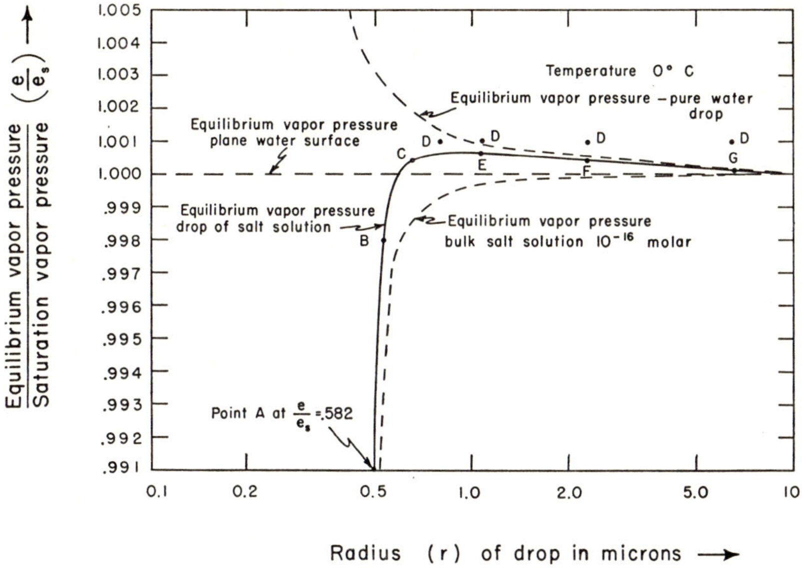

Fig. 7.2 Illustration of the growth of an aqueous salt solution in droplet form as a function of the relative humidity e/e_s. In terms of per cent, the maximum relative humidity needed for the process to go from point A to G is a relative humidity $> e/e_s \times 100 = 100.07$ per cent or 7/100 per cent supersaturation.

a size consistent with position E, the solution is so weak that the drop may be considered pure water. As long as the vapor pressure difference Δe is maintained, the drop will continue to grow at a rate which now can be calculated.

At no place in this discussion has the rate of growth from A to G been mentioned, since we have confined the discussion to equilibrium states. Experimental evidence indicates that the drops will reach their equilibrium sizes as quickly as the relative humidity can reasonably be varied, going from A to E under a constant humidity at D in less than 1 second.

† *Ambient* as in *ambient pressure, ambient temperature,* or *ambient density* is the adjective that refers to environmental conditions of pressure, temperature, or density, respectively. Since the conditions in the environment can be varied, we use the word *ambient* to suggest a temporary condition.

The equations governing the growth and evaporation of a droplet

Consider that we have a pure water droplet with radius the order of 1 micron. The ambient vapor pressure is e, and the saturation vapor pressure at the drop surface is e_s. A vapor pressure difference $e - e_s = \Delta e$ exists. When $\Delta e > 0$ the drop will grow and we have the condition specified for point E of Fig. 7.2 and discussed in the last section. If $\Delta e < 0$, then $e_s > e$, and the drop will evaporate until it reaches an equilibrium size given in Fig. 7.2 for $r < 1$ micron. Let us consider the derivation of the law for droplet growth.

In the absence of radiation, excessively strong electric fields, and a relative velocity between the water droplet and the air stream, a droplet grows by the diffusion of water vapor from the atmosphere to the droplet surface. Condensation follows and the droplet size increases. It has been found that the diffusion of water vapor is proportional to the concentration gradient of water vapor, a quantity nearly proportional to the gradient of vapor pressure. The rate of mass transferal by diffusion has, *under steady-state conditions,* a form analogous to Newton's law of cooling. The law, called *Fick's law of diffusion,* states that

$$\frac{dM}{dt} = -KA\,\frac{d\rho_w}{dx}\,. \tag{7.10}$$

dM/dt is the rate at which mass flows across a boundary under the driving effect of the concentration gradient $d\rho_w/dx$. ρ_w, the concentration of water vapor, is just the mass of water vapor in a unit volume of space. It is numerically equal to the product ρq, where ρ is the density of air and q is a meteorological quantity called the specific humidity. A is the area of the surface through which the mass of water vapor is diffused and is always measured normal to x. Thus, for spherical symmetry, A is the surface area of a sphere of radius x. For circular symmetry (neglecting the end effects), A is the surface of a cylinder and x is the radius of the cylinder. Similar reasoning holds for disks or flat plates. K is called the diffusivity, and as we are interested in water vapor diffusing in air, the diffusivity selected must be for this combination of gases. Under conditions of constant ambient temperature and relative humidity, i.e., constant ambient q, Eq. 7.10 may be integrated.

Under steady-state conditions, dM/dt is constant. For spherical symmetry, Eq. 7.10 may be written as

$$\frac{dM}{dt} = -4\pi r^2 K\,\frac{d\rho_w}{dr}\,. \tag{7.11}$$

At the surface of the drop, the variable r is equal to $a/2$, the drop radius. The density of water vapor is the saturated water vapor density at the temperature of the drop surface. This quantity will be denoted by ρ_{0w}. The ambient vapor density ρ_w must be measured at a position uninfluenced by the presence of the drop. This position is implied by $r = \infty$.† Therefore, the physical conditions are met if we integrate Eq. 7.11 between the limits shown in Eq. 7.12,

$$\frac{dM}{dt} \int_{a/2}^{\infty} \frac{dr}{r^2} = -4\pi K \int_{\rho_{0w}}^{\rho_w} d\rho_w, \tag{7.12}$$

obtaining after integration and substitution of limits

$$\frac{dM}{dt} = 2\pi K a(\rho_w - \rho_{0w}). \tag{7.13}$$

If we assume that all the water vapor is condensed as soon as it reaches the boundary at $r = a/2$, the drop must grow at a rate

$$\frac{dM}{dt} = D 4\pi r^2 \frac{dr}{dt} = \frac{D\pi a^2}{2} \frac{da}{dt}, \tag{7.14}$$

where D is the density of liquid water and $4\pi r^2\, dr$ is the change in the volume of the drop. The two a's in Eqs. 7.13 and 7.14 always signify the same position in space, so that we may equate Eq. 7.13 to Eq. 7.14. On simplifying, this operation leads to

$$a\, da = 4 \frac{K}{D} (\rho_w - \rho_{0w})\, dt, \tag{7.15}$$

the differential equation that expresses the rate of growth of the drop. If we set $a = a_0$ at time $t = 0$, the drop will grow to a diameter a in time t given by

$$\int_{a_0}^{a} a\, da = 4 \frac{K}{D} (\rho_w - \rho_{0w}) \int_0^t dt. \tag{7.16}$$

Integration and simplification of Eq. 7.16 show that

$$a^2 = a_0{}^2 + 8 \frac{K}{D} (\rho_w - \rho_{0w}) t. \tag{7.17}$$

a is the diameter of the drop. In the cgs system, $D = 1$ g/cm³ for liquid water and is often omitted when the growth of water droplets is considered.

The equations governing the growth of ice particles by diffusion can

† See remarks on infinity in Chapter 3. In this case, infinity is essentially any distance a few drop diameters from the drop itself.

also be developed, but because of the geometry of the crystals (plates, needles, or plane dendritic), the mathematical development is not so straightforward as for spherical water droplets. Fortunately, there is an electrical analogy for Eq. 7.10, so that in all generality we may write the equivalent of Eq. 7.10 as

$$\frac{dM}{dt} = D' \frac{dV}{dt} = 4\pi KC(\rho_w - \rho_{0w}). \tag{7.18}$$

C is the capacity of a condenser in air, a function only of the geometry of the condenser. The equivalence of the analogy can be seen from a study of Fig. 7.3. The surface of a condenser is a line of con-

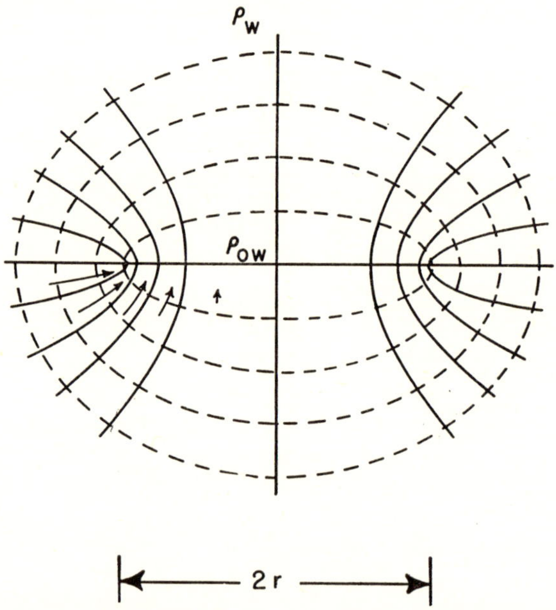

Fig. 7.3 The vapor density field of an infinitely thin circular plate of ice whose radius is r. The dashed lines are lines of constant vapor density with values lying between ρ_{0w} at the surface of the plate to ρ_w, the ambient vapor density, at an infinite distance from the sphere. The solid hyperbolic lines are streamlines which are lines enclosing the flow of vapor. The strength and direction of the flow is indicated by the vectors and the flow will be noticed to be strongest at the edges of the plate. The plane of the circular disk is normal to the page.

stant potential. Similarly, the surface of a crystal or a drop is a surface of constant vapor density ρ_{0w}. The condenser shapes the electric field, as the crystal or drop shapes the vapor density field, so that closed lines of constant potential or lines of constant vapor density can be drawn.

ρ_w, the ambient vapor density, corresponds to the zero potential line. Similarly, lines normal to the equipotential lines show the direction of the field, the line along which electric charge (in the electrical case) or water vapor (in the diffusion case) will flow. The rate of flow is proportional to the potential gradient or vapor density gradient. In the figure, notice that the vapor density gradient is strongest at the place where the curvature of the surface of deposition is greatest. This fact, in accord with observation, indicates that ice plates (approximated by circular disks) grow much more rapidly on the edge rather than at the flat surface of the disk. Table 7.2 indicates the values of C and D' to be used in computations involving Eq. 7.18.

Table 7.2

Various Types of Cloud Ice Crystals with the Constants to be Used in the Diffusion Equation, Eq. 7.18

(After Houghton)[13]

r is the radius of the particle, ϵ is the eccentricity, and L the length of the long axis of the ellipsoid.

Type	Assumed Shape	C	D', g/cm³
Plane dendritic	Disk	$2r\pi^{-1}$	0.5
Hexagonal plate	Disk	$2r\pi^{-1}$	0.8
Ice needle	Ellipsoid of revolution	$\epsilon L[\ln(1+\epsilon)(1-\epsilon)^{-1}]$	0.188
Sphere	Sphere	r	0.9

Type	b (Thickness)	M (Mass)
Plane dendritic	Constant = 0.0011 cm	$\pi D'r^2b$
Hexagonal plate	Constant = 0.0040 cm	$\pi D'r^2b$
Ice needle	$\frac{\text{Major axis}}{\text{Minor axis}} = \frac{L}{r} = 10$	$\frac{4}{3}\pi D'r^2L$
Sphere		$\frac{4}{3}\pi D'r^3$

Equation 7.18 is true only in the absence of a relative velocity between the evaporating (or growing) particle and the air (vapor field). Drops of any size tend to fall through the air under the influence of gravity. There is always a relative velocity between the free particle in the air and the air itself. This ventilating action causes an increase in the rate of deposition of vapor when $\rho_w > \rho_{0w}$. This action causes dM/dt to increase at a faster rate than would be computed from Eq. 7.18. On the other hand, when $\rho_w < \rho_{0w}$, evaporation of the drop is accelerated. Thus, a velocity factor of the form $[1 + f(v)]$ is indicated for the right-hand side of Eq. 7.18. $f(v)$ is the velocity factor,

which also depends on the shape of the particle and the viscosity of the air. Of course, $f(v) = 0$ when the velocity v is zero. One theoretical form of the correction, due to Frössling,[7] is

$$f(v) = \text{constant}\sqrt{Re}. \tag{7.19}$$

const. $= 0.276\left(\frac{K}{\mu}\rho\right)^{-1/3}$ and $Re = \frac{\rho v a}{\mu}$. v is the velocity of a drop of diameter a relative to air of density ρ and viscosity μ. Re is called the Reynolds number, whereas K is the diffusivity of water vapor in air. The method for the computation of fall velocities will be left until a later section.

In order to evaluate Eq. 7.18 or 7.13, it is necessary to find a means for determining ρ_{0w}, the saturation vapor density at the surface of the cloud particle. ρ_w, the ambient vapor density, is just the product of the relative humidity, measured, and the saturation vapor density which can be computed once the temperature of the air is known. Not so for ρ_{0w}, however. The temperature of the drop surface is not a constant independent of ambient relative humidity. The surface temperature shows a marked depression over ambient temperature when the ambient relative humidity is low. In the cases of droplet growth, the surface temperature is expected to be higher than ambient because of the condensation of water and its attendant release of latent heat of condensation to the drop.

ρ_{0w} can be evaluated through the following scheme. Energy in the form of heat can be brought to and absorbed by the water drop through conduction of heat by the air, radiation, or convective transport. This absorbed heat is used to change the state of the liquid to the vapor phase, the vapor in turn being diffused away from the drop. If $\sum H$ is the rate at which all forms of heat are being added to a drop, and L the latent heat of condensation (or sublimation of ice), during a steady state process

$$\sum H = L\frac{dM}{dt}. \tag{7.20}$$

The conduction of heat to the drop is the only term of the left-hand side of Eq. 7.20 that is of importance. The conduction of heat follows an equation analogous to (7.18), except that we utilize the molecular coefficient of heat transfer of air, κ_H equal to $\rho c_p \kappa$, instead of the diffusity K, and the temperature difference $(T_0 - T)$ instead of the vapor-density difference $(\rho_w - \rho_{0w})$. We write that

$$H = \frac{dQ}{dt} = 4\pi\rho c_p \kappa C(T_0 - T), \tag{7.21}$$

where T_0 is the surface temperature of the drop and T is the ambient temperature. Placing Eq. 7.21 equal to Eq. 7.18 multiplied by L satisfies Eq. 7.20 under the conditions specified above, and as a result

$$T_0 - T = \frac{K}{\kappa}\frac{L}{c_p}\frac{(\rho_w - \rho_{0w})}{\rho}. \tag{7.22}$$

It is to be observed that Eq. 7.22 is independent of the geometry of the cloud particle and is strictly true only for zero relative velocity of the particle in the air. The velocity correction for the heat transfer

Table 7.3

Kinematic Viscosity and Thermometric Conductivity of Air, and Diffusivity of Water Vapor in Air at 1000 Millibars Pressure. The Latent Heat of Vaporization and of Sublimation

(From Smithsonian Meteorological Tables)

Temperature, T °C	Kinematic Viscosity,† η cm² sec⁻¹	Thermometric Conductivity,‡ κ cm² sec⁻¹	Diffusivity of Water Vapor in Air, K cm² sec⁻¹	Latent Heat of Vaporization, L_v cal/g	Latent Heat of Sublimation, L_s cal/g
−20	0.117	0.165	0.197	609	678
−10	0.126	0.177	0.211	603	678
0	0.135	0.189	0.226	597	677
10	0.144	0.202	0.241	592	
20	0.153	0.215	0.257	586	
30	0.162	0.228	0.273	580	
40	0.172	0.242	0.289	575	

The specific heat at constant pressure is 0.240 cal per g per °C.

At pressures other than 1000 mb, the values of η, κ, and K are given by multiplying the tabular value by $1000/p$ where p is in millibars.

† See Eq. 7.28.

‡ Also called thermal diffusivity.

by diffusion and that for heat transfer by conduction are probably not the same. This point, which has not been investigated, seems to be an unimportant correction, since the velocity corrections to both sides of Eq. 7.20 are of about the same order of magnitude, in the same sense, and as a result tend to cancel. Hence, during the period of cloud growth when Eq. 7.18 is of importance, the velocity correction is small and may in many applications be ignored.

The evaluation of Eq. 7.22 must be accomplished through a "cut-and-try" procedure. ρ_w and T are determined by ambient conditions. ρ_{0w} is a function of T_0 and is known only in tabular form. Therefore, it is necessary to choose a value of T_0, find ρ_{0w} from tables of saturation vapor density, put the numbers in Eq. 7.22, and see if an equality results. One should note that L, K, and κ are also dependent on temperature but in a relatively insensitive fashion. The "cut-and-try" procedure must be followed until an equality results. At equality, the heat-balance equation has been satisfied in the steady state. In this way $(\rho_w - \rho_{0w})$ is determined for later substitution into the integrated form of (7.18), of which Eq. 7.17 is the example for the spherical case.†

Table 7.3 is a table of a number of useful constants used in evaluating Eqs. 7.18 and 7.22.

The growth of cloud droplets to raindrops by the accretion process

Supersaturations that exist in the air for any length of time are usually due to the cooling associated with rising air, and apparently do not exceed one-tenth of 1 per cent. Computations of the growth of drops by diffusion indicate that even for such high supersaturations it takes a one micron drop 1 second to grow to 10 microns, $\frac{1}{2}$ minute to grow to 50 microns, 2 minutes to 100 microns, 3 hours to 1 millimeter, and 1 day to 3 millimeters. As a result of these calculations, the diffusion process can never explain the formation of heavy rain from showers. The length of time required to form drops greater than 1 millimeter‡ in diameter precludes accepting such a theory. There are also other objections, one of which is that the equations show all drops growing by diffusion to tend toward a uniform size after the process has been in operation for a relatively long time. One consequence of the diffusion process, even when the restrictive time factor is ignored, is the unrealistic result that rainfall would be uniform in size rather than of the distribution of sizes usually observed in showers. A computation on the time available for supersaturation to be maintained through rising motion and adiabatic cooling of the air shows that air rising at the rate of 1 meter per second through 3 kilometers (about 10,000 feet) has about 50 minutes of transit time, not enough time for a cloud drop to attain 1 millimeter in diameter. Finally, the question may in all justice be asked how a large drop of less than 1 millimeter in diameter, but certainly much larger than the general size in the lower section of the cloud, falls through the cloud and grows

† Note that although Eq. 7.18 is written for an ice particle, changing D' to D and taking vapor densities in respect to water instead of ice makes Eq. 7.18 identical to Eq. 7.13 and after integration equal to Eq. 7.17.

‡ Shower rain has been observed to range from $\frac{1}{2}$ to 4 millimeters in size.

without benefit of supersaturation† and yet reaches the ground at normal raindrop size. The explanation for the growth of a large drop falling through a high concentration of small drops involves a study of fall velocities of water drops in air and the efficiency with which a large drop will collect and add to its own mass, by collision, smaller drops that are directly in its line of fall.

Let v be the velocity of fall relative to the air of a droplet of radius r. t is the time of fall. In a time dt the droplet will sweep out a volume equal to the product of its cross-sectional area and the distance $v\,dt$. If this volume of air contains n_i drops of water per unit volume having diameters a_i, the total mass of liquid water per unit volume of air is $\frac{\pi}{6} D \sum_i n_i a_i^3$ denoted by w. The mass of water geometrically intercepted by the falling drop is

$$w\pi r^2 v\,dt.$$

A fraction E of the intercepted water is added to the falling drop to increase its mass. E is called the collection efficiency because the ratio of the water actually collected to the water geometrically intercepted may vary between 0 and 1. Thus, the growth of a drop by the accretion of water is given by

$$dM = \pi \mathrm{E} w r^2 v\,dt. \tag{7.23}$$

But as

$$dM = 4\pi D r^2\,dr, \tag{7.24}$$

substitution of Eq. 7.24 in Eq. 7.23 predicts that the change in diameter, $r = a/2$, of a drop falling through a mass of water in the form of drops is

$$da = \frac{\mathrm{E}w}{2D} v\,dt = \frac{\mathrm{E}w}{2D} dz \tag{7.25}$$

where dz is the vertical distance of fall.

The evaluation of Eq. 7.25 under all but the simplest of assumptions is difficult. The difficulty lies in the aerodynamics of the problem for evaluating the collection efficiency. An introduction to the problem is given in the next two sections.

The Terminal Velocity of Freely Falling Drops. A spherical droplet of density D falling through a medium of density ρ is accelerated

† The raindrop presumably falls in regions of low upward velocity, where low or no supersaturation exists.

under the action of gravity with a force

$$f = \tfrac{4}{3}\pi(D - \rho)r^3 g. \tag{7.26}$$

Equation 7.26 presents the difference between the unimpeded fall of the drop in a vacuum ($\frac{4}{3}\pi D r^3 g$), and the buoyant force of the air ($\frac{4}{3}\pi \rho r^3 g$). r is the radius of the drop and g the acceleration of gravity. The accelerating force is opposed in any real fluid by a frictional force arising from the viscosity of the medium and the turbulence induced in the medium by the passage of the drop. A measure of the turbulence is given by a nondimensionless parameter that enters in the theory of the flow of fluids. This parameter, called the Reynolds number, is defined as

$$Re = \frac{2\rho r v}{\mu} \tag{7.27}$$

where v is the *relative velocity* of an object, whose principal cross-sectional dimension (the diameter of a sphere) is $2r$, to a fluid of density ρ and viscosity μ. The ratio

$$\eta = \mu/\rho \tag{7.28}$$

is a common combination, and is called the *kinematic viscosity* of the fluid. However, μ is the quantity experimentally measured. The cgs unit of viscosity is the poise of dimensions g cm^{-1} sec^{-1}. The kinematic viscosity has dimensions of cm^2 sec^{-1}.

The generalized resisting force on spheres has been found by theory and experiment to be of the form

$$f_r = 6\pi\mu r v \left(\frac{C_D Re}{24}\right). \tag{7.29}$$

C_D is called the *drag coefficient* and is the constant of proportionality between the measured force f_r and the physical quantities on the right-hand side of Eq. 7.29. For low Reynolds numbers, $Re < 1$, $C_D Re/24 = 1$, and Eq. 7.29 becomes

$$f_r = 6\pi\mu r v. \tag{7.30}$$

Equation 7.30 is called *Stokes' law*, for the man who first worked out the theory of the fall of small droplets in a viscous medium.

C_D is not a constant but is a function of the Reynolds number. However, since C_D varies only slowly with r and v, it is often useful to write Eq. 7.29 as

$$f_r = \frac{\pi}{2} C_D \rho r^2 v^2. \tag{7.31}$$

The aid of Eq. 7.27 for the Reynolds number was invoked to obtain (7.31). Thus, over limited ranges of r and v we may select an average value of C_D, and note that the resisting force for larger drops increases as $(rv)^2$ instead of (rv), the case for small drops obeying Stokes' law.

Freely falling drops accelerated by a force given by Eq. 7.26 are opposed by a constantly increasing resisting force given in general by Eq. 7.29. A point is reached where the two forces balance one another and equilibrium exists. At this point, the velocity reaches a maximum, called the terminal velocity. By equating (7.26) and (7.29) we find that the terminal velocity v_T becomes

$$v_T = \frac{2}{9}\frac{(D - \rho)}{\mu} g \left(\frac{24}{C_D Re}\right) r^2. \tag{7.32}$$

Equation 7.32 may be written in a variety of ways, but the definitions of drag coefficient derived from this equation are consistent with modern aerodynamic usage and identical with the drag coefficient as defined in the *Smithsonian Meteorological Tables*[(B2)] (sixth revised edition, 1951).

For the remainder of the section we shall confine our discussion to spheres of water or ice falling in air. $D \gg \rho$ that $D - \rho = D$ to all practical limits. When Stokes' law is followed, and it is valid for water drops in air for $r \leqq 40$ microns, Eq. 7.32 becomes

$$v_T = \frac{2}{9}\frac{Dg}{\mu} r^2 = \frac{1}{18}\frac{Dg}{\mu} a^2. \tag{7.33}$$

Calculation of v_T by Eq. 7.33 is straightforward and presents no difficulties. Once Stokes' law has become invalid, the calculation of v_T, necessarily by Eq. 7.32, becomes quite arduous. It is a cut-and-try solution because v_T is explicit in the definition of Re and implicit in the definition of C_D. Table 7.5 gives values of a, Re, and v_T.

The collection efficiency of drops

Let us consider the two-dimensional problem of a circle of diameter a (sphere or cylinder in three dimensions) moving through the air with a velocity relative to the air of v. In a time t, the drop sweeps through an area equal to vta. Any spherical particle initially in this area will be intercepted by the circle unless it can be acted upon by a sideward force that will move it outside the shaded area of Fig. 7.4. The air particles are forced around the circle in characteristic fashion, following the lines of flow, called streamlines, shown in Fig. 7.5. A spherical particle of diameter a_i and mass $M_i = (\pi/6)Da_i^3$ will tend to follow a path that is intermediate between the straight line (inertial) motion of Figs. 7.5 and 7.6 and the path of the air particles themselves. Viscosity provides the force tending to hold the particle to streamline flow. We shall neglect the fact that interception actually will occur

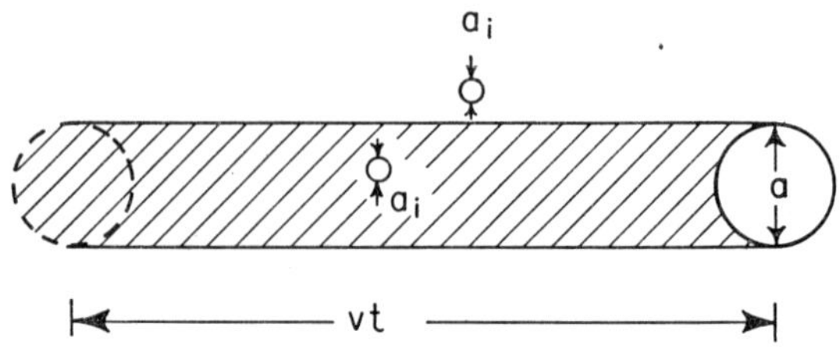

Fig. 7.4 In the absence of viscous forces, a drop of diameter a moving with velocity v will collide with any drop of diameter a_i if the small drop lies inside a volume $(\pi/4)a^2vt$. A section of this volume is shown by the cross-hatched lines. In two dimensions, the collision "volume" is the area vta.

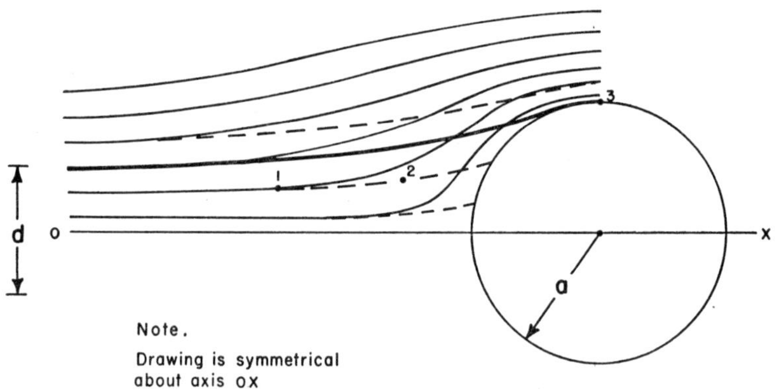

Fig. 7.5 The streamlines (solid lines) of the air flow and the trajectories (dashed lines) of particles in the air flow. All particles within the distance d will hit the cylinder (or sphere) of radius a. The limiting trajectory is shown as the heavy solid line. For a cylinder, the collection efficiency per unit length would be the ratio d/a. For a sphere, the collection efficiency would be in the ratio d^2/a^2. In both cases, the collection efficiency is in the ratio of areas. (See reference 24 for a complete discussion.)

in the area $vt(a + 2a_i)$ although the intercepted particle will not coalesce when the two diameters a and a_i are tangent (see point 3 of Fig. 7.5).

Suppose that we bring two successive positions in the trajectory of a particle, given by points 1 and 2, infinitesimally close together and examine the forces that must be acting to bring the particle at 1 to its next position at 2. The vector diagram is shown in Fig. 7.6. A particle at O, position 1, if acted on by no deviating force, would proceed along the vector OA and in a given unit of time dt it would arrive at A. Because the particle has mass, any attempt to change the motion of the particle will be opposed by a reaction force of inertia, $M_i \frac{d\mathbf{u}}{dt}$. In a fluid or a gas such as air, viscosity is a property that influences motion. For velocities low enough so that Stokes' law applies, this viscous force has components along the x and y axis of

$$f_{r_x} = 3\pi\mu v_x a_i \quad (7.34a)$$

$$f_{r_y} = 3\pi\mu v_y a_i. \quad (7.34b)$$

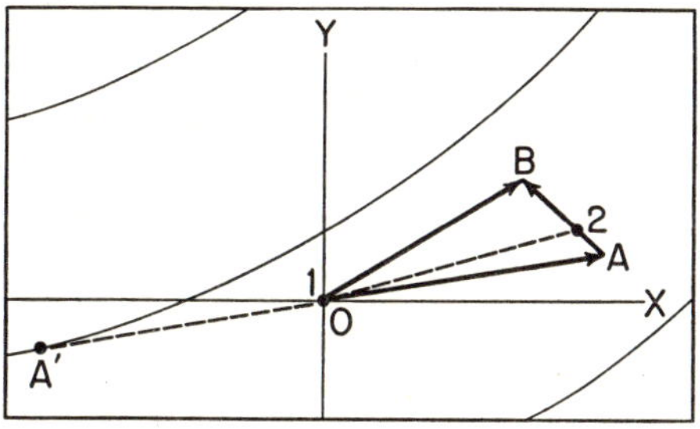

Fig. 7.6 OB is the vector representation of the air velocity U. This vector is tangent to a streamline. OA is the vector representation of the trajectory velocity of the particle at position 1. OA and OA' are equal. OA' is the trajectory or path of the particle just subsequent to its arrival at 1. Assuming unit time makes the velocity vector OA and the path vector OA' equal. AB denotes the vector difference between the path velocity and the streamline velocity. As constructed, the magnitude of the vector **AB** is

$$\mathbf{AB} = |-(\mathbf{u} - \mathbf{U})| = \sqrt{(u_x - U_x)^2 + (u_y - U_y)^2}.$$

The latter terms, $(u_x - U_x)$ and $(u_y - U_y)$, are obtained from Eqs. 7.35a and 7.35b.

The vector difference of the velocities associated with these forces is AB, and this vector is proportional to the deflecting force on the particle. The deflection forces the particle to point 2. The path of the particle in going from 1 to 2 is called the trajectory. The components of the inertial force are $M_i \frac{du_x}{dt}$ and $M_i \frac{du_y}{dt}$. From Fig. 7.6 and Eqs. 7.34a and 7.34b the deflecting forces are seen to be†

$$x \text{ component} \quad M_i \frac{du_x}{dt} = -3\pi\mu a_i(u_x - U_x) \quad (7.35a)$$

$$y \text{ component} \quad M_i \frac{du_y}{dt} = -3\pi\mu a_i(u_y - U_y). \quad (7.35b)$$

† U is the velocity of the fluid along a streamline. u is the velocity of the particle along a trajectory. The difference $(u - U)$ represents the difference in velocity between streamline and trajectory flow, i.e., particle velocity relative to

The solution of this set of differential equations will give the velocity of the particles as a function of time. If we set $u_x = dx/dt$ and $du_x/dt = d^2x/dt^2$ and use a similar notation for u_y and du_y/dt, the set of equations above can be written as

$$\frac{Da_i^2}{18\mu}\frac{d^2x}{dt^2} + \frac{dx}{dt} - U_x = 0 \tag{7.36a}$$

$$\frac{Da_i^2}{18\mu}\frac{d^2y}{dt^2} + \frac{dy}{dt} - U_y = 0 \tag{7.36b}$$

when we substitute for the mass of the particle its equivalent $(\pi/6)Da_i^3$. It is beyond the scope of this book to do other than indicate that the solution to this set of differential equations is two equations, one of which shows x as a function of t and the other shows y to be a function of t. Elimination of t between the two equations gives an equation in x and y, which describes the path of the particle as it nears the collecting circle of diameter a. The path, called the trajectory, of the particle may or may not hit the collecting circle. The streamlines and trajectories of typical particles are shown in Fig. 7.5. It will be noticed that not all the particles that were originally on an interception course will collide and be collected. The ratio of the number that are actually intercepted to the number in the area vta is the collection efficiency E. Of course, the problem is in reality a three-dimensional problem that is just an extension of the two-dimensional case. It is not difficult to see that the collection efficiency is a function of the geometry of the collector, because the streamline pattern depends on the shape of the collector, being somewhat different for spheres from what it is for cylinders, even though both may have a common circular cross section at one point.

Before leaving the subject of trajectories, it is instructive to find a parameter that is important in determining the collection efficiency. This parameter, a nondimensional quantity, is probably the most important single bit of information that comes out of the set of equations denoted by Eq. 7.36. We arrive at this parameter by choosing scale factors whose variation from system to system is important. For example, the velocity U_0 of the undisturbed stream is a measurable parameter of the system that is important in the theory. So, also, is the diameter a of the collector. The combination a/U_0 has the dimensions of time. Therefore, if we change to nondimensional parameters

---

the fluid stream. It is this quantity that is proportional to the acceleration of the particle along a trajectory. This is the interpretation given to v of Eqs. 7.34a and 7.34b to obtain Eqs. 7.35a and 7.35b.

χ, ψ, Ξ and τ by letting

$$x = a\chi; \qquad u_x = \frac{dx}{dt} = U_0 \frac{d\chi}{d\tau}; \quad \frac{du_x}{dt} = \frac{d^2x}{dt^2} = \frac{U_0^2}{a}\frac{d^2\chi}{d\tau^2}; \quad U_x = U_0\Xi_x$$

$$y = a\psi;$$

$$t = \frac{a}{U_0}\tau; \qquad u_y = \frac{dy}{dt} = U_0 \frac{d\psi}{d\tau}; \quad \frac{du_y}{dt} = \frac{d^2y}{dt^2} = \frac{U_0^2}{a}\frac{d^2\psi}{d\tau^2}; \quad U_y = U_0\Xi_y$$

we find that Eq. 7.36 takes the form

$$\Lambda \frac{d^2\chi}{d\tau^2} + \left(\frac{d\chi}{d\tau} - \Xi_x\right) = 0 \tag{7.37a}$$

$$\Lambda \frac{d^2\psi}{d\tau^2} + \left(\frac{d\psi}{d\tau} - \Xi_y\right) = 0 \tag{7.37b}$$

where Λ is a nondimensional parameter given by

$$\Lambda = \frac{DU_0 a_i^2}{18\mu a} \cdot \dagger \tag{7.38}$$

Physically, Λ is the ratio of the force necessary to stop a particle traveling with initial velocity U_0 in the distance a, to the fluid-resistance force opposing a particle having a velocity U_0.

It has been shown by responsible investigators(19,24) that E is a function of Λ only. Some theoretical curves by these investigators are shown in Figs. 7.7 and 7.8. Langmuir and Blodgett(19) suggest the equation

$$E = \frac{\Lambda^2}{(\Lambda + \frac{1}{2})^2} \tag{7.39}$$

as suitable for describing the collection efficiency of a sphere over a range of velocities given by $\Lambda > 0.2$. It has been found that no drops are collected on spheres when $\Lambda < \frac{1}{12}$ and on cylinders when $\Lambda < \frac{1}{8}$. Applying the criterion for no collection we find the largest size of the collector for which no drops are collected. If we know a, we can compute U_0 from *Stokes' law*.‡ Putting in the condition that $\Lambda_{\text{crit}} = \frac{1}{12}$ we obtain the table on page 229.

† When relative velocities are so high that Stokes' law must be modified by taking into account the correction factor $\frac{C_D Re}{24}$, a better parameter than Λ is $\Lambda' = \frac{24}{C_D Re}\Lambda$.

‡ For a stationary collector, U_0 of Eq. 7.38 equals v_T of Eq. 7.33 when drops are falling vertically in still air. For airfoils, U_0 is interpreted as the true air speed. For two falling drops, U_0 is interpreted as the relative velocity between the two drops. In Table 7.4, there is no relative velocity between the 14-micron drops, so that in reality they never come together. We interpret the table to mean that one drop is stationary while the other falls according to *Stokes' law*.

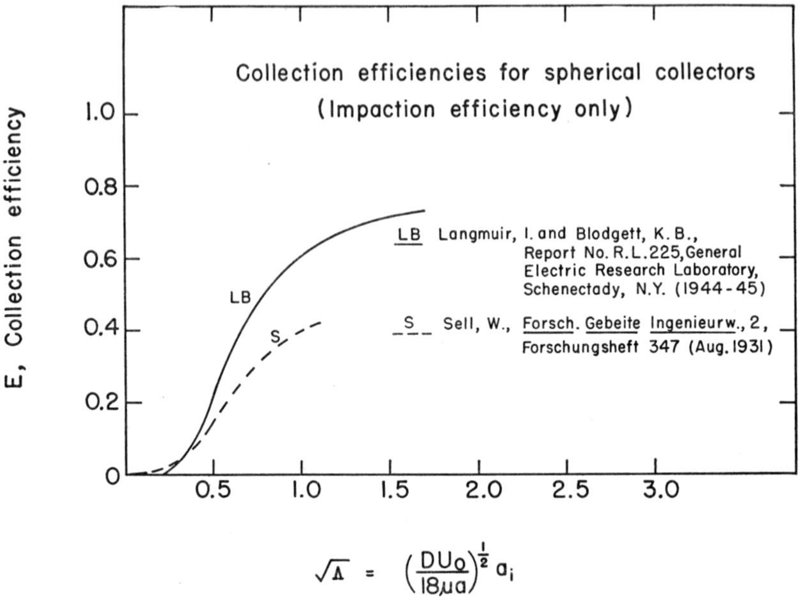

Fig. 7.7 The collection efficiency of spheres.(24) (See Eq. 7.38 for definition of Λ.)

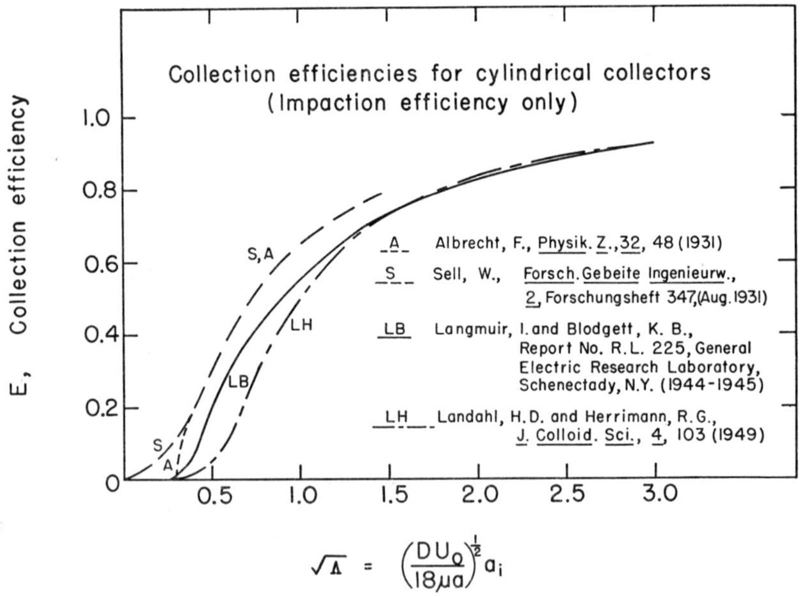

Fig. 7.8 The collection efficiency of cylinders.(24) (See Eq. 7.38 for definition of Λ.)

Table 7.4

Critical Diameter of Falling Drops below Which No Accretion Occurs, for Various Cloud-Drop Diameters (in Microns)†

(After Langmuir)

Cloud-Drop Diameter	Minimum Diameter of Falling Drop for Collection		
a_i	a	a_i	a
3	600	10	31
4	350	12	20
6	140	14	14
8	58		

† See note after Table 7.6.

A drop falling through air is not a good approximation to a sphere, except for drops less than about 1000 microns in diameter. Larger drops become flattened and are in a constant state of oscillation because of the increased resisting pressure of the air as the drops fall.(4) In fact, the oscillations become so violent for large drops that surface tension cannot keep them from breaking up. As a consequence, drops larger than 5 to 7 millimeters are not observed. In a subsequent chapter, some electrical consequences of this breakup action will be discussed. For the present, the effect of flattening changes both the fall velocity and collection efficiency of actual raindrops over spheres of equal mass. The collection efficiency of drops of diameter a falling through other drops of diameter a_i is given in Table 7.6. The diameter of the large drops is figured from an equivalent sphere having the same mass as the drop selected.

The calculation of the growth of drops by accretion

We now have enough information to calculate the growth of a falling raindrop by the accretion process. As one can see by the multitude of tables involved, Eq. 7.25 will have to be evaluated by numerical means because of the complexity of the function for the collection efficiency. For example, on integration, Eq. 7.25 becomes

$$a_2 - a_1 = \frac{\overline{\mathrm{E}}\overline{w}}{2D}(z_2 - z_1) \tag{7.40}$$

when average values of E and w are used as a drop falls from level z_1 to level z_2. $z_2 - z_1$ is considered a positive number. a_1 is the ini-

Table 7.5

A Short Table of Terminal Velocities, Reynolds Number, and Correction Terms to Stokes' Law† for Falling Raindrops

Diameter (microns)	$C_D\, Re/24$	v_T (cm/sec)	Re
20	1.00	1.24	0.017
50	1.00	7.72	0.268
80	1.00	20.0	1.11
200	1.68	72	9.61
400	2.98	162	4.32
800	5.94	327	175
1200	9.41	464	372
2000	18.7	649	866
2800	30.4	782	1461
4800	77.0	907	2905

† Stokes' law applies for spheres in air less than 80 microns in diameter.

Table 7.6

The Collection Efficiency E for Drops of Diameter a Falling through a Cloud of Smaller Drops of Diameter a_i

(After Langmuir and Blodgett)(19)

	a_i, microns							
a, microns	4	6	8	12	16	20	30	40
30					0*	0*	0*	0*
50				0.02*	0.22*	0.33*	0.50*	0.50*
80				0.13*	0.36*	0.48*	0.67*	0.74*
200			0.133	0.418	0.564	0.62*	0.739	0.82*
400		0.085	0.326	0.564	0.684	0.756	0.859	0.908
800	0.040	0.303	0.500	0.698	0.793	0.849	0.919	0.950
1200	0.121	0.355	0.530	0.731	0.827	0.876	0.939	0.963
2000	0.140	0.358	0.535	0.738	0.834	0.886	0.944	0.966
2800	0.168	0.360	0.534	0.735	0.840	0.890	0.950	0.970
4800	0.075	0.220	0.372	0.606	0.743	0.823	0.920	0.950

Note: Values in this table marked with * are computed by Dr. Raymond Wexler, using the Langmuir and Blodgett formula, with the relative velocity between the falling drops employed instead of a drop falling through a cloud of stationary particles. It is evident from Stokes' law that a 30-micron drop will not fall through a cloud of 40-micron drops because of its smaller fall velocity. At larger values of radius, the difference between relative velocity and the velocity used by Langmuir and Blodgett is negligible.

tial size of the drop, and a_2 is its final size. w is the liquid-water content in grams of water per cm^3 of air. D, the liquid-water density, is unity in the cgs system.

Suppose, for example, that we wish to find how large a drop will grow in falling 1 kilometer through a cloud containing 0.7 gram of liquid water per cubic meter of air. The size of the cloud particles will be considered to be 20 microns in diameter. The drop itself will have an initial diameter of 200 microns. Using the cgs system of units and Eq. 7.40, we find that

$$a_2 - 2 \times 10^{-2} = \frac{\bar{\mathrm{E}} \times 7 \times 10^{-7}}{2} (1 \times 10^5)$$

$$a_2 = 2 \times 10^{-2} + 3.5 \times 10^{-2}\bar{\mathrm{E}}.$$

For $a_1 = 200$ microns we find from Table 7.6 that $\bar{\mathrm{E}} = 0.62$ for a cloud of 20-micron drops. Substitution makes

$$a_2 = 2 \times 10^{-2} + 3.5 \times 10^{-2} \times 0.62 = 4.2 \times 10^{-2}\ \mathrm{cm} = 420\ \text{microns}.$$

However, a drop of 420 microns has a collection efficiency of 0.76. A collection efficiency of 0.62 was too low for the drop; a collection efficiency of 0.76 will be too high. However, with the latter collection efficiency, the final size is

$$a_2 = 2 \times 10^{-2} + 3.5 \times 10^{-2} \times 0.76 = 4.7 \times 10^{-2}\ \mathrm{cm} = 470\ \text{microns}.$$

It is evident that the drop will grow to somewhere between 420 and 470 microns in the 1-kilometer fall. For the best answer, let us take an interpolated value of $\bar{\mathrm{E}} = 0.69$ and obtain a final drop size of

$$a_2 = 2 \times 10^{-2} + 3.5 \times 10^{-2} \times 0.69 = 4.4 \times 10^{-2}\ \mathrm{cm} = 440\ \text{microns}.$$

In the next kilometer of fall, the initial diameter of the drop will be $a_1 = 440$ microns. Successive calculations of the same nature as shown will give the growth of a raindrop as it falls through a cloud several kilometers thick.

Evaporation of raindrops

Raindrops falling from clouds may partially evaporate before they hit the ground. This effect will occur provided that the air underneath the cloud has a reasonably low relative humidity. The equations for calculating the decrease in diameter and mass are the same as given in this chapter for the growth of raindrops by diffusion, with the velocity correction to the rate of diffusion taken into consideration because of the high rates of fall of large raindrops.

In the next chapter we shall consider theories of the growth of clouds and rain, using the principles set forth in this chapter.

References

1. Aufm Kampe, H. J., Weickmann, H. K., and Kedesdy, H. H., "Remarks on 'Electron-microscope study of snow-crystal nuclei,'" *J. Meteorol., 9,* 374–375 (1952).
2. Best, A. C., "Drop-size distribution in cloud and fog," *Quart. J. Roy. Meteorol. Soc., 77,* 418–426 (1951).
3. Best, A. C., "The evaporation of raindrops," *Quart. J. Roy. Meteorol. Soc., 78,* 200–225 (1952).
4. Blanchard, D. C., "The Behavior of Water Drops at Terminal Velocity in Air," *Trans. Am. Geophys. Union, 31,* 836–842 (1950).
5. Bowen, E. G., "The formation of rain by coalescence," *Australian J. Sci. Research, Ser. A, 3,* 193–213 (1950).
6. Davies, C. N., and Aylward, M., "The Trajectories of Heavy Solid Particles in a 2-dimension Jet of Ideal Fluid Impinging Normally upon a Plate," *Proc. Phys. Soc.* (B), *64,* 889–911 (1951).
7. Frössling, N., "Über die Verdunstung fallender Tropfen," *Gerlands Beitr. Geophys., 52,* 170–216 (1938).
8. Fuchs, N., "Über die Verdampfungsgeschwindigkeit kleiner Tröpfchen in einer Gasatmosphäre," *Physik. Z. Sowjetunion, 6,* 224–243 (1934).
9. Gunn, R., "Vertical Shaft for the Production of Thick Artificial Clouds and for the Study of Precipitation Mechanics," *J. Appl. Phys., 23,* 1–5 (1952).
10. Gunn, R., and Kinzer, G. D., "The terminal velocity of fall for water droplets in stagnant air," *J. Meteorol., 6,* 243–248 (1949).
11. Gunn, K., and Hitschfeld, W., "A laboratory investigation of the coalescence between large and small water-drops," *J. Meteorol., 8,* 7–16 (1951).
12. Houghton, H. G., "On the Physics of Clouds and Precipitation," *Compendium of Meteorology,* American Meteorological Society, Boston, 165–181 (1951).
13. Houghton, H. G., "A preliminary quantitative analysis of precipitation mechanisms," *J. Meteorol., 7,* 363–369 (1950).
14. Houghton, H. G., and Radford, W. H., "On the Local Dissipation of Natural Fog," *Papers in Phys. Oceanog. Meteorol.,* Mass. Inst. of Technology and Woods Hole Oceanographic Inst., Vol. VI, No. 3 (1938).
15. Howell, W. E., "The growth of cloud in uniformly cooled air," *J. Meteorol., 6,* 96–104 (1949).
16. Junge, C., "Nuclei of Atmospheric Condensation," *Compendium of Meteorology,* American Meteorological Society, Boston, 182–191 (1951).
17. Kinzer, G. D., and Gunn, R., "The evaporation, temperature, and thermal relaxation time of freely falling waterdrops," *J. Meteorol., 8,* 71–83 (1951).
18. Kumai, M., "Electron-microscope study of snow-crystal nuclei," *J. Meteorol., 8,* 151–156 (1951).
19. Langmuir, I., "The production of rain by a chain reaction in cumulus clouds at temperatures above freezing," *J. Meteorol., 5,* 175–192 (1948).
20. Ludlum, F. H., "The Production of Showers by the Coalescence of Cloud Drops," *Quart. J. Roy. Meteorol. Soc., 77,* 402–417 (1951).
21. Mason, B. J. and Ludlum, F. H., "The Microphysics of Clouds," *Reports of the Physical Society on Progress in Physics, 14,* 147–195 (1950).
22. Möller, F., "Thermodynamics of Clouds," *Compendium of Meteorology,* American Meteorological Society, Boston, 199–206 (1951).

23. Neiburger, M., and Wurtele, M. G., "On the nature and size of particles in haze, fog, and stratus of the Los Angeles region," *Chem. Rev., 44,* 321–335 (1949).
24. Ranz, W. E., "The Impaction of Aerosol Particles on Cylindrical and Spherical Collectors," *Tech. Report No. 3,* Engineering Experiment Station, University of Illinois (1951). A.E.C. Report SO-1004 (declassified).
25. Rodebush, W. H., "Nuclei in Evaporation and Condensation," *Chem. Rev., 44,* 269–276 (1949).
26. Simpson, G. C., "On the Formation of Cloud and Rain," *Quart. J. Roy. Meteorol. Soc., 67,* 99–134 (1941).
27. Woodcock, A. H., "Condensation nuclei and precipitation," *J. Meteorol., 7,* 161–162 (1950).
28. Woodcock, A. H., "Atmospheric salt particles and raindrops," *J. Meteorol., 9,* 200–212 (1952).
29. Wright, H. L., "Sea salt nuclei," *Quart. J. Roy. Meteorol. Soc., 66,* 3–12 (1940).

Source Books

B1. Haurwitz, B., *Dynamic Meteorology,* Chapter III, McGraw-Hill Book Co., New York (1941).
B2. List, R. J., *Smithsonian Meteorological Tables,* Smithsonian Miscellaneous Collections, Vol. 114, Washington, D.C. (1951).

Problems

7.1 The reduction in vapor pressure resulting from dissolving M moles of electrolyte in M_0 moles of water is given as

$$\frac{e - e_s}{e_s} = -\frac{iM}{iM + M_0}.$$

e is the equilibrium vapor pressure of the solution and e_s is the saturation vapor pressure of water. i is called the van't Hoff factor and is related to the degree of dissociation of the solute in water. For nonelectrolytes $i = 1$, and the above equation is called Raoult's law.

a. Show that this equation can be reduced to Eq. 7.2. State the approximations necessary and express the constant in proper units.

b. Compute the mole fraction of sodium chloride in water when 10^{-15} gram is dissolved in a water drop $\frac{1}{2}$ micron in diameter. The mole fraction is equal to $M/(M + M_0)$. The molecular weights of sodium chloride and water are 58.5 and 18.0, respectively.

c. Find the reduction in the saturation vapor pressure resulting from the presence of the dissolved salt for a drop whose temperature is 0° C. i has the following values as the mole fraction varies.

$\frac{M}{M + M_0}$	0	0.0013	0.0036	0.0108
i	2.00	1.90	1.82	1.85

7.2 Find the equilibrium vapor pressure over a pure water drop of $\frac{1}{2}$-micron diameter when the drop is at a temperature of 0° C. What is the equilibrium vapor pressure and relative humidity of a $\frac{1}{2}$-micron drop at 0° C that contains 10^{-15} gram of NaCl when both the curvature and ionic effects are considered?

At 0° C, the surface tension of water is 75.6 dynes per centimeter and the saturation vapor pressure is 6.11×10^3 dynes/cm^2.

7.3 In thunderstorms, it is assumed that electric fields as high as 4000 volts per centimeter may exist before lightning occurs. The maximum charge Q that a spherical drop of radius r can attain in an electric field E[14] is

$$Q = Er^2\left(1 + 2\frac{\epsilon - 1}{\epsilon + 1}\right).$$

E is measured in statvolts per centimeter (see Table 9.1 for conversion units) and r in centimeters. For a water droplet $\epsilon = 81$. (*a*) From the Thomson equation for the equilibrium vapor pressure over a curved surface, find the importance of the electric charge under extreme thunderstorm electric fields relative to the curvature effect for a 20-micron diameter drop whose temperature is 0° C. (*b*) Is there any condition under which the electric charge on a water droplet in a cloud can affect seriously the equilibrium vapor pressure? (*c*) Calculate how long it would take a 20-micron drop to evaporate in an electric field of 4000 volts per centimeter. The ambient temperature and relative humidity are maintained at exactly 0° C and 100 per cent, respectively.

7.4 For spheres, the drag coefficient C_D varies with Reynolds number *Re* in the following fashion.

Re	134	175	220	269	372	483
C_D	0.926	0.815	0.729	0.671	0.607	0.570

Find the terminal velocity and Reynolds number of a 1-millimeter raindrop falling in air where the pressure and temperature are 900 millibars and 2° C, respectively.

7.5 A 1-millimeter raindrop falls through a cloud having a liquid-water content of 0.6 gram of water per cubic meter of air. How large will the drop have grown in falling 500 meters through this cloud? The water content is distributed so that 0.1 g/m^3 is contained in 8-micron drops, 0.3 g/m^3 in 14-micron drops, and 0.2 g/m^3 in 20-micron drops.

7.6 A water drop and an ice crystal exist together in a cloud where the ambient temperature is −10° C. The mass of each particle is assumed to be 4×10^{-9} gram. The water droplet is spherical while the ice particle can be assumed to be disk 1 micron thick. The density of the ice will be taken to be 0.8 g/cm^3. On the assumption that the cloud is at 99.9 per cent of water saturation, how large will the ice crystal have grown in the time required for the water drop to evaporate?

7.7 A water droplet falls out of the base of a cloud into an isothermal atmosphere of 5° C and relative humidity of 70 per cent. How large will the raindrop be when it hits the ground if it was 2 millimeters in equivalent spherical diameter when it left the cloud base 800 feet off the ground? The fall velocity of raindrops is

Equivalent spherical diameter, mm	2.0	1.8	1.6	1.4	1.2
Fall velocity, m/sec	6.49	6.09	5.65	5.17	4.64

Neglect the effect of fall velocity on evaporation rate and assume the vertical air velocity to be negligibly small.

CHAPTER 8

Natural and Artificially Stimulated Precipitation, Icing of Aircraft, and Radar Meteorology

The results of any theories of precipitation that may be advanced must be consistent with the physical laws of growth of droplets and other cloud particles discussed in Chapter 7. In general, the conclusions of Chapter 7 are summarized in the statement that the formation of precipitation that will reach the ground is favored by thick clouds, low cloud bases, and a drop-size distribution such that there are enough large drops to grow through coalescence with numerous smaller droplets. As a result, the large drops may grow to the size of a precipitation element, usually a raindrop.

The problem of arriving at a mechanism for explaining the formation in a cloud of a drop-size distribution of the proper magnitude is a major impediment to a completely satisfactory theory of precipitation. Let us examine the facts and separate them from hypotheses.

We are able to calculate by the methods outlined in the last chapter the growth of a relatively large drop falling through a cloud of droplets smaller than itself. The calculation is made more difficult if the smaller droplets are not uniform in size, but it is still possible and has been successfully attempted. For reasonable values for the sizes of drops, we must turn to empirical determinations.

All experimental evidence indicates that cloud elements are not uniform in size, but vary from 1 to 100 microns in diameter with a few drops per unit volume that are much larger. The most frequent size ranges from 10 to 20 microns in diameter. The techniques for measuring the size of cloud particles are somewhat less than satisfactory.

The most common measuring technique is to expose a flat plate, covered with a thin film of oil or other nonwetting compound, from an airplane or blimp flying through a cloud. In the short interval of time that the plate is exposed broadside to the air stream, a few cloud particles will adhere to it, essentially keeping their spherical shape. The particles are immediately photomicrographed, to be measured and counted from the photographic film at a more convenient time. After the sizing is done, the results can be displayed as a measured size distribution in clouds. Corrections to the raw data so measured have to be made in order to render the data comparable to the distribution of cloud particles that actually existed in the cloud in the absence of the collector.

A flat plate exposed to the air will divert the air around it in a manner analogous to that described in the discussion of the collection

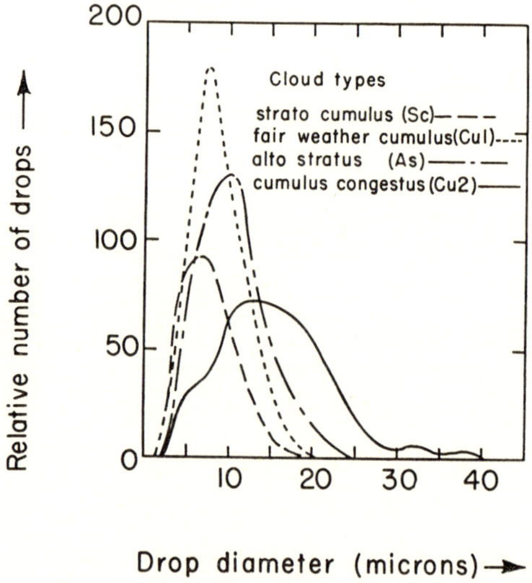

Fig. 8.1 (*After Diem.*) Average distribution of drop sizes in clouds. Present research indicates that there may be a large number of droplets in clouds with diameters < 5 microns, making these curves nonrepresentative below this value. (See Aufm Kampe, Ref. 1, Chapter 3.)

efficiencies of cylinders and spheres in the last chapter. As would be expected, the glass plate has a poor collection efficiency for the small particles, so that the drops collected by the plates are heavily biased toward the larger drops. The raw data from the plates, therefore, must be corrected for the variation in collection efficiencies of the plates for various drop sizes. The unreliability of the correction factor, plus the fact that no very small drops may be collected at all, plus the ever-present danger of the evaporation of the small droplets, makes data on cloud-drop sizes under 5 microns in diameter untrustworthy. Even when checks on the total liquid-water content of clouds by independent techniques are made, the errors involved are at least of the order of the extremely small mass involved in drops of less than 5 microns. For drops greater than 5 microns, the slide techniques are quite satisfactory. Figure 8.1 shows some average size distributions in different generic cloud types measured by M. Diem.(10) Figure 8.2 shows some typical size distributions by the same investigator. The figure indicates the wide spread of the individual data from which averages are drawn. These figures illustrate another case where average data and typical data may be quite different. For correct results, average data, when used, must also be highly typical.

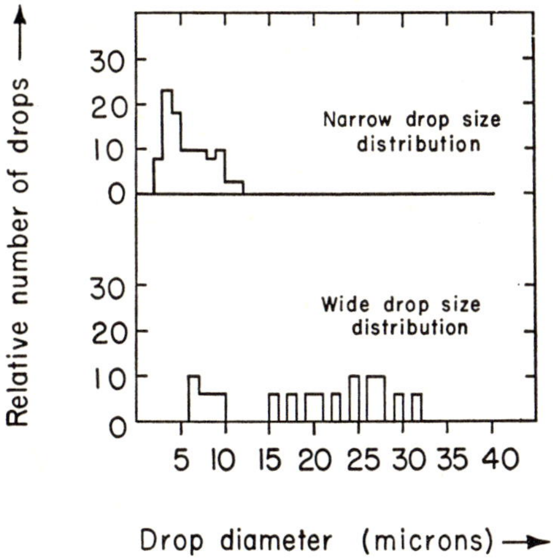

Fig. 8.2 (*After Diem.*) Two cases of drop-size distributions measured by the slide technique from an aircraft. Both clouds were altostratus. The narrow distribution was observed in a cloud whose temperature was +4° C and the wide distribution in an undercooled cloud at about −6° C. A glory was observed with the latter, which may have meant the presence of smaller drops than were observed.

Another popular way of measuring drop-size distributions is through using a series of rotating multicylinders(15,C1) under conditions when undercooled liquid-water droplets exist in clouds. The multicylinder principle has also been extended to measuring drops at temperatures above 0° C. Briefly, the method says that if a number of cylinders of different diameters are simultaneously exposed to the same cloud, the difference in collection efficiency of the various cylinders will cause differences in the mass of liquid which will hit and be collected by each cylinder in a unit time. Under icing conditions, the water will freeze. Later, the ice can be melted off and weighed, and the mass of water collected by each cylinder determined. The cylinders are rotated to preserve the cylindrical shape during the freezing and collection of the water.

In the multicylinder method, the shape of the distribution must be assumed. It was found from data similar to Diem's that a Gaussian type of distribution could be assumed when the volume of a drop was plotted against the frequency of occurrence of the various drop diameters. The mean and the standard deviation of the distribution are the only unknown parameters in the normalized Gaussian curve. If, then, a normal distribution of cloud drops is assumed, the mass of water that will be collected by each cylinder can be computed from a knowledge of the collection efficiencies. Variation of the mean and standard deviation of the distribution varies the relative amounts of water collected by each cylinder. This theoretical information can be tabulated or graphed. Provided that the theoretical assumptions are correct, a measurement of the water collected by the series of cylinders determines a curve that will match closely one of the theoretical curves. This procedure automatically determines the best fit by a Gaussian distribution to the actual distribution of cloud particles measured.

Other methods, such as direct photography, can be used to measure the size of snowflakes and ice particles in the atmosphere. Since the particles are first caught on plates, it is necessary to design apparatus that will decelerate the ice or snow particles to speeds slow enough to prevent shattering on impact.

All the experimental evidence, therefore, points to a nonuniform drop-size distribution from cloud to cloud. The observations do not explain how the distributions arise.

One of the most plausible theories† for a drop-size distribution in clouds states that rising air has a wide range of velocities over a small horizontal extent, implying quite rightly that most instruments measuring vertical velocities in the free atmosphere average over a horizontal area much greater than the individual vertical velocities extend. The wide range of vertical velocities leads to the maintenance of significantly different relative humidities. Under the conditions stated, application of the law of droplet growth by diffusion leads to significantly different sizes. The theory then requires a fine-grained horizontal turbulence to exist, still on a scale not readily measured, that will mix the various sizes of particles.

However, even by this theory it is difficult to explain the relatively few droplets that are ever so much larger than the general run of sizes. It is these few large droplets that eventually grow large enough to fall to the ground as rain.

In an attempt to explain the existence of the large drops, a theory

† Arenberg, D. L., "Turbulence as the Major Factor in the Growth of Cloud Drops," *Bull. Am. Meteorol. Soc.*, *20*, 444–448 (1939).

involving the coexistence of ice and undercooled water in a cloud as a starting point for the big drops was proposed. The theory, called the Bergeron-Findeisen theory† after its originators, presents the following argument.

It is known that at temperatures greater than 0° C, the melting point of water, clouds are predominantly composed of liquid-water drops. If no temperature in the cloud is lower than 0° C, the cloud must be composed entirely of water.

At cloud temperatures lower than 0° C, four conditions may and do exist. For very cold temperatures (and it has been the aim of much current research to say how cold), the cloud will be composed entirely of ice particles. For temperatures intermediate between these very cold temperatures and 0° C, ice (including snow) may exist alone, undercooled water may exist alone, or ice particles and subcooled water particles may exist together. The latter condition occurs in extratropical regions with great frequency and is the important case for the proof of the Bergeron-Findeisen theory.

When ice and water coexist in the atmosphere at temperatures below 0° C, thermodynamics states that the equilibrium vapor pressure is greater over water than it is over ice. The curvature effect is neglected and is justified for spheres greater than 1 micron in diameter. Figure 8.3 graphically illustrates the statement. The figure shows the equilibrium vapor pressure of both ice and water as a function of temperature. It is possible then to choose a vapor pressure (see point A of the figure) where the actual vapor pressure is greater than the equilibrium vapor pressure over ice, yet less than the equilibrium vapor pressure over water. Thus, in a cloud it is possible to have at one and the same time supersaturated vapor over the ice particles and subsaturated vapor over the undercooled water droplets. Application of the diffusion equations, Chapter 7, shows that the ice crystals grow and the water droplets evaporate. Although point A may be at any position between ice and water saturation in a cloud, evidence indicates that most clouds are nearer water than ice saturation and that even clouds that are entirely composed of ice crystals do not form until water saturation is reached.

The Bergeron-Findeisen theory is completed by the statement that the large ice particles eventually fall to a region of the cloud greater than 0° C, and in melting form a water droplet much larger than its neighbors in this portion of the cloud. The process of accretion does the rest in forming a raindrop.

† Bergeron, T., "On the Physics of Cloud and Precipitation," *P. V. Météorol. Un. géod. géophys. int.*, Part 2, 156–178 (1935).

Although the Bergeron-Findeisen theory does not explain the rain from clouds that do not have temperatures below 0° C (a common observation in the tropics), the theory is believed to be valid for most extratropical rain and for thundershower rain in the tropics.

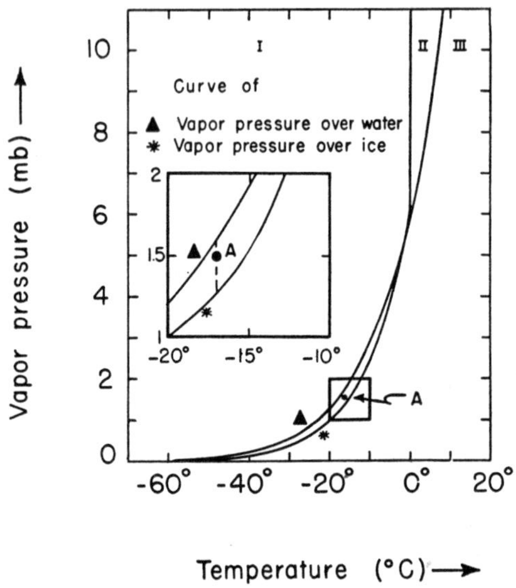

Fig. 8.3 Saturation vapor pressure over a water surface and over an ice surface. In this phase diagram, region I is ice; II is liquid water; and III is water vapor.

This theory, together with the icing problem in aircraft, has stimulated research that aims to explain the relative amounts of ice and water that will exist in a cloud under various conditions of temperature and drop-size distribution. A knowledge of the undercooling properties of water becomes essential when a quantitative answer is required.

Undercooling of water

Observations of undercooling (supercooling is the common term) of a liquid have been confined in the past to large masses of the liquid. Such observations, when made on large masses of water, indicated that water undercooled only 3 to 5° C below its melting point before the onset of freezing. N. E. Dorsey,[12] of the National Bureau of Standards, in the 1930's pioneered the study of the undercooling of liquid water in small samples. The samples of water, which were of the order of 1 or 2 grams mass, were sealed in glass vials so that experiments could be and were repeated on the same samples after the lapse of months and even years. The water in these samples was drawn from such varying sources as laboratory taps and ponds to the other limit of very carefully distilled water. The containers for each sample were scrupulously cleaned before each sample was sealed in.

Dorsey's results showed samples that undercooled to —20° C in many cases, certainly less than —10° to —15° C for a large portion of the samples. The freezing temperatures for a given sample were reproducible within a degree or two even after periods of months. While in the undercooled state, the water could be agitated and poured from one end of the vial to another without causing the sample to freeze. Dorsey found, however, no predictive value of the freezing point based on chemical purity of his samples, because there were cases when pond water froze at temperatures as low as or lower than those for distilled water.

Since the late 1930's a large amount of experimental work has been done by meteorologists on the freezing of water droplets. The droplets are in the cloud- and small-rain-droplet range of diameters. The work of V. J. Schaefer[(C8)] with his cold chamber has been highly popularized and is of great scientific value. Other men, such as Cwilong[(C5)] in England, Weickmann[(C5)] in Germany, and Levine[(19)] in this country have made important contributions to the literature on the experimentally determined freezing temperatures of liquid-water droplets ranging from 5 to 2000 microns in diameter, and masses of 6.5×10^{-11} to 4.2×10^{-3} gram, respectively. The results on chemically pure water show that the smallest water droplets consistently approach a freezing temperature of —40° C. The rise in freezing temperature is a rapidly changing function of size until a diameter of about 500 microns (mass of 6.5×10^{-5} gram) is reached. The freezing temperatures for the larger drops (masses $> 10^{-4}$ gram) then become comparable with Dorsey's results for larger masses. Figure 8.4 shows the variation of the course of the freezing temperature for water with mass, and indicates the source of the observations.

Figure 8.4 indicates that the choice of a freezing temperature for an unknown sample of water can be made only on a statistical basis and is relatively independent of mass above 10^{-4} gram. For smaller masses, the statistical choice has an increasingly smaller range of error and seems to indicate that there is some natural limit beyond which a sample of water cannot be cooled. This evidence supports the theory of heteorogeneous freezing nuclei as being active for all but the freezing of the very smallest drops. The probability that a foreign nucleus of the proper configuration may exist in a sample of water increases as the mass of the sample increases. In support of this, if two samples of water with significantly different freezing temperatures are mixed together, the mixture freezes at the higher temperature.

The chemical and crystallographic composition of the freezing nuclei has been investigated with some success. The assumption guiding the investigations is that droplets first form as undercooled water and then freeze. It does not seem likely that the condensation nuclei and the

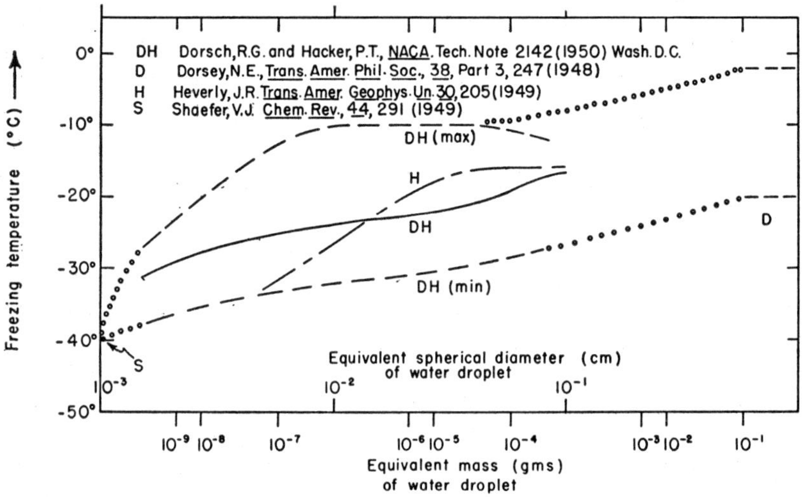

Fig. 8.4 The freezing temperatures of water as observed experimentally. The dashed lines indicate the range of observed freezing temperatures and the solid line is the average value. The dotted lines indicate the trend of the curves outside observed regions.

freezing nuclei are the same. The first provides a surface on which the molecules of water vapor can be adsorbed, holding the molecules until a sufficient number can combine to form a droplet. The activity of the freezing nuclei, on the other hand, stems from the fact that the nucleus has a crystalline structure very similar to that of ice. Water molecules can be found and held to this crystal nucleus with the minimum of energy exchange between liquid and solid as a molecule or group of molecules releases in the form of latent heat the energy that has kept molecules separated from their neighbors. The molecules possess in the liquid state a mobility not found in the crystal state. That such a situation exists is shown by X-ray diffraction patterns of undercooled water.[11] These data indicate that groups of water molecules tend to cluster in icelike arrangement, the extent of the clustering increasing as the temperature of the water is decreased.

The condition in a liquid that groups of molecules move as a unit and preserve within this unit a crystalline structure similar to the crystal pattern of its solid phase is known as *short-range* order.

A study of crystals having a structure similar to that of water indicates no exact match, water being of a unique structure. Silver iodide, cadmium iodide, lead iodide, and quartz have similar dimensions (crystal-lattice constants) and are indicated in Table 8.1. Figure 8.5

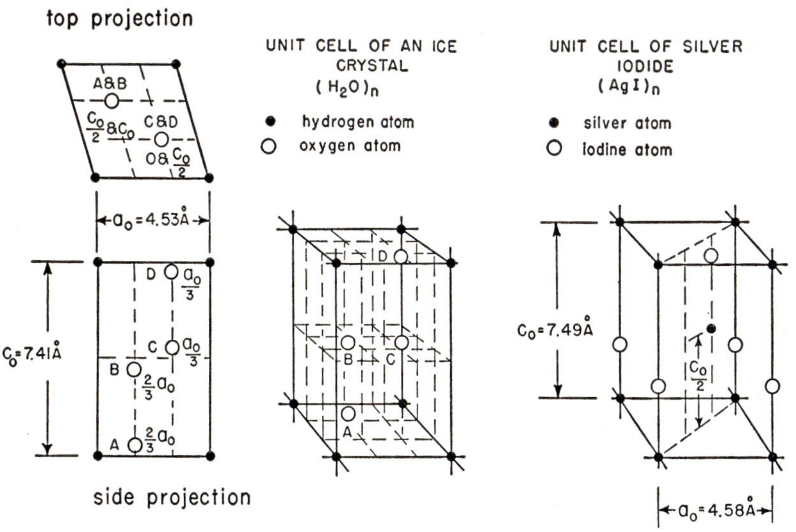

Fig. 8.5 An illustration of the lattice structure of ice and silver iodide. The data were obtained from Wyckoff.(B3)

indicates where these lattice constants are measured. All of the iodides but not quartz are effective in preventing a large magnitude of undercooling. Experiments[13] indicate that silver iodide inhibits undercooling of water below —3° C. The use of these undercooling inhibitors, especially silver iodide, has been useful in changing undercooled clouds to ice, inducing by artificial means the precipitation mechanism called for by the Bergeron-Findeisen theory. It is not known, however, whether the crystals indicated in Table 8.1 or other nuclei as yet undiscovered are in the atmosphere in sufficient quantities to act as natural freezing nuclei.

In the vicinity of —40° C, all water samples in excess of a mass of 10^{-11} gram are frozen, suggesting that at this temperature homogeneous nucleation occurs in the absence of prior heterogeneous nucleation. The temperature at which the clusters of water molecules act as their

Table 8.1

A Table of the Structure of a Unit Crystal and Its Lattice Constants for Selected Crystals

(See Fig. 8.5 for explanation of a_0 and c_0.)

Substance	Chemical Structure	Crystal Symmetry	Lattice Constants in angstrom units a_0	c_0
Water (ice I)	H_2O	Hexagonal	4.53 Å	7.41 Å
Silver iodide	AgI	Hexagonal	4.58	7.49
Cadmium iodide	CdI_2	Hexagonal	4.24	6.84
Lead iodide	PbI_2	Hexagonal	4.54	6.86
Quartz	SiO_2	Hexagonal	4.90	5.39

own nuclei, i.e., a homogeneous nucleation process, is often called the *spontaneous freezing temperature of water*. More experimental and theoretical evidence is necessary before the true spontaneous freezing temperature for water can be established.

In summary, there is abundant evidence that little if any undercooled water exists in clouds at temperatures lower than —40° C. Clouds whose temperatures are above —15° C and less than 0° C are very likely to be predominantly of undercooled water, especially if rain is not reported as falling from them.

The Icing of Aircraft. [15,34,C3,C7] Clouds of undercooled water present a hazard to the operation of aircraft and, to a lesser extent, other forms of transportation. The undercooled water, on striking a surface such as an aircraft wing, freezes to that surface. The continuation of this process for any length of time can cause buildups of ice on the leading edges of wings and control surfaces that so distort the airflow pattern that lift is decreased below that required to sustain the aircraft in flight. Air inlets to carburetors, the first stage of air compressors in jet turbines, and Pitot tubes for instruments can also become so clogged with ice that the airflow is reduced to the extent of causing malfunctioning. Icing can also obscure vision through windshields, freeze landing gear in wheel-wells, load wires to the breaking point, and in general cause dangerous operating conditions. Fortunately, the application of heat to the icing surface at the start of the icing condition, or preferably just before its onset, can keep the ice accretion below dangerous limits. The wings and control surfaces of aircraft present a special problem. In some circumstances, the heat available for supply to the leading edge of the wing in the hot-wing type of aircraft may be insufficient to prevent serious icing. Similarly, the mechanical action of de-icing boots, an alternate method of wing-ice removal, may not be adequate under severe icing conditions. It is then necessary to recognize the factors that lead to dangerous icing and when they are meteorological in nature to learn to avoid or minimize them as much as possible.

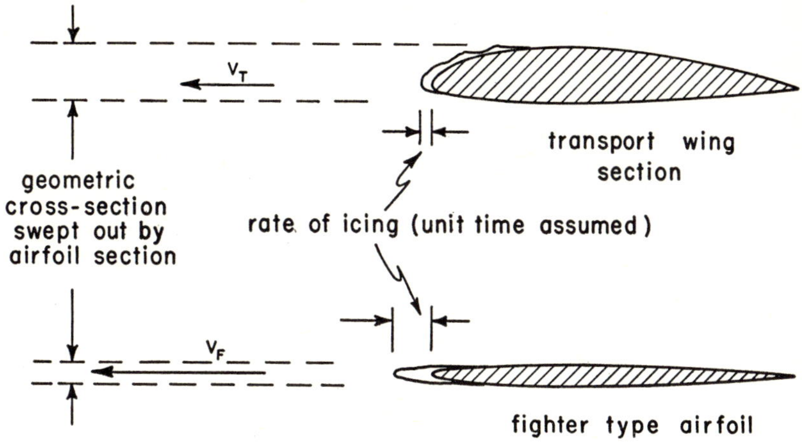

Fig. 8.6 Schematic drawing of the relative icing rates of thick- and thin-wing airplanes whose relative air speed is denoted by the horizontal arrows marked v_T and v_F.

In Chapter 7, under the section discussing collection efficiencies of surfaces, we have developed the concepts necessary for an understanding of wing icing in aircraft. Figure 7.4 may be referred to and used in conjunction with Fig. 8.6. Figure 8.6 shows two airfoils and delineates the icing rate schematically. Borrowing from Eq. 7.23, we see that the mass accumulation of ice on the *leading* edge of a wing is given by

$$dM = \mathrm{E}Awv\,dt, \tag{8.1}$$

where as before w is the liquid-water content of the cloud in units of mass per unit volume, v is now the airspeed† of the aircraft, A is the cross-sectional area measured normal to the air flow at the leading edge of the wing, E is the collection efficiency of the wing, t is the time, and M is the mass of ice collected. The mass of ice collected is the product of the density of ice D', the cross-sectional area A, and the length of buildout of ice into the airstream L. The icing rate dL/dt from Eq. 8.1 becomes

$$\frac{dL}{dt} = \mathrm{E}\frac{w\,v}{D'}, \tag{8.2}$$

† Airspeed is the speed of the aircraft relative to the air, not to the ground.

because

$$dM = D'A\,dL. \tag{8.3}$$

Evaluation of Eq. 8.2 becomes difficult because the quantities entering into this equation may vary over a wide range. The collection efficiencies for wings are approximated by analogous collection efficiencies for cylinders that have the same general flow pattern as around the leading edge of a wing. In lieu of measurements on individual clouds, only average values can be taken. Average values for E, w, and D' are given in Table 8.2, but individual cases may vary widely from these normals.

Table 8.2

Average Values of E for the Wings of Airplanes in the 250-Mph Class

Average values of w for various condensation states and D' for various ice forms are given.

	E	w	D'
	(dimensionless)	g/m^3	g/cm^3
Raindrops	1.0	0.5 to 5.0	Glaze ice 0.9
Drizzle	0.6 to 0.9	0.5 to 2.0	Rime plus glaze 0.6 to 0.9
Cloud drops	0.4 to 0.6	0.2 to 1.0	Rime ice 0.6

A word of explanation as to how the values for D' were arrived at follows. Only when the droplets arrive at the surface in such profusion and at such high temperatures that the surface on which they impinge cannot conduct heat away from them fast enough to freeze them competely before the next drop hits does D' represent the density of ice given in physical tables. The semiliquid state of the upper layer under these conditions causes all the spaces between drops to fill in, giving a solid ice mass of maximum density. Such ice is called *glaze* ice.

If, however, there are relatively few and/or small droplets at cold temperatures impinging on the wing, each drop freezes as soon as it hits, thereby preserving its nearly spherical shape. Air spaces result between the individual ice particles (as with BB shot in a box), resulting in a friable mass of ice of low density. Ice with these characteristics is called *rime* ice.

Under any individual icing condition, both rime and glaze ice may be observed. For example, even if an aircraft entered a pure rime-ice situation and icing began, the rime ice could take on glaze ice characteristics after an interval of time, simply because the drops which initially were impinging on a good conducting surface (the metal wing) after a while impinge on a poor conducting surface (the rime ice), a condition more favorable for the formation of glaze ice.

A study of collection efficiencies† shows that the collection efficiency E decreases with increasing speed. The variation of the operating range of speeds in an aircraft is not great. An increase in speed will increase the icing rate because of the factor v in Eq. 8.2, but will decrease the total ice collected in a given path because of the decrease in E and the constancy of the integral of $v\,dt$. However, the change in E with increase in speed is so small, and the other dangers due to high speed in the meteorological situations accompanying icing conditions may be so large, that it can only be said that a pilot should leave icing conditions as rapidly as is consistent with safe flying. Equation 8.2 does indicate through the relative constancy of E that if an airplane is forced to cruise in an icing zone for a given amount of time, e.g., holding at a range station before letting down for a landing, the icing rate and hence the total accumulation of ice are minimized by slow flying. This problem is very real for jet airplanes, which have thin wings and high cruising speeds. The thin wings have a high collection efficiency compared to the thick wings of slower transport aircraft.

Although most serious icing conditions occur in clouds where liquid water is initially present, piston-type aircraft of low horsepower can experience loss of power and even engine failure in clear air through the formation of carburetor ice. A section of a carburetor is shown in Fig. 8.7, with the icing region indicated in the vicinity of the butterfly valve. This valve, controlled by the throttle, determines the power output of the engine by controlling the volume of fuel and air that ultimately passes into the cylinders to be burned. Ice may reduce the flow in the region near the butterfly valve, causing a consequent loss of power. In extreme cases, the valve can be frozen in one position with disastrous results. Carburetor icing may occur in cloud-free air through the following process.

Moist air at an ambient temperature above freezing and an ambient relative humidity below 100 per cent enters the throat of the carburetor. Gasoline is sprayed from the main discharge nozzle (see Fig. 8.7) in a fine spray. The spray evaporates into the air, the air supplying an amount of heat equivalent to the total latent heat of vaporization of the gasoline spray. The air can supply the required heat only by a reduction in its temperature. Therefore an equilib-

† I. Langmuir (see reference at end of Chapter 7).

rium condition is built up where the spray continues to be evaporated at the expense of a reduction in the temperature of the fuel-dry air-water vapor mixture. At equilibrium, the temperature of the entire mixture plus the walls of the carburetor may be reduced below 0° C. The relative humidity may also reach saturation, with any further increase to supersaturation being precluded by the condensation of

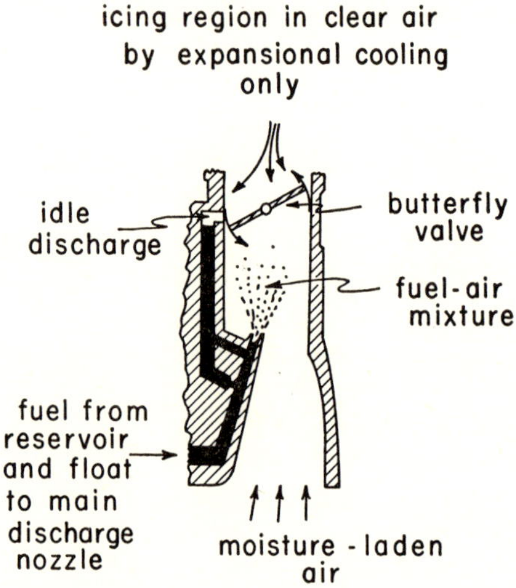

Fig. 8.7 Schematic drawing of a carburetor for a low-powered aircraft engine showing the site of carburetor ice.

the water vapor. The liquid water condensed in turn freezes to ice at the site of the butterfly valve. In order to suck the gasoline into the air stream, the main discharge nozzle is located in a region where the air is expanding, giving a lower exit pressure with an accompanying lower temperature. Studies of the thermal effects have led to conclusions concerning the occurrence of carburetor ice, which are summarized by Fig. 8.8 for an average carburetor in a low-powered engine. As would be expected, low temperature and high relative humidity favor carburetor icing. Fortunately, carburetor ice can be removed by heat, either by external carburetor heat in the simple carburetors, or by careful design such as is characteristic of modern carburetors installed in high-horsepower engines. These high-pressure carburetors keep the vapor in a sufficiently compressed state and hence at high temperature in the critical carburetor icing sites to prevent the formation of ice.

The subject of aircraft icing is but one of the many problems with which cloud physics is concerned. The entire field has been aided by information obtained from the behavior of microwaves in clouds and rain. Radar principles are used. Since we wish to utilize these data as background information for any precipitation mechanism that may be advanced, we shall investigate at this point the principles governing radar meteorology with a view toward ascertaining the type and reliability of the data available.

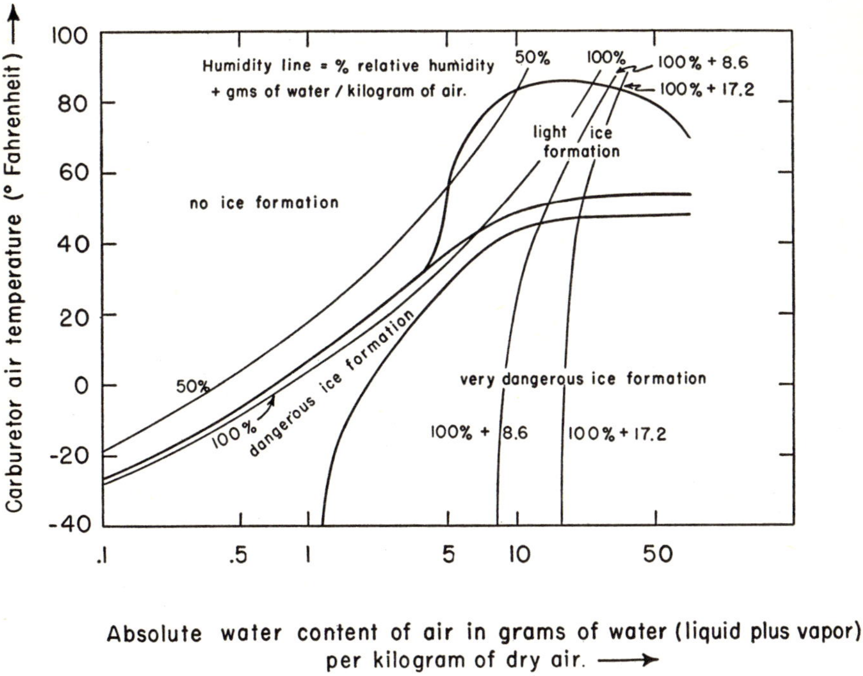

Fig. 8.8 The temperature and moisture conditions favorable for carburetor ice formation in low horsepower aircraft engines. (*After V. J. Skoglund,* "Icing of Carburetor Air Induction Systems of Airplanes and Engines," *J. Aeronaut. Sci., 8,* 437–464 [1941].)

Radar Meteorology. (27,B1,C4,C9) Since the end of World War II, radar has become a useful tool in the detection of areas of rainfall. An analysis of the strength of the signal returned from a storm area can give information on the rainwater content of clouds, the size of drops, the large-scale turbulence in the cloud, and the thermodynamic phase of the cloud elements.† Certain aspects of the problem of micro-

† The information on the topics mentioned is in many cases only semi-quantitative and on a scale that is of limited meteorological value.

wave propagation are discussed in Chapter 1 and especially in Chapter 2. In this section, we shall endeavor to discuss some of the problems of evaluating the data that a radar set is capable of giving. We shall start by discussing the pertinent factors of radar propagation and will neglect, unless specifically mentioned, refraction and absorption effects of the atmosphere.

A radar emits a pulsed, highly directional signal into the atmosphere. A cross section of the beam may show an elliptical shape, through whose surface flows one-half the power radiated by the antenna. The periphery of the ellipse is the locus of the half-power points and this locus defines the beam. The radar radiates energy in short, evenly spaced pulses, and acts as a receiver of the pulses after they are reflected from a distant target. In a typical radar,† the pulse duration may be 1.5 microseconds, and the pulses repeated at the rate of 400 per second. The figures indicate that the radar is emitting energy only 1/1667 of the time and acts as a receiver the other 1666/1667 of the time. Since microwaves travel with the speed of light, the same example indicates that each pulse is 0.45 kilometer long. For the 10.7-centimeter radar described, each pulse contains 4200 complete periods of vibration. Furthermore, a single pulse can travel a maximum distance of 375 kilometers (235 miles) to a target and then return before the next pulse is emitted by the radar. Thus, as the radar measures the range of a target by measuring the time that it takes a pulse to reach a target and return before the next pulse is emitted, 235 miles represents the maximum range of the radar at a pulse repetition rate of 400 seconds^{-1}. Allowing a finite time for the radar to change from source to detector reduces the effective range below the theoretical to a maximum of 120 miles.

It is also instructive to compute the volume of space occupied by a single radar pulse as it travels through space. A target cannot be located with any more precision than to say it lies within a volume given by one-half the volume of the radar pulse at the range of the target. The beam width of this radar is 4 degrees both horizontally and vertically, making the cross-sectional area of the beam equal to 12.5 square miles at a range of 57 miles. One-half the volume of space occupied by the radar beam becomes 1.75 cubic miles. Since a showering cumulus cloud may occupy an effective radar volume of 100 cubic miles, the radar even at this distance is sampling relatively small sections of the cloud. The geometry of the radar beam and a

† SCR 615-B. S-band radar putting out a peak radio-frequency power of 750 kilowatts.

drawing of the oscilloscope presentation of the reflected microwaves are shown by Fig. 8.9.

Let us consider a radar that emits E_T units of radiant microwave energy into the atmosphere. E_T is so defined that E_T multiplied by the effective area of the antenna A_T is equal to one-half the radiant

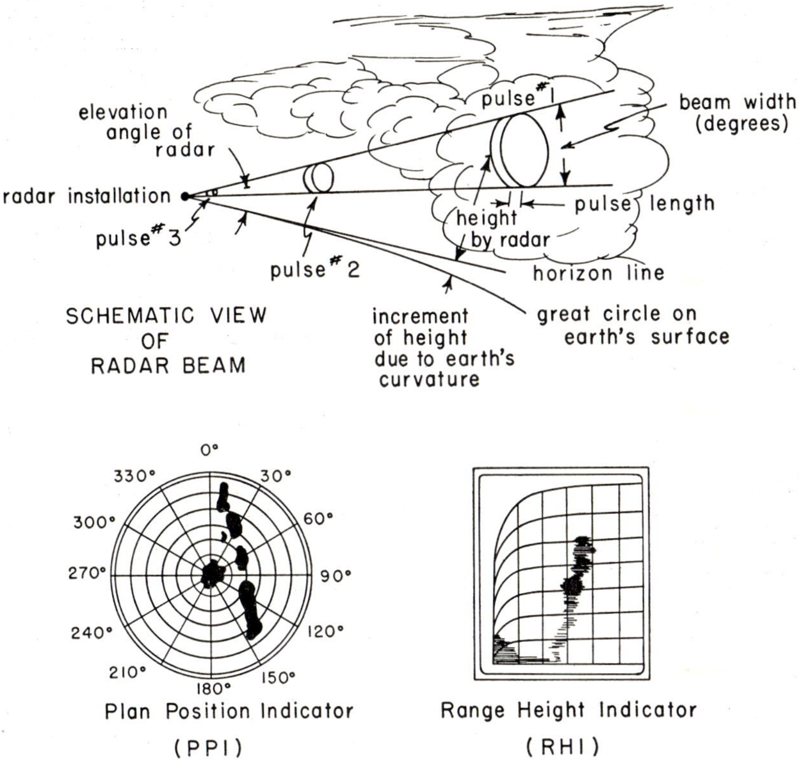

Fig. 8.9 Radar-scope presentation with a schematic of the radar beam illuminating a cloud. For practical measurements, only a single pulse would be propagated in the atmosphere at any one time, i.e., if pulse No. 1 were present, pulses No. 2 and No. 3 would be absent. The detector is unable to distinguish one pulse from another. The dark areas on the PPI and RHI are precipitation areas.

flux emitted by the antenna. This flux is contained in a beam whose solid angle is ω. We shall assume that there is a constant efficiency with which radiant energy is converted to power generated by or detected in the radar set proper.

Because of the constant beam width ω, the energy density decreases with increasing distance from the radar. If a target r units of distance from the radar is illuminated by the beam, the energy density incident on the target has decreased to

$$E_0 = \frac{A_T}{\omega}\frac{E_T}{r^2}. \tag{8.4}$$

The cross-sectional area of the beam at distance r is just $r^2\omega$. E_0, the flux density incident on the target, can be considered as parallel radiation if r is large. If scattering particles, such as rain or cloud droplets, are present, a certain part of the incident energy will be scattered back toward and detected at the radar as a flux of density E_s. Of course, the flux scattered by a single particle will not be detected, but if a large number of particles exist in the region illuminated by the radar the total flux will be large enough to be detected.

From scattering theory one can compute the flux scattered by a single particle illuminated by a plane wave whose energy density measured at the particle is E_0. Calling the flux scattered from a single particle F_s, we define and can compute from theory a quantity having the dimensions of area,

$$K_s' = \frac{F_s}{E_0} = 4\pi\frac{I_s}{E_0} = 4\pi r^2\frac{E_s}{E_0}. \tag{8.5}$$

E_s is measured r units from the scattering particle and is a flux density. When E_s is measured at the site of the radar antenna after having been "reflected" from the target, E_s is the back-scattered flux density and K_s' is called the *back-scattering cross section.*

In general, when a radar illuminates a cloud the beam cross-sectional area is entirely filled with cloud particles. As a radar emits radiation in pulses of length l, a volume of space equal to $l\omega r^2$ is illuminated by a radar beam at any distance r from the radar. If n scatterers per unit volume exist in this section of the cloud, the number of particles that scatter the incident radiation is $n(lw/2)r^2$. The $\frac{1}{2}$ factor enters because the pulse may be scattered from any point lying between $r - (l/2)$ and $r + (l/2)$ units distant from the radar. On the average, the effective length of the pulse in determining the scattering volume is $l/2$ instead of l. The total flux detected at the radar is just $(l\omega/2)nr^2E_s$. Denoting this quantity by E, we find with the help of Eq. 8.5 that

$$E = \frac{l\omega}{2}nr^2E_s = \frac{l\omega}{8\pi}nK_s'E_0, \tag{8.6a}$$

and with the help of Eq. 8.4 that

$$\frac{E}{E_T} = \left(\frac{lA_T}{8\pi}\right)\frac{nK_s'}{r^2} = \left(\frac{lA_T}{8\pi}\right)\frac{1}{r^2}\sum_i n_iK_s{}'_i. \tag{8.6b}$$

The ratio E/E_T is measured at the site of the radar itself. It is the ratio of the total back-scattered flux detected to the flux that was emitted initially. r is the range of a target which consists of a concentration of n particles per unit volume whose individual back-scattering cross section is K_s'. $(lA_T/8\pi)$ is an instrumental constant. If the scattering volume is not completely filled with scatterers, the $1/r^2$ relationship will break down. The ratio E/E_T will then become proportional to a variation in range lying between $1/r^2$ and $1/r^4$. The summation sign indicates that more than one size of scatterer may be present in the scattering volume.

For a cloud much greater in extent than the radar pulse illuminating it, we may write

$$\frac{E}{E_T} = \frac{C}{r^2} e^{-\sigma x} \sum_i n_i K_s{}'_i \tag{8.7}$$

as the governing equation relating the ratio of the detected to the emitted power or flux density. C is an instrumental constant equal to $(lA_T/8\pi)$.

We have taken attenuation of the beam into account by including the $e^{-\sigma x}$ factor. The back-scattering coefficient may be expressed for the Rayleigh region $(a/\lambda) < 0.03$ for water spheres, <0.015 for ice spheres), as

$$\frac{K_s'}{4\pi} = \frac{\pi^4}{4\lambda^4} \left\{ \frac{(m^2 - 1)^2 + m^2 k^2[2(m^2 + 1) + m^2 k^2]}{(m^2 + 2)^2 + m^2 k^2[2(m^2 - 2) + m^2 k^2]} \right\} a^6. \tag{8.8a}$$

The absorption coefficient per wavelength is k. For a nonabsorbing sphere, Eq. 8.8a reduces to

$$K_s' = \frac{\pi^5}{\lambda^4} \left(\frac{m^2 - 1}{m^2 + 2} \right)^2 a^6. \tag{8.8b}$$

The consequences of Eq. 8.8 have been explained in Chapter 2. Some raindrops are so large that the more general Mie theory of scattering should be used. Similarly, ice crystals and snowflakes are hardly spheres or small compared to the wavelength of the incident radiation. For these cases, the Mie theory for other geometries than that of the sphere should be used. Some progress is being made toward a more exact form of the scattering coefficient from the Mie theory to supplant pure Rayleigh scattering.

The absorption factor $e^{-\sigma x}$ is expressed as the product of a distance x and extinction coefficient σ, the latter being dependent on the wavelength of the radar set. x is not in general comparable with r, the range of the target from the radar, but is much less. The attenuation of the radar energy in the atmosphere is of significance primarily in the saturated cloud air, increased by the absorption and scattering of the energy by the water droplets in the clouds. Thus, x is of the order of twice the penetration distance of the radar beam into the cloud and σ is the sum of the scattering and absorption coefficients, the latter coefficient being due mainly to the water vapor rather than to the liquid water droplets. We shall neglect the absorption in cloud-free air.† Thus, if a radar beam passes through a considerable length of cloud, no return may be detected from the distant portion of the cloud because the power return falls below the sensitivity threshold of the radar.

We are now in a position to examine the methods used and assumptions made in measuring meteorological parameters by radar. A most important parameter of interest to meteorologists and capable of measurement by radar is the rainwater content of clouds. Measurements have been made both by ground and airborne radar.

An inspection of Eqs. 8.6 and 8.7 shows that the radar signal

$$\frac{E}{E_T} \sim \sum_{i=1}^{i=\infty} n_i a_i^6$$

when all other variables are held constant. Primarily, the sixth-power dependence on diameter means that the power return from cloud drops is significantly less than the power return from raindrops. For an approximate computation we may consider that a cloud contains 500 drops per cubic centimeter, 10 microns in diameter. In the same cloud, the rainwater content may be of the order of 2 drops per liter, 1 millimeter in diameter. The ratio of the effect due to the rainwater in respect to the cloud-drop effect is, for this typical example, $\dfrac{2 \times 10^{-3}(10^{-1})^6}{5 \times 10^2(10^{-3})^6} = 4 \times 10^6$. Cgs units are used in the computation. It is clear from this example that the principal return is from the rainwater in clouds. In fact, clouds without rain in them are usually not detected on a radar set. The mass varies as $\sum n_i a_i^3$ so the ratio of mass of rainwater to cloud-water content for the same conditions is $\dfrac{2 \times 10^{-3}(10^{-1})^3}{5 \times 10^2(10^{-3})^3} = \dfrac{4}{1}$. Thus, the liquid-water content computed from a radar signal is actually the rainwater content, which in turn is of the order of four-fifths of the total liquid-water content of the cloud.

Since the mass of a single drop of water is $(\pi/6)Da^3$, we find that the radar signal observed should be

$$\frac{E}{E_T} \sim \sum n_i M_i^2 = \sum \frac{(m_i)^2}{n_i}. \tag{8.9}$$

† Oxygen gas is a significant absorber of microwaves.

M_i is the mass of the individual droplet and n_i is the number of drops per unit volume in a class Δa_i units wide. m_i is the mass of liquid water in a unit volume of air and is equal to $n_i M_i$. Under the assumption that the raindrops are nearly uniform in size and concentration,† Eq. 8.9 may be written as

$$\frac{E}{E_T} = Am^b \tag{8.10}$$

where A and b are considered as constants to be determined experimentally. On taking logarithms of both sides,

$$\log \frac{E}{E_T} = b \log m + \log A. \tag{8.11}$$

Thus simultaneous measurements of E/E_T and m plotted as $\log E/E_T$ versus $\log m$ will yield the value of b from the slope of the best straight-line fit to the data. A is found from the intercept of the straight line with the axis of $\log E/E_T$. Values of b found from experiment lie between 1.44 and 1.72 and are independent of the units in which m is measured. One lack of exact agreement arises from the difficulty in measuring directly

$$m = \frac{\pi n}{6} D \sum_i \frac{n_i}{n} a_i^3, \tag{8.12}$$

where a_i is the mean diameter and n_i the number of drops per unit volume in a class interval Δa_i wide. n is the total population of drops in a unit volume. D as usual is the density of water. Instead of m the rate of rainfall is measured, a quantity directly proportional to m.

Suppose that a flat collecting surface (such as blotter paper with a dye sensitive to water) is exposed approximately normal to the rain for a short interval of time t. Here again is the same problem of collection of drops discussed in the last chapter. If a collection efficiency of unity for the flat plate is assumed, it collects a mass of water in the volume Avt of

$$M = nAt \frac{\pi}{6} D \sum_i \frac{n_i}{n} v_i a_i^3. \tag{8.13}$$

A is the area of the plate, n is the total number of drops per unit volume measured by the plate, n_i/n is the fraction of drops in the ith class, and v_i is the relative fall velocity of the drops. For raindrops v_i is dependent on size and therefore needs a class subscript. The rainfall

† This statement imposes a stringent restriction on Eq. 8.10, and only experiment will determine the validity of the assumption.

rate is the depth of rain accumulation per unit of time. In terms of precipitable water w, the rainfall rate is

$$\frac{dw}{dt} = \frac{M}{DAt} = \frac{\pi}{6} n \sum_i \frac{n_i}{n} v_i a_i^3. \tag{8.14}$$

From Eq. 8.14, we are able to deduce m as a function of the rainfall rate, dw/dt. Assuming that an average velocity $\bar{v}$ can be assigned to the falling drops, we find that m should be equal to $M/A\bar{v}t$ and that as a result of the definitions given by Eqs. 8.12 and 8.14

$$m = \left(\frac{D}{\bar{v}}\right) \frac{dw}{dt}. \tag{8.15}$$

Substitution of Eq. 8.15 into either Eq. 8.10 or Eq. 8.11 shows that the power returned to the radar should be approximately equal to the square of the rainfall rate. These relationships can be checked by simultaneous measurements of rainfall rate, radar signal, and drop-size distribution. dw/dt can be measured by an extremely sensitive recording rain gauge. Both $\sum_i \frac{n_i}{n} a_i^3$ and $\sum_i \frac{n_i}{n} v_i a_i^3$ can be measured by exposing filter paper to the rain and the size of the spots can be converted to the original diameter of the raindrops. The terminal velocity of raindrops has been both theoretically determined and empirically observed and is available in meteorological tables. A measurement of the distribution of raindrops determines both summations.

Thus the square of the rainfall rate appears to be nearly proportional to the square of the mass concentration which in turn by Eq. 8.12 is proportional to $\left[n \sum_i \frac{n_i}{n} a_i^3\right]^2$. However, in general

$$\frac{1}{n}\left[n \sum_i \frac{n_i}{n} a_i^3\right]^2 \neq n \sum_i \frac{n_i}{n} a_i^6 = n \sum_i \left(\frac{n}{n_i}\right)\left(\frac{n_i}{n} a_i^3\right)^2 \sim \frac{E}{E_T}$$

even if some average value of n/n_i were included on the right-hand side of the last equation to allow this factor to be taken outside the summation sign. There are a number of cross-product terms between the various classes arising from the notation of the left-hand side of the equation that are absent from the right-hand side. As the radar return is proportional to $n \sum_i \frac{n_i}{n} a_i^6$, plus the addition errors involved in making an exact measure of the rainfall rate, it is not surprising that the exponent b of Eq. 8.10 is not 2. However, and this is most important, there still remains an approximate relationship between the strength of the radar signal and the rainwater content of a cloud. Radar thus becomes an important tool in determining the spatial distribution of the rainwater content of a cloud and by implication the total liquid-water content of clouds. Figure 8.10 shows the intensity of an airborne radar signal as a function of height in a precipitating cumulus cloud

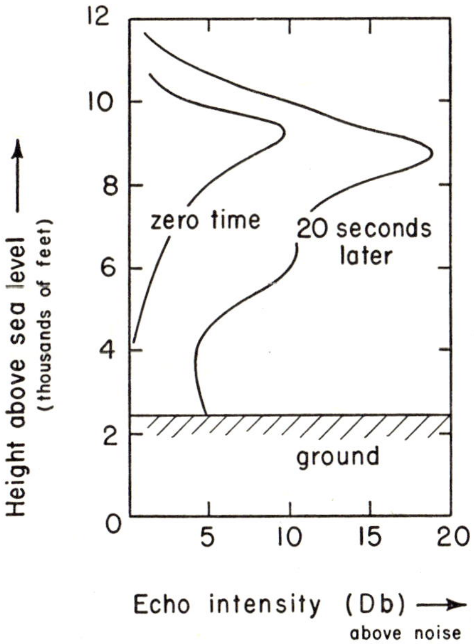

Fig. 8.10 (*After Bowen.*) The variation of a typical radar signal returned from a cloud as a function of height and of time. Since radar return is roughly proportional to the square of the liquid-water content, this curve is also typical of the variation of liquid-water content in a cloud.

which had not reached the freezing level. The shape of the curves indicates that the mass of rainwater must increase with height. This result corroborates other direct measurements of liquid-water content, mainly from nonprecipitating clouds. Typical liquid-water contents of nonprecipitating clouds range from 0.2 to 0.3 gram per meter^3 at the bottom to 0.7 to 1 gram per meter^3 toward the top of the cloud. If precipitation is falling from the cloud, this amount may be increased by a factor of 2 or 3.

Mapping the rainwater content of clouds represents a second use of radar in cloud-physics investigations. The theory of mapping by radar arises from the fact that, like all instruments, a radar is not an infinitely sensitive instrument and will not detect and indicate a signal below a threshold value E_{min}. The minimum detectable energy may be thought of as the product of three factors, all dependent on the operating characteristics of the radar. The three factors are E_T, the power radiated by the radar; G, the receiver gain or amplification of the returned signal to a point where it can be displayed on a scope and measured; and ϵ, a constant equal to the ratio of the minimum detectable power to the radiated power at the highest receiver gain setting. Thus, we see from Eqs. 8.7, 8.8*b*, and 8.10 that

$$\epsilon = \text{const.}\ \frac{Gm^b}{r^2}\, e^{-\sigma x}. \tag{8.16}$$

In those cases where the variation of the attenuation with range can be neglected

$$m = \text{const.}\ \left(\frac{r^2}{G}\right)^{1/b}. \tag{8.17}$$

The constants in Eqs. 8.16 and 8.17 are not the same.

We interpret Eq. 8.17 to mean that at a gain setting G, the mass of liquid water that will just give a detectable signal varies as $r^{2/b}$, which is approximately directly as the range. This information is used as follows.

At a given gain setting a plan position indicator (PPI) indicates the rain areas to be concentrated in easily recognized and well delineated cells as shown in Fig. 8.9. The outline of these cells represents the line of minimum detectable power. The variation of points on this line from the center of the PPI is the distance r of Eq. 8.17 and can be measured. From these values of r, corresponding values of m can be ascertained.

If the gain be turned down to a lower value, the areas of radar return on the PPI are diminished until it is possible to turn the gain down to such a low value that the radar echoes finally disappear. From each gain setting, values of rainwater content are evaluated as above. The values of m can be plotted on a chart of the area and contours of rainwater content of the various cells drawn for a given height. Results are shown in Fig. 8.11.

Radar indicates that clouds are cellular in nature, and that even regions of synoptically steady rain show a well-defined variation in intensity over the area scanned by a radar set. It is possible to trace the formation and dissipation of individual cells and it is found that the life of a rain cell may vary from 20 minutes to 3 hours, with an average life of about 1 hour. A convective shower cloud, which may be visually a thunderstorm extending over 100 square miles, appears to a radar as a group of cells undergoing a constant change in intensity.

The foregoing is a brief discussion of the theory of measurement of two important uses of radar in cloud-physics application rather than in routine storm-detection work. A little thought will convince one that radar measurements are subject to corrections reminiscent of those in optics. Microwaves have line-of-sight transmission, but refraction and reflection errors can give distance errors on the radar. The radar measures distance by measuring the length of time required for a pulse to be emitted and returned to the radar. Errors in scope interpretation may arise from this system. In particular, refraction may channel the microwave to a target further away from the radar

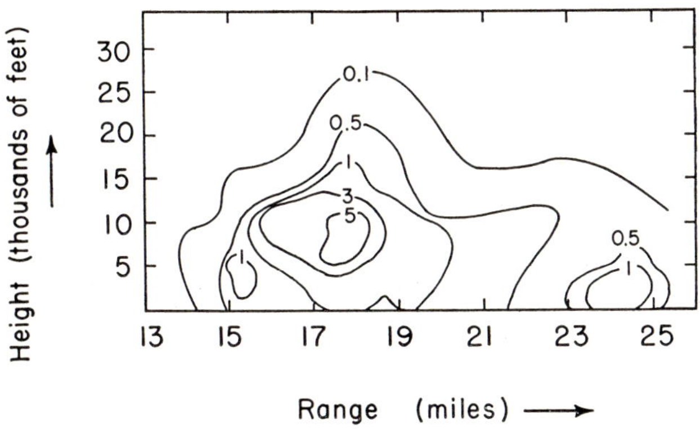

Fig. 8.11 (*After R. C. Langille.*) Illustration of the contours of rainwater content of clouds as measured by a 10-centimeter radar. The numbers on the lines of constant rainwater content are the number of grams of rainwater per cubic meter of air.

than one-half the pulse repetition rate, making the total time between emission and detection greater than the period between pulses. The radar, being unable to discriminate between the sequence of individual pulses, indicates the target at a position consistent with the last pulse emitted rather than at a position consistent with the pulse that was reflected from the target.

The line-of-sight transmission feature also means that a radar scanning horizontally will show on the PPI echoes from a level in the cloud that increases in the same sense as the distance of the cloud from the radar. The curvature of the earth accounts for this effect, because the PPI is a display of the signal from targets in a flat disk or cone opening upward, with the center of disk or apex of cone tangent to the earth's surface at the ground radar installation. At a distance of 100 miles, the minimum height of a detected cloud segment in the absence of refraction is 6700 feet. If the beam is pointed 2 degrees above the horizontal, the portion of a cloud seen at 100 miles is at least 8500 feet from the ground.

The range height indicator (RHI) shows a vertical section of a precipitating cloud along a given azimuth. The vertical distribution of the rainwater content can be inferred from the intensity of the radar echo. In the vertical, however, the temperature in the cloud is decreasing with height so that at temperatures considerably below 0° C, the reflecting particles are large ice crystals or snowflakes. The radar echo from these solid hydrometeors is in part due to their lower index of refraction compared to liquid water (discussed in Chapter 2) and in part due to their geometry, orientation, and of course numbers. The shape factor may increase the reflecting powers by an appreciable amount, so that it is often difficult to distinguish snow from rain purely on the intensity of the signal detected.

At the elevation of the melting isotherm (0° C) and for a kilometer or two below it, snow melts to rain. When this is the case, the radar signal from this region shows a manifold increase over the return from either the snow or water region of the cloud. Observed on the fluorescent RHI oscilloscope, the signal appears as a bright band,[4] giving this name to the region. The bright band occurs because this section of the cloud is the optimum region for a high concentration of particles of high index of refraction arising from the melting snow, with the additional favorable factor of nonspherical particles. The bright band is therefore a convenient and accurate (to within a few degrees) means of determining the 0° C isotherm in a cloud under study.

An examination of the information derivable from microwave propagation in the atmosphere, especially in clouds and precipitation areas, concludes the experimental and observational data on the structure and composition of clouds that we shall discuss. We shall now be concerned with how these fragments of information can be fitted together into a logical, comprehensive, and useful theory of the precipitation process.

General Theory of Precipitation. The mechanics of the growth of a cloud from tiny nuclei of condensation to the raindrops, snowflakes, or hailstones that ultimately fall to earth is fairly well understood by most meteorologists. It is accepted that clouds are formed by air maintained at a slight supersaturation condensing on nuclei and subsequently growing to cloud-droplet size in accordance with known laws. There is further agreement that cloud particles are not all of the same size. The process by which some cloud particles become significantly larger than others is not understood. It is at this point that a theory of precipitation satisfying in all details is lacking. Various theories have been advanced to describe the precipitation process once an initial distribution of cloud particle sizes has been postulated. Two such theories are discussed later, the details of the first having been worked out in a quantitative fashion by H. G. Houghton and the second by E. G. Bowen. Both processes become identical once the precipitation particle becomes a water droplet enough larger than its neighbors to have a fall velocity relative to other cloud particles and also to the ground. Further growth of the droplets then occurs by collision and coalescence with other cloud droplets until the droplet, now of raindrop size, falls out of the base or side of the cloud. Such a collection mechanism, a function of the drop's collection efficiency, will always occur when a droplet falls through other smaller droplets. The collection of many drops by one is known as the accretion process. Finally, there will be evaporation of the raindrop after falling from the cloud into unsaturated air, with the evaporation continuing until the drop hits the ground as a raindrop, or until the drop completely evaporates in the air.

Let us examine in more detail the mechanism of growth of a cloud particle, somehow favored over its neighbors to grow to raindrop size. In the first discussion we shall be guided by the results of Houghton's calculations, based substantially on but a significant extension of the Bergeron-Findeisen theory of colloidal instability in ice or ice-water clouds. The precipitation mechanism postulated follows as a logical consequence of the observation that most clouds, even ice clouds, form at water saturation. A corollary of this observation is that the air in young, growing clouds of ice crystals is highly supersaturated in respect to the ice crystals themselves.

Clouds containing a mixture of undercooled water droplets and ice crystals form an unstable colloidal suspension of water in two phases dispersed throughout a continuum of air. Because of the difference in saturation vapor pressure between ice and water, cloud-water drops evaporate into and water vapor condenses from the air onto ice crystals. The tendency is to create a single ice phase of large particles. The latter, which are rapidly formed, have a fall velocity relative both to their neighboring ice particles and undercooled droplets and to the ground. As the ice particles become larger, the diffusion process becomes less important in the growth process than is the capture of undercooled water particles by the falling ice particles. By the time the accretion process is well established, the melting level in the atmosphere is reached. The ice crystals melt, forming large water droplets of the same mass as the corresponding ice crystals. These water droplets continue to grow by collision and coalescence as long as they remain in the cloud. When the drops leave the cloud base, evaporation begins, continuing until they reach the ground as rain. The evaporational step usually does not decrease the size of the drops much in heavy rain falling from clouds with low bases. This precipitation mechanism is very important in all regions of the earth.

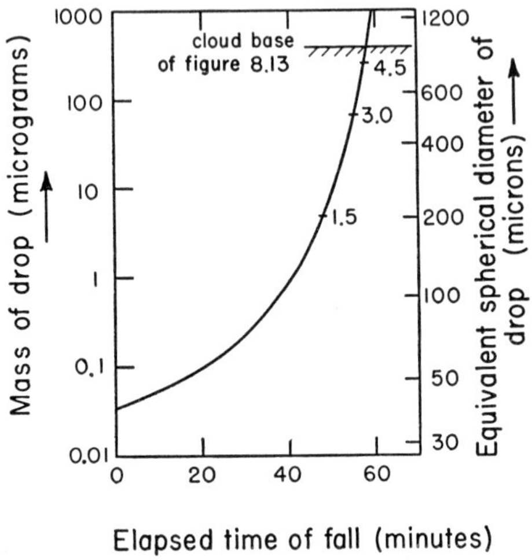

Fig. 8.12 (*After H. G. Houghton.*) Growth of an ice crystal by sublimation and of the subsequent water drop (formed after melting) by the accretion process as a function of time. The crystal initially has a mass of 0.05 microgram. It is assumed to fall through a cloud of droplets whose mean diameter is 24 microns, with a drop spectrum of sizes ranging from 4 to 42 microns in diameter. The liquid-water content of the cloud is 1 gram per meter^3. The numbers entered near the curve at intervals are the distance fallen through the cloud in thousands of feet under conditions of zero vertical velocity.

Figure 8.12, illustrating the theory just described, must be translated into an actual distribution of rainwater content of a volume of air with height. Upon the assumption that the cloud replenishes by further condensation of water vapor the moisture leaving the cloud as rain, so that the growth of one drop is representative of the average process of a large number of drops, Fig. 8.13 has been drawn. The information obtained from Fig. 8.12 has been utilized to derive two different curves of height versus time. The difference in the vertical air velocities assumed in the cloud makes the difference in the initial height of the drop.

The conclusions of H. G. Houghton† on the validity of the ice crystal theory are quoted.

† H. G. Houghton, *J. Meteorol.*, *7*, 363 (1950).

"It is clear that the ice-crystal mechanism is much more rapid than the accretion process in the early stages of the growth of precipitation elements. Thus, if ice crystals are present, it seems certain that the precipitation will be initiated by the ice-crystal process rather than by accretion. When the larger cloud elements have a mass of the order of ten micrograms or more, it appears that the accretion process is more rapid than the ice-crystal mechanism.

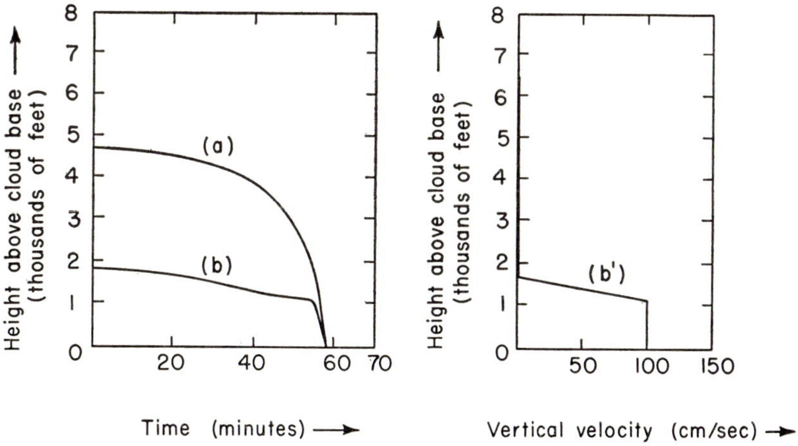

Fig. 8.13 Curves of particle height above cloud base as a function of time for two different vertical velocity distributions. Curve *a* is the height-time profile for zero vertical velocity of the air. Curve *b* is the height-time profile for the vertical distribution of air velocity, curve *b′* of the right-hand figure. Both curves *a* and *b* are consistent with mass-time curve of Fig. 8.12. Note how an *upward* air velocity favors the growth of drops in a relatively shallow layer of cloud.

"The ice-crystal process is most rapid when the atmosphere is saturated with respect to water at −15° C and the crystals are of dendritic form. Under these optimum conditions, single crystals of about 1000 micrograms, equal in mass to 1.2-mm drops, might be formed. Under more typical conditions, it appears unlikely that ice crystals of mass greater than about 500 micrograms will form. This is in general agreement with such observations as are available.

"The rate of growth of a particle by the accretion process is dependent on the mass of the particle, the liquid-water content of the cloud, the median cloud drop diameter and the breadth of the cloud drop-size distribution curve. A true comparison with the optimum rate of growth of ice crystals cannot be made because it is not possible to set definite upper limits on these parameters. It is believed that the Diem nimbostratus distribution, with a liquid-water content of 2 g m^{-3}, is a reasonable approximation to an optimum. On this basis, it seems reasonable to expect a growth of rain drops of, say, 2-mm diameter in about one hour while falling about 2.5 km. Judging from the available data on drop size and liquid-water content, most of the clouds of middle latitudes could not be expected to release drops larger than drizzle by the accretion process.

"It may be concluded that, in typical mid-latitude conditions, neither the ice-crystal process nor the accretion process, acting alone, is likely to produce rain drops of 1 to 5 mm diameter. In deep clouds of high liquid-water content, the accretion process may lead to rain drops of the above size. If it turns out that tropical clouds are composed of much larger drops than those of mid-latitudes, the accretion theory is capable of explaining the release of rather large rain-drops from clouds 2 to 3 km deep.

"Much, if not most, of the middle-latitude precipitation falls from clouds which extend above the freezing level. The writer believes that most of the precipitation from such clouds is initiated by the ice-crystal effect, but that the growth beyond a mass of about 100–500 micrograms results from clumping of the crystals to form snow flakes. Alternatively, the precipitation may form as graupel, which is presumed to grow both by sublimation and accretion. Only in those ways can the observed sizes of precipitation elements be brought into accordance with the theory.

"It is not believed that the uncertainties of the computations are sufficient to invalidate the general conclusions presented above, but caution should be exercised in using the exact numerical values given in this paper. Additional information is needed on collisions between small drops, collisions between ice crystals, the velocity coefficients of snow flakes, the dimensions and masses of ice crystals, and the drop-size and liquid-water content of tropical clouds."

A second precipitation mechanism is required to explain the factors entering into the production of rain from clouds that do not extend to the freezing level and hence must be composed entirely of water droplets. The restriction to any theory that the cloud temperature be above 0° C is severe, because it limits the depth of cloud to about 3 kilometers in summer in extratropical regions, higher of course in the tropics. The following theory for rain from nonfreezing clouds is attributable to E. G. Bowen† and is the result of intensive work on cloud physics in Australia.

The theory requires a few drops in a cloud that through an unspeci-

† E. G. Bowen, *Australian J. Sci. Res., Ser. A, Phys. Sci.,* *3,* 193 (1950).

fied process are sufficiently larger than their neighbors to have a fall velocity relative to the surrounding smaller droplets. For computational purposes the cloud was considered to consist of drops 20 microns in diameter. By Stokes' law the fall velocity of these drops is so small relative to the upward air velocity U_z in the cloud that no relative velocity between the small droplets and the air is assumed. A few large drops which will eventually grow to raindrop size will not be carried up with the air current as rapidly as the small drops because of a velocity v_T relative to the air stream. Their rate of fall in respect to the ground is $v_T - U_z$, where v_T can be calculated from Eq. 7.32. v_T is a variable quantity, increasing with an increase in the size of the drop.

The theory states that initially some large drops exist along with many smaller ones at the base of a cloud. Through the base of the cloud a uniform upward velocity exists U_z units large, e.g., 1 meter per second. The drops at the base of the cloud begin to move upward, the small ones at the uniform velocity U_z and the large ones at the velocity $v_T - U_z$. Because of the relative velocity v_T between the larger and the smaller drops, collisions occur and by the accretion process, the large drops grow larger. This process as it continues means that as the large drops grow v_T increases. At some point in the cloud

$$v_T - U_z = 0 \tag{8.18}$$

and the large drops can rise no farther. In fact, as the relative velocity v_T continues to exist even at the point of maximum height, the drop continues to grow by accretion, still further increasing v_T. From this point on $v_T - U_z > 0$ meaning that a drop begins to fall relative to the earth. The drop continues to fall and increase in size until it falls out of the base of the cloud. Such a process has been tested by numerical example, and some results, which are presented as Fig. 8.14, are very reasonable answers.

A further test of the theory lies in the prediction that the liquid water content increases with height, attaining a maximum liquid water content in the region where $v_T - U_z = 0$. If N is the total number of water drops of rain drop size passing through a unit horizontal plane in unit time, the number of raindrops in a unit volume of air is

$$n = \frac{N}{(v_T - U_z)} \tag{8.19}$$

because by definition the product of area and time is unity. Now, n times the mass of a raindrop is the mass of rainwater per unit volume

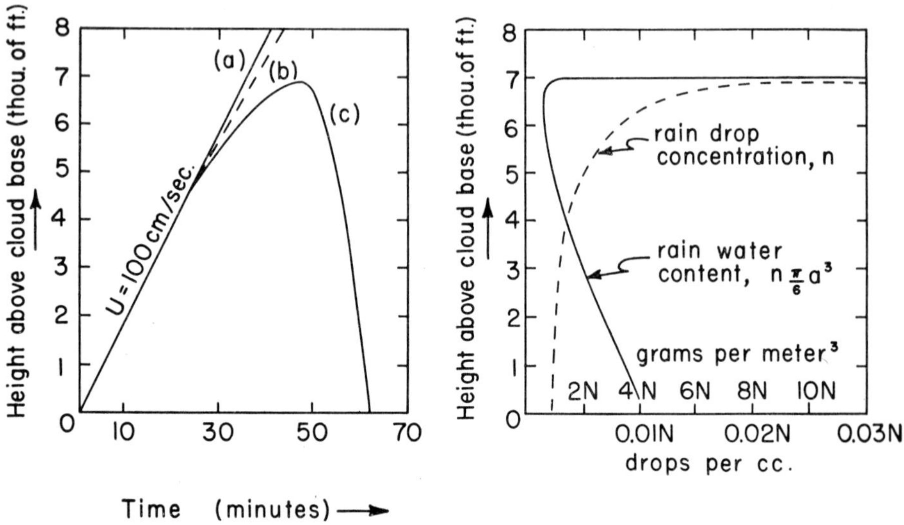

Fig. 8.14 (*After Bowen.*) The results of the coalescence theory when a drop of diameter $a = 20$ microns moves upward through a cloud of 20-micron drops (liquid-water content of 1 gram per meter^3) with an air velocity in the vertical of $U_z = 100$ centimeters per second. N is the number of drops crossing unit area in unit time and n is the number of drops per unit volume. Curve a is the motion of the air, b the motion of the cloud drops, and c the motion of the cloud particle growing by coalescence until it falls out the base of the cloud 62 minutes later as a drop 1.53 millimeters in diameter. Higher vertical air velocities result in particles rising higher in the cloud, falling from the base as larger drops, and having shorter travel time in the cloud during the growing time. For vertical velocities smaller than the 100 centimeters per second illustrated, the converse is true: namely, rising to lower heights in the cloud, falling from the base as a smaller drop, and having a longer growth time in the cloud. The theoretical results are in good agreement with observation.

of air—m of Eq. 8.12. From Eqs. 8.19 and 8.12,

$$m = \frac{\pi}{6} \frac{ND}{(v_T - U_z)} \sum_i \frac{n_i}{n} a_i^3.\dagger \tag{8.20}$$

Earlier in this chapter, we found experimentally that the rainfall rate is proportional to the strength of a radar echo (see Eq. 8.15). From Eq. 8.20 the rainfall rate should increase as $v_T - U_z$ decreases, provided that the numerator remains small and finite. That such is

† Strictly, $(v_T - U_z)$ should be written as $[(v_T)_i - U_z]$ and taken inside the summation sign. However, large raindrops have nearly uniform velocities, and it simplifies the argument to assume $(v_T - U_z)$ as an average value for the array of drops considered.

the case has been established by numerous investigations. Therefore, we reason that the radar echo should increase with height, and to prevent a too rapid increase of the radar signal with height we expect that the liquid water content also increases with height, reaching a maximum liquid water content where $v_T - U_z$ is a minimum value. Of course, any cloud is a complexity of drop sizes and velocities so that the average value of $v_T - U_z$ never goes to zero. Figure 8.10 substantiates the foregoing statements.

A summary of the results of this theory is quoted directly from Bowen's† original paper.

"Since the great majority of clouds do not produce rain, it is appropriate to conclude by considering briefly the conditions which need to be met before rain can form by the condensation-coalescence process. Apart from the basic requirement that coalescence shall occur as a result of collisions between cloud droplets, the following five conditions are necessary:

"(1.) A distribution of cloud droplet sizes must exist so that the droplets have an opportunity of falling relative to one another and coming into collision. It is probable that this condition is, in fact, met in the great majority of clouds.

"(2.) The width of the cloud must be such that drops which grow by coalescence will not be carried out of the cloud by a wind shear or another similar factor. This requirement is common to almost any theory of rain formation.

"Coming now to the conditions which arise from the theory as developed in the present paper:

"(3.) The cloud droplets must attain a certain minimum size before collisions are frequent enough to give a reasonable number of raindrops. The present state of knowledge on the collision process is not good enough to define this limit at all accurately.

"(4.) The vertical air current in a cloud must be maintained long enough for the growth process to be completed. It is a matter of observation that many convective clouds do not meet this requirement, going through their whole cycle of growth and dissipation in a shorter time than that required for raindrops to form by coalescence.

"(5.) For a given upward air velocity the depth of the cloud must be greater than that required for the drops which grow by coalescence to come into equilibrium in the upward air current.

"If any of these conditions are not met, then rain is unlikely to form by the condensation-coalescence process."

† E. G. Bowen, *Australian J. Sci. Res., Ser. A, Phys. Sci., 3,* 193 (1950).

The stimulation of precipitation by artificial means[5,14,25,29]

The imagination of the world was aroused when shortly after the close of World War II it was announced that man had caused the clouds to rain by the simple expedient of dropping Dry Ice into them. Since this announcement was made, much Dry Ice and later silver iodide have been dropped into clouds of all descriptions, with varying amounts of success. Much has been written on the validity of the results, and only recently have any large-scale controlled experiments been performed.

The results to date indicate that man under extremely favorable conditions has under his control the necessary factors to start precipitation. He does not have at present any control on the duration of the precipitation and/or the amount of precipitation if any that reaches the ground. However, again under favorable local conditions, rain may be started by man and then continue through natural processes with the result that enough rain falls to the ground to be economically important.

It appears that in order for substantial amounts of rain to be produced by "cloud seeding," as it is called, the atmosphere must be in such a metastable state of equilibrium that once rain is started, the water that falls from the cloud is replenished by natural convective means. Otherwise, the cloud rains or snows out, and a hole is left in it; the cloud droplets grow to raindrop size, fall, and are not replaced. This effect has been produced in undercooled clouds when the atmosphere has been meteorologically stable, so that precipitation once started has not continued. The rain or snow that falls from a cloud under these circumstances usually evaporates before reaching the ground, because a combination of high cloud bases and low relative humidity may exist between these cloud bases and the ground.

The principal success in "rainmaking" has been the changing of an undercooled cloud in a metastable state of colloidal equilibrium to a state of colloidal instability. The change is a result of the production of a large number of ice crystals in the volume of cloud air occupied by the undercooled water droplets. Once accomplished, the conditions for the Bergeron-Findeisen theory hold and the physics of rain formation is the same as for natural rain. Since many clouds exist in nature with both undercooled water and ice crystals present and yet do not rain, the same condition may be a final state of "seeded" clouds.

Theory shows that the one "rainmaking" tool man has at his disposal is the ability to cause supercooled water to freeze. One very cheap and simple way is to cause local portions of the cloud to become so cold that the temperature drops below the natural freezing point of supercooled water condensed initially from the water vapor in the air. The freezing of the supercooled droplets already present in the cloud is probably not important in this thermal effect. The local cooling of clouds is accomplished by dropping Dry Ice (solid carbon dioxide which vaporizes at a temperature of $-78.5°$ C) into the undercooled cloud. The pellets of Dry Ice as they drop sublime, the latent heat of sublimation being furnished by the air in the environment of the falling pellet. The local cooling of the air causes additional water vapor to condense out as a host of tiny ice crystals† in the regions where the air temperature has been reduced below about $-40°$ C. Once the ice particles are produced and the pellet passes, the particles remain in the air, whose temperature returns to the ambient temperature (but still below 0° C), and grow by diffusion in accordance with the theory of ice-crystal growth.

A second method which has also been successful in producing precipitation is the dropping of finely divided silver iodide crystals (diameter of 10^{-6} centimeter) into clouds. The concentration of silver iodide is so high that there will be a collision and coalescence between the silver iodide particles and a substantial fraction of the undercooled drops in each unit volume of air. The experimental evidence indicates that a silver iodide particle cannot exist in an undercooled waterdrop at drop temperatures more than a few degrees below 0° C. Hence, silver iodide, because of its icelike structure, causes these drops to freeze, inducing a state of colloidal instability in the cloud.

A third method for artificially stimulating precipitation follows the dictates of the theory introduced by Bowen and described in this chapter as a condensation-coalescence process. In conclusion (3), page 267, it is suggested that some clouds may not rain because they lack a sufficient number of relatively large drops to begin the process envisioned in the theory. To remedy the situation, experiments[7] have been conducted to see whether rain can be started when a large number of drops (mean size about 50 microns in diameter) are sprayed into a nonprecipitating cloud. Eleven experiments in situations selected to give an adequate control were carried out in the autumn of 1949. In only one seeding out of the eleven were no effects observed. In seven cases the clouds were less than 5000 feet thick and

† It has been estimated by V. J. Schaefer that 1 gram of Dry Ice will produce the order of 10^{16} ice crystals. If distributed uniformly, 10^{16} ice crystals represent 1000 per cubic centimeter in a volume of cloud air 100 meters × 100 meters × 1000 meters.

in six of these trials only virga† was observed. In the four experiments where the clouds were more than 5000 feet thick, a considerable fall of rain or hail was observed. No effects were observed on other clouds of the same structure in the vicinity. Computation and observation agreed that the thicker clouds should begin to rain at a longer interval of time after seeding than the thinner clouds and that the rain shower should be heavier and composed of larger drops. There is undoubtedly some breaking of large drops into smaller ones to augment the rain process. The time of the beginning of the rainfall was as little as 20 to 30 minutes after seeding for clouds 2000 to 4000 feet thick (rain fell as virga) to 30 to 60 minutes for clouds 6000 to 8000 feet thick (rain and hail).

In summary, there is evidence that cloud seeding will modify clouds and in carefully selected cases will cause rain to fall on the ground in measurable amounts. Evidence has been accumulated to indicate that in order to obtain an economically important amount of rain on the ground, the clouds should be at least 5000 feet thick and the height of the bases should not exceed the cloud thickness. For supercooled clouds, the tops should have a temperature of at least —5° C and should contain no measurable amounts of ice crystals. If measurable amounts of ice crystals are present (as there will be for temperatures below —15 or —20° C) and the cloud is not raining, additional seeding to form more ice crystals will probably not help.

It seems only fair to add that the number of days when the conditions above are met in a given locality and natural rain does not occur may not exceed a day or two a month.

References

1. Atlas, D., and Banks, H. C., "The interpretation of microwave reflections from rainfall," *J. Meteorol., 8,* 271–282 (1951).
2. Aufm Kampe, H. J., and Weickmann, H. K., "The effectiveness of natural and artificial aerosols as freezing nuclei," *J. Meteorol., 8,* 283–288 (1951).
3. Aufm Kampe, H. J., and Weickmann, H. K., "Trabert's formula and the determination of water content in clouds," *J. Meteorol., 9,* 167–171 (1952).
4. Austin, P. M., and Bemis, A. C., "A quantitative study of the bright band in radar precipitation echoes," *J. Meteorol., 7,* 145–151 (1950).
5. Bergeron, T., "The Problem of Artificial Control of Rainfall on the Globe," *Tellus, 1,* 32, No. 1 (February, 1949); 15, No. 3 (August, 1949).
6. Bowen, E. G., "Radar observations of rain and their relation to mechanisms of rain formation," *J. Atm. and Terrest. Phys. 1,* 125–140 (1951).

† Virga is the streamerlike band of precipitation falling from the bases of clouds. Since the precipitation evaporates before reaching the ground, the streamer has a lower limit.

7. Bowen, E. G., "A new method of stimulating convective clouds to produce rainfall and hail," *Quart. J. Roy. Meteorol. Soc., 78,* 37–45 (1952).
8. Byers, H. C. (chairman), "Report of the Committee on Cloud Physics," *Trans. Am. Geophys. Union, 32,* 760–764 (1951).
9. Coons, R. D., Gentry, R. C., and Gunn, R., "First Partial Report on the Artificial Production of Precipitation," U. S. Weather Bureau Research Paper No. 30 (August, 1948), Washington, D.C.
10. Diem, M., "Messungen der Grösse von Wolkenelementen," *Meteor. Rundschau, 1,* 261–273 (1948).
11. Dorsch, R. G., and Boyd, B., "X-ray Diffraction Study of the Internal Structure of Supercooled Water," *Natl. Advisory Comm. Aeronaut.,* Tech. Note 2532 (October, 1951), Washington, D.C.
12. Dorsey, N. E., "The Freezing of Supercooled Water," *Trans. Am. Phil. Soc., 38,* Part 3, 247–328 (1948).
13. Hosler, C. L., "On the crystallization of supercooled clouds," *J. Meteorol., 8,* 326–331 (1951).
14. Houghton, H. G., "An Appraisal of Cloud Seeding as a Means of Increasing Precipitation," *Bull. Am. Meteorol. Soc., 32,* 39–46 (1951).
15. Howell, W. E., "A Comparison of Icing Conditions on Mount Washington with Those Encountered in Flight," *Trans. Am. Geophys. Union, 32,* 179–188 (1951).
16. Kerker, M., Langleben, P., and Gunn, K. L. S., "Scattering of microwaves by a melting spherical ice particle," *J. Meteorol., 8,* 424 (1951).
17. Langille, R. C., "The Scattering of Ten-Centimetre Radio Waves by Rain," *J. Geophys. Research, 55,* 51–52 (1950).
18. Langille, R. C., and Thain, R. S., "Some Quantitative Measurements of Three-Centimeter Radar Echoes from Falling Snow," *Can. J. Phys., 29,* 482–490 (1951).
19. Levine, J., "Statistical Explanation of Spontaneous Freezing of Water Droplets," *Natl. Advisory Comm. Aeronaut.,* Tech. Note 2234 (December, 1950), Washington, D.C.
20. Ludlum, F. H., "The production of showers by the growth of ice particles," *Quart. J. Roy. Meteorol. Soc., 78,* 543–553 (1952).
21. Marshall, J. S., Langille, R. C., and Palmer, W. McK., "Measurement of Rainfall by Radar," *J. Meteorol., 4,* 186–192 (1947).
22. Marshall, J. S., and Gunn, K. L. S., "Measurement of Snow Parameters by Radar," *J. Meteorol., 9,* 322–327 (1952).
23. Mason, B. J., "The nature of ice-forming nuclei in the atmosphere," *Quart. J. Roy. Meteorol. Soc., 76,* 59–74 (1950).
24. Mason, B. J., "The spontaneous crystallization of supercooled water," *Quart. J. Roy. Meteorol. Soc., 78,* 22–27 (1952).
25. Orr, J. L., Fraser, D., and Pettit, K. G., "Canadian Experiments on Artificially Inducing Precipitation," *Bull. Am. Meteorol. Soc., 31,* 56–59 (1950).
26. Rankine, A. O., "Experimental Studies in Thermal Convection" (33rd Guthrie Lecture), *Proc. Phys. Soc.,* (B), *63,* 225–257 (1950).
27. Ryde, J. W., "The attenuation and radar echoes produced at centimetre wavelengths by various meteorological phenomena," *Meteorological Factors in Radio Wave Propagation,* London, The Physical Society, 169–188 (1946).
28. Schaefer, V. J., "The formation of ice crystals in the laboratory and the atmosphere," *Chem. Rev., 44,* 291–320 (1949).

29. Schaefer, V. J., "The effects produced by seeding supercooled clouds with dry ice and silver iodide," *Centenary Proc. Roy. Meteorol. Soc.*, 42–50 (1950).
30. Smith, E. J., "Observations of rain from non-freezing clouds," *Quart. J. Roy. Meteorol. Soc.*, *77*, 33–43 (1951).
31. Turnbull, D., "Formation of Crystal Nuclei in Liquid Metals," *J. Appl. Phys.*, *21*, 1022–1028 (1950).
32. Vonnegut, B., "Nucleation of supercooled water clouds by silver iodide smokes," *Chem. Rev.*, *44*, 277–289 (1949).
33. Wexler, R., "Theory of the radar upper band," *Quart. J. Roy. Meteorol. Soc.*, *78*, 372–376 (1952).
34. "Meteorology and the operation of jet aircraft," Discussion in *Quart. J. Roy. Meteorol. Soc.*, *78*, 427–456 (1952).

Source Books

B1. Ridenour, L. N., "Radar System Engineering," *Mass. Inst. of Technology Rad. Lab. Series, 1* (1947) (Chapters 1 through 3), McGraw-Hill Book Co., New York.
B2. Seitz, F., *The Modern Theory of Solids,* McGraw-Hill Book Co., New York (1940).
B3. Wyckoff, R. W. G., "The Structure of Crystals," *Amer. Chem. Soc. Monograph Series,* Book Dept. of Chemical Catalogue Co., New York (1931).

The *Compendium of Meteorology,* Boston, American Meteorological Society (1951), has several articles on topics discussed in this chapter. These articles have excellent bibliographies up to 1950. The following names and titles are from this book.

C1. Bemis, A. C., "Aircraft Meteorological Instruments," 1223–1231.
C2. Coons, R. D., and Gunn, R., "Relation of Artificial Cloud-Modification to the Production of Precipitation," 235–241.
C3. Lewis, W., "Meteorological Aspects of Aircraft Icing," 1197–1203.
C4. Ligda, M. G. H., "Radar Storm Observation," 1265–1282.
C5. Ludlum, F. H., "The Physics of Ice Clouds and Mixed Clouds," 192–198.
C6. Nakaya, U., "The Formation of Ice Crystals," 207–220.
C7. Rodert, L. A., "Physical and Operational Aspects of Aircraft Icing," 1190–1196.
C8. Schaefer, V. J., "Snow and Its Relationship to Experimental Meteorology," 221–234.
C9. Wexler, R., "Theory and Observation of Radar Storm Detection," 1283–1289.

Problems

8.1 The rate of formation of ice nuclei in supercooled water per unit volume per unit time has the form (homogeneous nucleation)

$$J = \frac{nkT}{h} e^{-\left(\frac{A+F_c}{kT}\right)}.$$

J is the specified rate, n is the number of molecules per unit volume in the liquid phase, h and k are respectively the Planck and Boltzmann constant, and T is the absolute temperature. A is called the free energy of activation for self-diffusion of the liquid molecules, and F_c is the free energy of formation of a nucleus.

(*a*) Express J as a function of diameter a, assuming that n refers to a spherical drop. (*b*) Assume that $J = 1$ ice nuclei are formed per second per drop. Plot $(A + F_c)$ as a function of drop diameter for $T = 0°$ C; $-10°$ C; $-20°$ C; $-30°$ C; and $-40°$ C. (*c*) Plot on the graph resulting from part *b* the values of the average freezing temperature as a function of drop diameter given in Fig. 8.4. (*d*) Plot $(A + F_c)$ as a function of T.

8.2 Theory specifies that the form of F_c (see previous problem) should be

$$F_c = \tfrac{1}{3}\sigma' g r_c^2$$

and

$$r_c = \frac{2\sigma'}{D' L_F \ln \dfrac{T_0}{T}}.$$

σ' is the specific free energy of the solid-liquid interface. r_c is the critical radius for the formation of a stable ice nucleus. g is a geometric factor such that $g r_c^2$ represents the total surface area of the nucleus. D' is the density of ice and L_F is the latent heat of fusion. T_0 is the melting temperature of ice and T is the ambient temperature of the supercooled liquid. Values of these quantities have been found to be (see McDonald, J. E., "Theoretical Cloud Physics Studies," Final Report Project NR 082 093 Office of Naval Research Contract Nonr 757, Task Order (00), Iowa State College, Ames, Iowa, January 31, 1953)

T	σ', erg cm^{-2}	L_F, joules g^{-1}	D', g cm^{-3}
0° C	10.0	333.6	0.918
−10° C	9.6	311.9	0.919
−20° C	9.1	288.0	0.921
−30° C	8.5	263.7	0.922
−40° C	7.7	235.7	0.924

(*a*) Compute F_c as a function of temperature assuming that $g = 23$, the shape factor for a hexagonal column. (*b*) Find A of Prob. 8.1 as a function of temperature.

8.3 A, the free energy of activation for self-diffusion of the liquid-water molecules, has the form

$$\mu = Ce^{\frac{A}{kT}}.$$

μ is the viscosity of water and C is a constant. At 0° C, $A = 3.5 \times 10^{-13}$ erg per molecule, and $\mu = 0.018$ poise.

a. Determine C.

b. Determine A from the following table of μ as a function of T and extrapolate the values of A to $-40°$ C.

T, °C	−2°	−4°	−6°	−8°	−10°
μ (poise)	0.019	0.021	0.022	0.024	0.026

c. At what temperatures are the values of A from this problem and A from Prob. 8.2 satisfied? What is the physical significance of these temperatures?

8.4 Show that the back-scattering coefficient per unit cross-sectional area of the intercepting sphere $\dfrac{K_s'}{\pi(a/2)^2}$ can be written as $4\alpha^4\left(\dfrac{m^2 - 1}{m^2 + 2}\right)^2$. Show that the scattering area coefficient K_s can be written as $\frac{8}{3}\alpha^4\left(\frac{m^2-1}{m^2+2}\right)^2$. Explain how the difference in numerical coefficients arises. Rayleigh scattering is assumed.

8.5 It has been found experimentally that the attenuation coefficient in rain, referred to the base e, has a value $\sigma = 0.0125$ per mile of rain per millimeter of rainfall per hour. The quantity $Z = \sum_i n_i a_i^6$ enters the radar equation by considering scattering of microwave energy from rain to be Rayleigh scattering. A relationship between Z and the rainfall rate $\frac{dw}{dt}$ has been found experimentally to be

$$Z = 220\left(\frac{dw}{dt}\right)^{1.60}$$

when Z is expressed in millimeter^6/meter^3 and $\frac{dw}{dt}$ in millimeters of rain/hour.

(*a*) How many drops 2.5 mm in diameter are required for an observed rainfall rate of 50 mm/hr? What is the average rainwater mass in a unit volume? What is the value of the attenuation coefficient per mile of rain? (*b*) At a certain receiver gain setting, the minimum detectable power represents a liquid-water content of 1 g/m^3 from the front edge of a rain cloud situated 30 miles from the radar. At this same gain setting and azimuth the rear edge of the storm is observed to be 32 miles from the radar. What is the liquid-water content of this section of the cloud if the average rainfall rate from the cloud is measured as 50 mm/hr? Consider the front edge of the cloud as seen by radar to coincide with the front edge of the cloud seen visually.

8.6 *a*. What is the liquid-water content of a cloud containing 2000 drops 2.5 mm in diameter in a cubic meter of cloud? Express the answer per cubic meter.

b. Defining the quantity Z as

$$Z = \sum_i n_i a_i^6,$$

find Z. n_i is the number of drops per unit volume and a_i is the diameter of the drops.

c. Find the number of drops in each class interval if the liquid-water content of part *a* is distributed such that 0.1, 0.2, 0.3, 0.4 parts of the liquid water are contained in drops of 1, 2, 3, and 4 mm diameter respectively.

d. Find both Z and the volume mean diameter for the raindrops of part *c*.

e. Compare the ratio of Z from part *d* to Z found from part *b*. Compare in similar fashion the volume mean diameters. Comment on the significance of the ratios found.

8.7 Consider that Fig. 7.8 may be used to find the collection efficiency of a wing whose thickest dimension is D.

Find the icing rate of a wing whose $D = 1$ foot when traveling 400 miles per hour at a pressure level of 700 millibars (about 10,000 feet). The ambient temperature is $-10°$ C. Assume that the liquid-water content of the cloud is 2 g/m^3, one-half of which is made up of 20-micron cloud drops and one-half of which is composed of 1-mm raindrops. Assume the average density of the ice formed as 0.8 g/cm^3.

CHAPTER 9 Atmospheric Electricity

The subject of atmospheric electricity relates to the bulk electrical properties of the earth's gaseous envelope. The vertical extent of the study is conveniently confined between two natural equipotential surfaces, the earth and the ionosphere. The latter is a region of ionized gas, highly conducting, and existing at altitudes in excess of 90 kilometers. Experiment and theory show it to be impossible for local concentrations of charge in the horizontal to build up in either the ionosphere or the earth without almost immediate dissipation (of the order of a few microseconds). The characteristics just described for the atmosphere make the earth and the extreme lower boundary of the ionosphere, the *conducting layer,* a formal analogy to the two surfaces of a spherical capacitor, with the space between conducting surfaces filled by a poorly conducting gas. Once the analogy has been accepted, the concepts of charge, electrical force, current, and resistance apply to the atmosphere and can be evaluated in terms of well-known units.

It was discovered in the latter part of the nineteenth century that the atmosphere, even on cloudless days, at all times possessed both a measurable electric field† and an electric current flowing from the atmosphere toward the earth. Experiment has shown the normal or fair-weather electric field of the earth to be such that the atmosphere is positively charged with respect to the ground. Local variations of the electric field exist, especially in the vicinity of thunderstorms. In fact, the direction of the electric field is usually reversed near thunderstorms. However, as thunderstorms have been estimated to occupy less than 1 per cent of the earth's surface, the fair-weather field is by far the normal electric state of the atmosphere. It has been estimated that normally the ionosphere is at a potential of the order of 360,000 volts positive with respect to the ground. The electric field is nonlinear and of the order of 130 volts per meter near the ground, about 4 volts per meter at 10 to 12 kilometers. However, by the proper

† Electric field is defined in Eq. 9.1 and current in Eq. 9.3.

assumptions, the electric field can be linearized and the large-scale electrical aspects of the atmosphere discussed in terms of a simple equivalent electric circuit having a capacitance shunted by a resistance. Measurements indicate the electric current is relatively independent of height. We shall discuss how these values were determined and their significance.

The primary source of current in the atmosphere is the ion, which may carry either a net positive or negative charge of electricity. Millikan in his famous oil-drop experiment determined that the quantity of electricity was not infinitely subdivisible, but existed in integral multiples of 4.803×10^{-10} electrostatic unit of charge (esu of charge). This basic unit of charge will be denoted by e. A charged body may have Ne units of charge where N is equal to 1, 2, 3, 4 . . . . Denoting the product Ne by Q, an experimental law of electrostatics states that such a charge Q exerts on a test charge Q_0 placed r units of distance away a mechanical force of

$$f = Q_0 E_r = \frac{Q_0 Q}{\epsilon r^2} . \tag{9.1}$$

This is the familiar inverse-square law of Coulomb that always occurs for the isotropic propagation of lines of flux. When E_r is the electric field intensity in statvolts per centimeter, f is the mechanical force in dynes and Q_0 is expressed in statcoulombs. ϵ, the dielectric constant for air, is unity in these units.† Equation 9.1 may be considered a defining equation for electric field intensity.

It has been found that an elementary charge does not exist for any length of time either as a free electron or as a proton with its correspondingly small mass. Rather, this small electrified particle attaches itself to a larger mass, such as a molecule, a dust particle, or a cloud droplet. As a consequence, a particle of large mass may acquire an electric charge and be set in motion by the force exerted on it by the earth's electric field. The motion of these charged particles, called ions, constitutes a current of electricity.

In accordance with the concepts of flux derived from previous chapters, we deduce that the flux of Ne units of charge through a cross-sectional area A in time t is related to the total charge Q in the volume Avt as

$$Q = Ne = nevAt = qvAt. \tag{9.2}$$

† Definitions of electrical units will be in the electrostatic system of units (esu). However, practical units will be used in the general discussion. Table 9.1 is a conversion table between the quantities to be used in the formulas and those quantities used in the general discussion.

Here, n is the number of charges per unit volume, ne equals q, the amount of charge per unit volume, and v is the average speed of this charge through the cross section normal to the stream of charges. Differentiation of Eq. 9.2 in respect to time with v constant defines the current

$$\mathrm{i} = \frac{dQ}{dt} = \mathrm{q}vA. \tag{9.3}$$

The current density is defined as

$$\mathrm{j} = \frac{\mathrm{i}}{A} = \mathrm{q}v = nev \tag{9.4}$$

and is a standard unit used in measurements of ionic current. Since ions can be either positively or negatively charged, two streams of ions will be set up in an electric field, one a negative stream and the other a positive stream. Since each stream travels toward opposite ends of the electric field, each stream contributes to the current density. Hence, we shall interpret the current density as the sum of the two ionic currents,

$$\mathrm{j} = \mathrm{j}_+ + \mathrm{j}_- . \tag{9.5}$$

The electric-field intensity is by definition the gradient of the electric potential. By this definition, the vertical component of the electric field is

$$\mathrm{E}_z = \frac{\partial \mathrm{V}}{\partial z} . \tag{9.6}$$

V is the electric potential and $\partial \mathrm{V}/\partial z$ is the difference in potential through a distance z. In the atmosphere z, the height of the potential surface, is positive in the upward direction and V also becomes more positive in the upward direction. Therefore, $\partial \mathrm{V}/\partial z$ is normally a positive quantity. The *ground will always be considered the surface of zero potential*, so that V also indicates the *magnitude* of the potential difference in respect to the earth. The electric-field intensity E_z is considered positive when oriented parallel to the direction that a free positive charge would move under the stress of this field. The atmosphere is considered to have a positive field when the atmosphere is at a positive potential above the ground potential. Ohm's law, in the beginning empirically derived, is stated as

$$\mathrm{V} = \mathrm{iR}. \tag{9.7}$$

Ohm's law is now considered a fundamental law in electrical theory, and relates the current i that flows in a circuit when an electric potential difference of V units is applied to a section of the electrical circuit under consideration. Note that we have considered one end of the circuit to be at zero potential, making V equal to the magnitude of the potential difference. The constant of proportionality is the resistance R.

Differentiation in respect to z of Eq. 9.7 shows with the aid of Eq. 9.6 that

$$\frac{\partial V}{\partial z} = E_z = i\frac{\partial R}{\partial z} = \frac{i}{A}\left(A\frac{\Delta R}{\Delta z}\right) \tag{9.8}$$

when we consider that the current i is a constant throughout the potential drop. The term $(\Delta R/\Delta z)A$ is called the specific resistance. Its reciprocal,

$$\lambda = \frac{1}{A}\frac{\Delta z}{\Delta R}, \tag{9.9}$$

is called the specific conductivity.

Substitution of Eqs. 9.9 and 9.4 into Eq. 9.8 makes

$$j_z = \lambda E_z. \tag{9.10}$$

Equation 9.10 is an alternate form of Ohm's law that is particularly adaptable to measurements of electrical quantities in gases.

With the equations that we have developed, we have the necessary relationships to determine the magnitude of the *conduction current* from atmosphere to earth. Since this current is continuous, the problem then becomes one of finding a supply current of the same magnitude as the conduction current in order to complete the electrical circuit between earth and ionosphere. If such a supply current could not be found, the law of the conservation of energy would be violated. We shall study each of the currents separately and begin by examining the causes of the fair weather electric field, the magnitude of the conduction current, and the methods of measurement of these quantities.

The measurement of the earth's electric field

If an uncharged conductor is placed in an electric field, the conductor, though remaining electrically neutral, has its charge redistributed along its surface in a way consistent with the law of like electrical charges repelling and unlike electrical charges attracting one another. The equivalent mathematical statement of this experimentally observed phenomenon in air ($\epsilon = 1$) is

$$E = 4\pi\sigma.\dagger \tag{9.11}$$

Stated in words, the magnitude of the charge per unit area σ is directly proportional to the strength of the electric field. E is directed normal to the area containing the charge. We may illustrate the meaning of Eq. 9.11 by Fig. 9.1.

In Fig. 9.1*a*, we show an electric field between the two plates of a condenser. If the plates are maintained at a difference in potential of V units, Eq. 9.6 indicates that a field E_z exists between them. The direction of the field is given by the direction in which a free positive charge would move if placed in the field. The field is assumed to be in the direction indicated and its strength is indicated by the spacing of the lines of force (arrow-tipped lines). Each line emanating from a unit charge indicates 4π lines of flux.

Let an uncharged body be placed in this field. Figure 9.1*b* indicates that the field will be distorted in the neighborhood of the body itself, but that neither the strength nor direction of the field at a great distance from the body will be affected appreciably. The body has been deliberately drawn in teardrop shape to indicate that the electric field is strongest in the region of greatest curvature. By Eq. 9.11, the surface charge per unit area is also greatest at the point of greatest curvature through the direct relationship between E and σ. In every case, it will be noticed that like charges tend to move as far away from their neighbors as possible and yet remain bound to the conductor. The lines of flux terminate on and are normal to the surface containing the charge.

If, while still in the electric field, a conductor is connected directly to one of the plates of the condenser, the charge on the body will become the same as the charge on the plate to which it is connected, because both plate and body are now the same conductor and as a result an equipotential surface. This state, shown by Fig. 9.1*c*, can arise from Fig. 9.1*b* only by the movement of charge through the wire

† Equation 9.11 is usually stated as

$$\int_A \mathbf{E}\cdot\mathbf{n}\, dA = \frac{4\pi}{\epsilon}\int_A \sigma\, dA = \frac{4\pi}{\epsilon} Q.$$

$\mathbf{E}\cdot\mathbf{n}$ is the vector expression for the magnitude of the electric field vector normal to and directed outward from an incremental surface of area dA. $\int_A$ is the integration over the entire surface. $\mathbf{n}$ is a unit normal vector. In this form Eq. 9.11 is called Gauss's theorem.

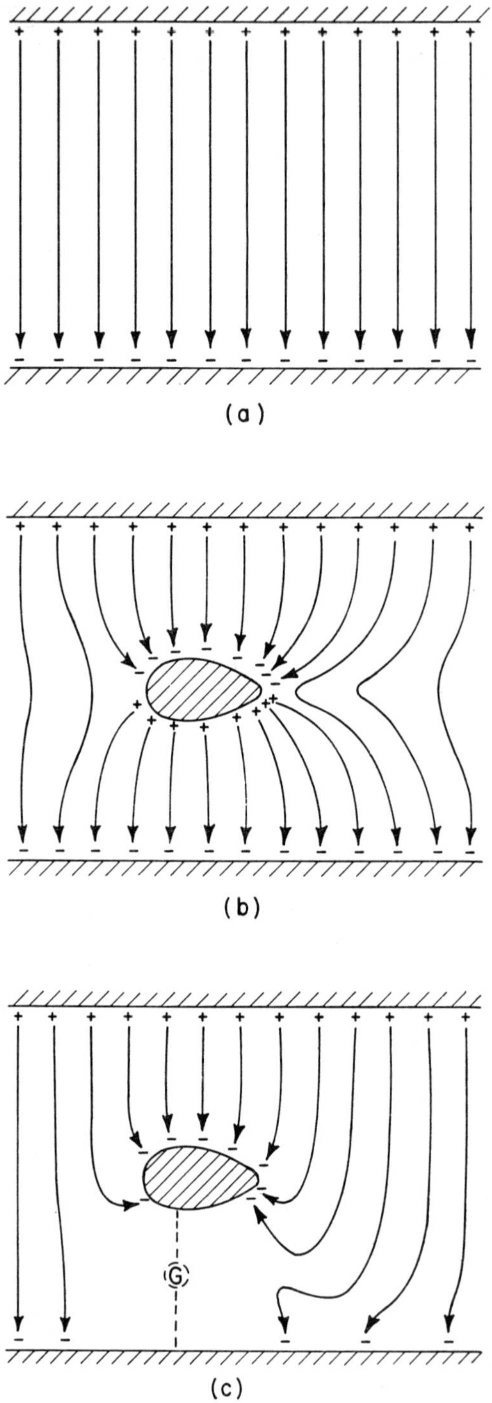

Fig. 9.1 Schematic representation of the electric field (shown by arrow-tipped solid lines) between the positively and negatively charged plates of a parallel plate condenser. (*a*) the undisturbed field; (*b*) the field as disturbed by an uncharged body; and (*c*) the field as disturbed by the body after it had been connected to one plate of the condenser, schematically shown by dashed lines to indicate a wire and galvanometer Ⓖ.

at the moment the wire was connected to both body and plate. In other words, an instantaneous current flowed when the connection was first made. This current theoretically can be detected by a sensitive galvanometer used as an ammeter. The strength of this current varies with the amount of separation of charge by the electric field, and hence is an indication of the magnitude of the electric field. Means other than the one just indicated may be used to equalize the potential between the body and one plate of the condenser.

Consider, for example, a metallic tank of water in the electric field of the earth. The tank, which is initially charged to the earth's potential (grounded), is then insulated from the earth's surface. The earth's surface, a good conductor of electricity, is one plate of a condenser and a conducting layer high in the atmosphere is the other. An electric field exists between them and is strongest near the ground.

Suppose that a pipe protrudes from the tank through which water from the tank can be discharged as a fine spray. Figure 9.2*a* shows the initial conditions of charge distribution and field before the water flows. This figure is analogous to the charged body of Fig. 9.1*c*. The electric field is most distorted near the mouth of the pipe, and hence the charge density per unit area is greatest there. The entire tank is at ground potential.

Water is now allowed to flow from the pipe in a fine spray. The water immediately prior to discharge forms a portion of the conducting surface of the tank, and gains a negative charge per unit area of σ. This charge is then forcibly torn from the tank as the stream of water breaks up into a spray of droplets. A fine spray is desired because of the large surface to volume ratio of the water, carrying away a maximum amount of charge for a given flow rate. By this mechanism, the water droplets carry electric charge to the ground, causing the tank to tend toward a more positively charged condition. Finally, an equilibrium condition is built up in which the tank is at the same potential as the adjacent air and the conditions of Fig. 9.2*b*, analogous to those of Fig. 9.1*b*, are reached, and prevail from this point on. The difference in potential between the water dropper and the earth can be measured by a quadrant electrometer, opposing plates of which are maintained at high potential in respect to ground, with the needle of the electrometer being connected to the reservoir. The water dropper just described is a potential equalizer and is credited to Lord Kelvin. The water dropper, though not sensitive to rapid changes in the earth's electric field, will charge to the potential of the air from ground potential in a matter of $\frac{1}{2}$ minute. In common with the other instruments used for electric-field measurements, corrections must be applied to indicated readings to obtain the true field because of the distortion in the earth's field by the instrument and its housing.

A second type of instrument used for electric-field measurements is an insulated wire about 6 feet long suspended horizontally between two posts about a meter above the ground. This conductor is coated

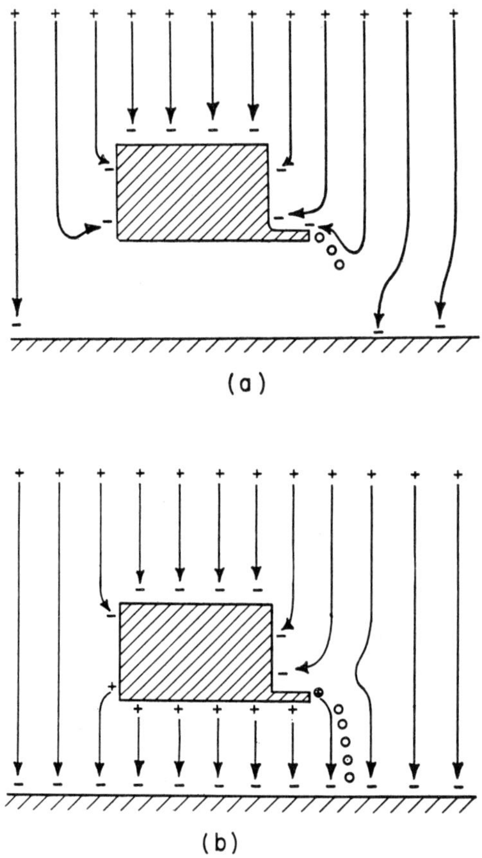

Fig. 9.2 Schematic drawing of the lines of flux of the water dropper as an instrument for determining the electric-field strength of the atmosphere. (*a*) lines of flux before the water has started to flow, and (*b*) lines of flux after equilibrium has been established at a constant water flow rate from outlet pipe.

with an electrolytic deposit of polonium, a source of α radiation having a half-life of 136 days. The polonium increases the ionization of the air in the immediate vicinity of the wire. Ions flow to the wire until the wire acquires a potential equal to that of the surrounding air, at which time the flow of ions ceases. The difference in potential between the atmosphere at the height of the wire and ground is then measured. By definition, this measurement indicates the magnitude of the vertical component of the electric field.

A third method of electric-field measurement makes use of Eq. 9.11. In principle, the method makes use of a plate set flush with the ground and insulated from it. The plate is arranged so that a grounded cover can be placed over it at intervals, thus causing the charge on the plate to flow to ground. When unshielded, the plate becomes charged by the atmosphere to a surface density σ, which through Eq. 9.11 is $E_z/4\pi$. An electroscope is the most direct method of observing σ. Refinements of this instrument include the *field mill*, which is a device for periodi-

Table 9.1

Conversion Factors between Electrostatic and Practical Electrical Units

Cgs mechanical units are used with the electrostatic units.

	Electrostatic Unit (equals)	Practical Unit
Potential	1 statvolt	300 volts
Charge	3×10^9 statcoulombs	1 coulomb
Current	3×10^9 statcoulombs per second	1 ampere
Resistance	9×10^{11} esu of resistance	1 ohm
Power	$1 \text{ statvolt} \times 1 \dfrac{\text{statcoulomb}}{\text{second}} \times 10^7$	1 watt
Capacity	9×10^{11} centimeters of capacity	1 farad

For example, an electric field of 130 volts per meter is

$$130 \frac{\text{volts}}{\text{meter}} = 0.433 \frac{\text{statvolts}}{\text{meter}}$$

$$= 4.33 \times 10^{-3} \frac{\text{statvolts}}{\text{cm}} = 3.45 \times 10^{-4} \frac{\text{statcoulombs}}{\text{cm}^2}.$$

$$3.45 \times 10^{-4} \frac{\text{statcoulombs}}{\text{cm}^2} = 1.15 \times 10^{-13} \frac{\text{coulombs}}{\text{cm}^2}$$

$$= 7.19 \times 10^5 \frac{\text{elementary electric charges}}{\text{cm}^2}.$$

The relationship $\sigma = E/4\pi$ has been used to convert electric field to surface charge density.

cally shielding the instrument from the earth's field. The alternating current that results has the advantages of easy amplification and rectification by electronic circuitry so that a continuous record of magnitude and size of the field can be obtained.

The results from the instruments described and others indicate that the vertical component of the earth's electric field has an average value at Kew Observatory outside London, England, of 1.3 volts per centimeter, with the sign of the field such that the atmosphere is positively charged in respect to the earth. The electric field under these conditions is called positive. The sign of the electric field can change, especially in the neighborhood of thunderstorms, and may reach measurable values up to 300 volts per centimeter. Higher values undoubtedly exist in extremely localized areas for short periods of time, especially before lightning flashes.

A table of electrostatic versus practical units is included as an aid for computations. All formulas dealing with static charges or used in converting electrical forces to mechanical energies or torques are expressed most naturally in electrostatic (esu) units. Flow of current is expressed in practical electrical units; i.e., electromagnetic units (emu) with appropriate powers of ten to make the units of practical size.

The measurement of electric charge in the atmosphere

The electric current and the flow of ions causing the current in the atmosphere are related through Eqs. 9.3, 9.4, and 9.5. The practical measurement of these quantities requires the equations to be rewritten in terms of instrumental quantities. We shall modify Eq. 9.4 by considering only ions of a single charge and then generalize the result through the use of Eq. 9.5.

If Eq. 9.4 be multiplied and divided by the electric field, the equation can be written as

$$\mathrm{j} = ne\frac{v}{\mathrm{E}}\,\mathrm{E}. \tag{9.12}$$

A comparison of Eqs. 9.12 and 9.10 shows that the specific conductivity

$$\lambda = ne\frac{v}{\mathrm{E}} = ne\kappa. \tag{9.13}$$

κ, the velocity of an ion in a unit electric field, is called the mobility of an ion. The mobility of ions of various sizes and charges has been the subject of measurement. The mobility of all but the largest (Langevin) ions is deduced from the rate of diffusion of a large number of the same type of ions in a known electric field at a fixed air density. The ions range in size from molecular diameters to dust-particle size. Larger particles, because of their low ratio of charge to mass, have negligible mobilities. It is only when the particles approach microscopic size that application of Stokes' law will give the mobility of ions in an electric field. For particles of the order of molecules, viscosity as applied through Stokes' law has no meaning, leading one to define the mobility as the average movement of a large number of ions, i.e., the diffusion process. The following is a table of ionic mobilities.

Table 9.2

A Table of Ionic Mobilities for Various Types of Ions

Name	Electronic Charge, units of elementary charge e	Range of Mass, grams	Mobility, $cm^2\ sec^{-1}\ volts^{-1}$
Small ion	1	10^{-21}	> 1.25
Intermediate ion	1	10^{-19} to 10^{-16}	Commonly 0.1 to 0.02
Large ion	$\gtrless 1$	$> 10^{-15}$	$< 10^{-3}$

For small ions, the negative ions have of the order of 1.3 times the mobility of their corresponding positive ions.

It is doubtful whether the smallest ions exist in any large numbers in a normal as opposed to a laboratory-clean atmosphere. The smallest ions quickly attach themselves to larger neutral atmospheric particles to form larger ions. The large ion so formed has as a result the charge of the small ions but essentially the mass of the atmospheric particle.

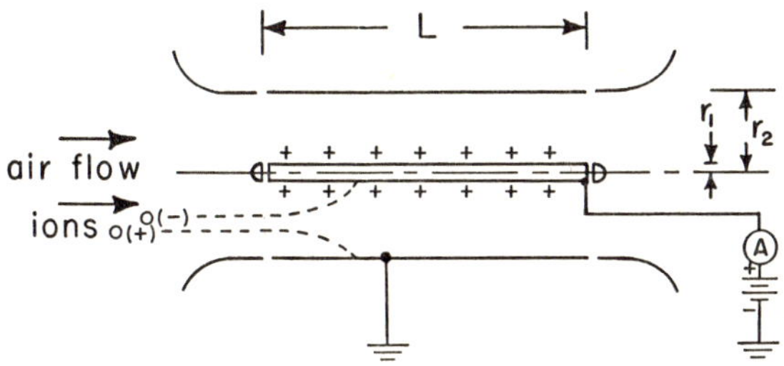

Fig. 9.3 Schematic drawing of an ion collector. Central plate is charged by battery to positive (or negative) potential in respect to ground. Ions hitting central plate (dotted path of negative ion shown) set up a steady current in the microammeter Ⓐ.

The measurement of n, the number of ions in a unit volume of air, requires in addition to mobility data, a knowledge of the magnitude and direction of the electric field between the two plates of the condenser used to collect these ions. A knowledge of an electrical unit called the capacitance of a condenser is needed.

Suppose that, for concreteness, we intend to pass a volume of air containing ions between the concentric cylindrical plates of a condenser (see Fig. 9.3). The outer cylinder of radius r_2 is kept at ground potential and the inner cylinder of radius r_1 is kept at a potential V volts above ground. An electric field

$$\frac{\partial V}{\partial r} = E_r = 4\pi\sigma \tag{9.14}$$

then exists between the two plates of the condenser. The electric field, which is directed radially between the plates, exerts a force on the ions which causes them to move along a radius toward the condenser plate having the appropriate electrical sign to attract the ions. The ions, on impinging on the condenser plate, tend to neutralize the existing charge. Since a constant charge on the plates of the condenser is maintained, a current must flow to the condenser plates in the external circuit, which is equal to the ionic current between the plates of the condensers. Reversing the direction of the field will cause a reversal of the direction of ion flow. Ionic currents of both polarities can be measured. By Eq. 9.5, the total ionic current is the numerical sum of the two separate currents. The problem therefore becomes one of determining E_r from Eq. 9.14. To do this, the geometry of the system, the capacitance, must be understood and evaluated.

Writing

$$\sigma = \frac{Q}{A} \tag{9.15}$$

for a conductor, substituting into and integrating Eq. 9.14 between the 0 and V levels of potential, expresses the potential difference between the two plates of the condenser† as

$$V = 4\pi Q \int_{r_1}^{r_2} \frac{dr}{A} = \frac{Q}{C} \cdot \tag{9.16}$$

The part of Eq. 9.16 expressed by

$$C = \frac{1}{4\pi \int_{r_1}^{r_2} \frac{dr}{A}} \tag{9.17}$$

is a function only of the geometry of the condenser plates, and is called the capacitance of a condenser. For the cylindrical condenser chosen, the surface area of the condenser which must at all times be normal to the field is

$$A = 2\pi r L, \tag{9.18}$$

† *Capacitor* is preferred in electrical-engineering nomenclature rather than *condenser*. Because of the wealth of literature referring to *condenser*, it has been used here.

a function only of the radial distance r. L is the length of the condenser and is a constant.

Integration of Eq. 9.17 under the conditions of Eq. 9.18 shows the capacitance of a cylindrical capacitor as

$$\mathrm{C} = \frac{1}{4\pi \int_{r_1}^{r_2} \frac{1}{2\pi L} \frac{dr}{r}} = \frac{L}{2 \ln r_2/r_1} \cdot \tag{9.19}$$

Equation 9.17 can be used to find the capacitance of condensers with other than cylindrical symmetry.

By the proper collection of terms, we can arrive at an expression that will give ionic current in terms of mobility, the volume transport of air per unit time $\Delta V/\Delta t$, electric potential, and capacitance.

Equations 9.12 and 9.13 together give

$$\mathrm{j} = ne\kappa \mathrm{E}_r = 4\pi\sigma ne\kappa \tag{9.20}$$

when $4\pi\sigma$ is substituted for E_r from (9.14).

Equations 9.15 and 9.16 together indicate that

$$\sigma = \frac{\mathrm{CV}}{A} \tag{9.21}$$

so that Eqs. 9.20 and 9.21 together give

$$\mathrm{j} = 4\pi ne\kappa \frac{\mathrm{CV}}{A} \cdot \tag{9.22}$$

However

$$\mathrm{j}A = \mathrm{i} \tag{9.23}$$

and for a steady current

$$ne = \frac{\mathrm{Q}}{V} = \frac{\mathrm{i}}{V} \int_0^t dt = \mathrm{i}\frac{t}{V} \cdot \tag{9.24}$$

Equation 9.22 reduces to

$$\kappa = \frac{1}{4\pi \mathrm{CV}} \frac{V}{t} \tag{9.25}$$

with the aid of Eqs. 9.23 and 9.24. V/t is interpreted as $\Delta V/\Delta t$, the volume of air swept through the condenser in unit time. Equation 9.25 indicates that for a constant volume flow of air through a condenser of known capacitance, the potential V can be adjusted so that all ions of mobility greater than or equal to κ will be collected by the condenser. For those ions whose mobility κ' is less than κ, a fraction κ'/κ will be deposited and will contribute to the current. The number of ions collected will be indicated by the current i obtained from Eq. 9.24, which is

$$n = \frac{i}{\frac{V}{t} e}. \qquad (9.26)$$

Equations 9.25 and 9.26 are used in the following way.

Under fixed conditions of C and V/t (the usual case), Eq. 9.25 shows that the lowest mobility that can be collected by our instrument depends only on the voltage applied between the plates of the instrument. At the same time, Eq. 9.26 shows that the strength of the

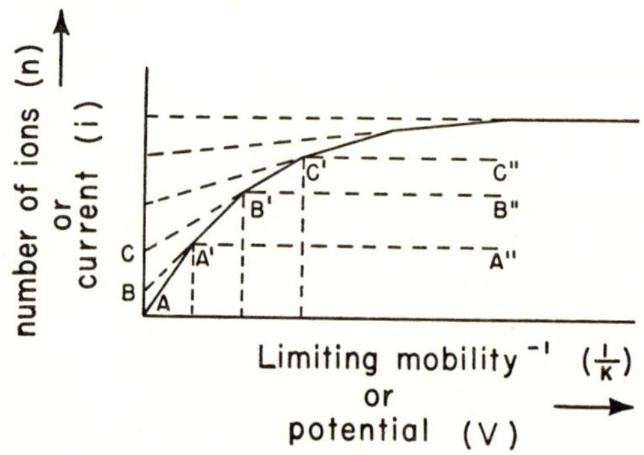

Fig. 9.4 Solid curve of straight segments is a plot of ionic current versus potential difference between condenser plates, or a plot of number of ions versus limiting mobility of the ions. If only one type of ion were present, the current would rise from A to A' and remain constant to A'', even though the potential were increased. A' shows the number and limiting mobility of the ions collected. For an ion of lower mobility, the path would be B, B', B''. The number is given by $B' - B$ and the limiting mobility by point B'. The combination of the two ions gives path A, A', B'. Similarly for C, C', C'', etc.

electric current depends only on the number of such ions present. Thus, if no ions were present with mobilities greater than the mobility specified by Eq. 9.25, no current would flow even though a potential V were applied across the plates. The current measured is the current of negative ions if the central plate is positively charged in respect to ground, whereas the current of positive ions is measured if the central plate is negatively charged. The total current is the sum of the two ionic currents.

From a measurement of V versus i, the corresponding quantities κ and n can be obtained. A plot of V versus i is shown in Fig. 9.4. The same figure is also a plot of $1/\kappa$ versus n. From curves that show the number of ions in a given range of mobility, there can be determined not only the conduction current of the atmosphere but also the number of ions of each sign present in various regions of the earth.

Measurements of ion concentrations show wide variations in space, time, and elevation. Over the oceans at sea level, the large ion count is expected to average 100 to 500 ions of each sign. Over the land the total large ion count is extremely variable, being of the order of 1000 to 2000 for the country and as high as 20,000 to 30,000 in cities. The small ion concentration is of the order of 400 to 500 over oceans, 100 in industrial areas, and 1000 in country air. All values are expressed in numbers per cubic centimeter. It has been found that at all levels there are more positive than negative ions, causing the air to have a positive space charge. At sea level the excess may run as high as 100 to 1000 per cubic centimeter.

Less is known about the vertical distribution of ions. It is known[C6] that the total ion concentration increases from the ground to about 6 kilometers after which it begins to decrease. On the average, the rate of increase is about 500 ions per cubic centimeter per kilometer elevation, provided that the measurements are made in fair weather away from cities. The explanation of the increase and subsequent decrease of ion concentration with elevation naturally leads to a discussion of the production and destruction of ions, ionizing radiations, and ionic equilibrium.

Ionic equilibrium and ionizing radiations in the atmosphere

Conductivity, current density, and electric-field measurements in the atmosphere indicate that the conductivity increases as a function of height. Concurrently, the vertical component of the electric field shows a decrease with height. As a result, the current density *is nearly* constant and the total current *is* constant with height. The *columnar resistance,* a quantity frequently used in this work, is defined as

$$\mathrm{R}(z) = \int_0^z \frac{dz}{\lambda} \tag{9.27}$$

and increases with height. These quantities are plotted schematically as Fig. 9.5.

The increase in the conductivity with elevation implies through Eq. 9.13 an increase in the product $ne\kappa$. Measurements show that the number of ions n (which on assuming one electronic charge per ion gives the product ne) increases to an altitude of about 6 kilometers and then decreases to a value nearly one-half this amount at 14 kilometers. The only explanation for the increase in conductivity above 6 kilometers becomes one in which it is postulated that the mobility of the ions increases with altitude above 6 kilometers to compensate for the decreased numbers of ions. This picture is consistent with the experimental evidence that the large ions are concentrated in the lower troposphere, whereas the smaller ions of higher mobility exist in the upper troposphere.

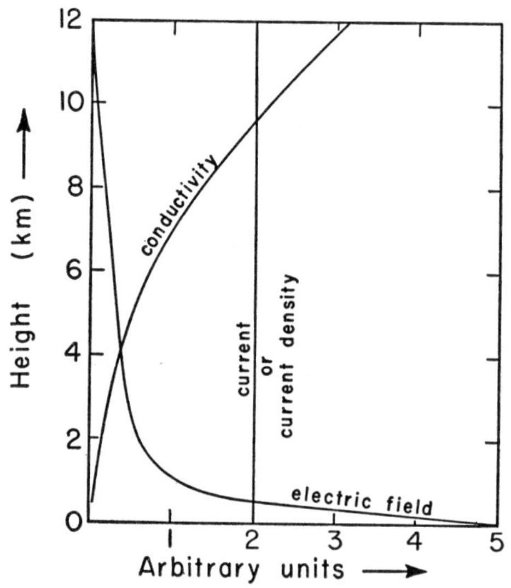

Fig. 9.5 Schematic representation of the trend of the various parameters of the atmosphere's fair weather electrical state as a function of elevation.

We have further stated in an earlier paragraph that large ions are formed by the capture of smaller ions and experimental evidence indicates that near the earth these small ions have on the average an extremely short life. In order for the values of conductivity to fit the observed data on ion counts it is necessary to find a source of ions near the ground and another in the high atmosphere whose rates of production of ions are equal to the rates of destruction of the ions by collision with other ions or by neutralization on contact with the ground. Otherwise, the ion content of the atmosphere would not be in a steady state (an equilibrium concentration of ions), and this is contrary to observation.

It has been established[C5] that the source of ionizing radiation near the ground is radioactive substances in the earth's crust, notably radium and thorium and their two isotopes, the gases radon and thoron. Radium and thorium exist in chemical combination with other substances as salts so that the effect of their ionizing radiations is confined to the air close to the ground. Their radioactive gases, radon and thoron, are exhaled from the ground in minute concentrations and can be carried by turbulence to higher levels before an ionizing radiation is emitted.

In the upper atmosphere, the ionizing radiations are cosmic rays from outer space. These high-energy particles, whose origin in space has not been definitely determined, possess so much energy that they can ionize particles which in turn possess enough energy to ionize other particles forming cosmic ray showers of ionized particles. The showers are produced by relatively few cosmic rays. It has been found that the maximum ionization by cosmic rays occurs at about 12 kilometers. Cosmic-ray ionization becomes less at lower altitudes because of the fewer cosmic rays that penetrate to this level, and less at higher levels because of the fewer number of air molecules in a given path to be ionized (less dense air). However, at the high levels all the ionized particles are small ions. Because of the lower air density and longer mean free paths, these ions possess much higher mobilities than would be measured for similar small ions at sea level.

In any process of ionization a negative ion must be formed from an electrically neutral particle at the same time as a positive ion is produced. Because of this pairing, the strength of ionizing radiations is usually expressed in the number of ion pairs formed per cubic centimeter per second. The quantity is usually indicated by the symbol I. The average rate of ion production by the various ionizing agents forms Tables 9.3 and 9.4.

Table 9.3

A Table of Values for Ionizing Radiations at Sea Level

$$\text{Units of } I = \frac{\text{Ion pairs}}{\text{Cm}^3\text{ sec}}$$

Element	Rays Emitted	Half-life	Ion Pairs Formed at Sea Level: Over Land	Ion Pairs Formed at Sea Level: Over Sea
Radon	α	3.8 days		
Thoron	α	55 sec	5 I	0
Actinon	α	3.9 sec		
Radium	α, β, γ	$\frac{1}{2}$ hr to 22 yr		
Thorium	α, β, γ	1 sec to 1 hr	3 I	0
Actinium	α, β, γ	1 sec to $\frac{1}{2}$ hr		
Cosmic rays	....	....	2 I	2 I
		Total	10 I	2 I

Table 9.4

Typical Variation with Height of Ion-Pair Formation by Cosmic Rays

From data of Bowen, Millikan, and Neher in *Compendium of Meteorology.*(C1)

Height, Kilometers	Station Madras, India Magnetic Latitude (3° N)	Station Omaha, U.S. Magnetic Latitude (51° N)
0	2 I	3 I
6	10 I	15 I
12	20 I	44 I
18	10 I	32 I
24	3 I	13 I

The fair-weather electric field of the atmosphere

Figure 9.5 of the previous section showed the idealized variations of the various electrical parameters with elevation. In any individual set of measurements, one might expect near the ground large variations of the electric-field intensity, yet a relatively constant current density. This observation indicates that the electric-field intensity and the conductivity vary in an inverse relationship. Above 1 kilometer, on the average, there is little variation in the electric field except in the vicinity of electrical storms. At 9 kilometers the electric field is only of the order of 0.05 volt per centimeter. The conductivity increases at such a rate that it appears that the current density is a constant independent of height. In any case, an integration of the curve for the variation of the electric field with height from the ground to the conducting layer will yield the potential difference between conducting layer and ground. A similar integration of the reciprocal of the conductivity curve gives the columnar resistance (see Eq. 9.27).

The columnar resistance between earth and conducting layer, when multiplied by the total area of the earth, shows the atmosphere to have a total resistance of 200 ohms. The current density at the surface of the earth multiplied by the area of the earth indicates an ionic conduction current of 1800 amperes for the entire earth. The corresponding potential difference between earth and conducting layer is 360,000 volts, a value consistent with the difference in potential calculated from the distribution of electric-field intensity with elevation.

The need for a supply current

The discussion of the fair-weather electric field and the flow of ions has proceeded on the tacit assumption that there is in the atmosphere a source of charge to maintain the conduction current and a sink to prevent charge from collecting at one location. In other words, a current equal to but oppositely directed to the conduction current has been inferred. If this latter current, called the supply current, were absent, and only a sink for charge were present, one can estimate the length of time for the conduction current of 1800 amperes to decrease to a value as close to zero as is desired. Should this end state be attained, it would signify a world-wide electric neutrality of the entire atmosphere and the absence of an electric field, a condition not observed.

The problem of calculating the time for the atmosphere to become electrically neutral in the absence of an electric field can be solved by considering that the earth and the conducting layer form two surfaces of a giant condenser. A difference in potential of 360,000 volts exists between these surfaces. The resistance of the atmosphere is 200 ohms, and a current of 1800 amperes normally flows through this resistance. One may then ask how long it will take for the current to fall to a small fraction, say one per cent, of its initial value should the potential difference of 360,000 volts be instantaneously removed.

This problem as stated is an elementary problem in transient currents in electrical networks. We may denote the atmosphere as the equivalent electric circuit shown in Fig. 9.11, where we have a resistor of R_2 units in magnitude shunted by a capacitor of C units. A potential of V units across the resistor and capacitor causes a direct current of i units to flow. Should a switch be opened instantaneously, the potential difference across the resistance-capacitance circuit becomes zero and the current through the resistor falls toward zero, because the energy stored in the electric field of the capacitor must be dissipated as heat in the resistor. As charge stored on the plates of the capacitor must eventually be dissipated through the resistor an instantaneous current i flows through this resistor, indicating that an instantaneous potential difference, iR_2, exists across the resistor. An equal and opposite potential difference must appear across the capacitor because the sum of the potential differences in the circuit must add to zero once the switch is opened. If we define a displacement current i equal in magnitude to the instantaneous current flowing in the resistor, we may rewrite Eq. 9.16 as

$$V = \frac{1}{C}\int_0^t i\,dt. \qquad (9.28)$$

On summing the potential drops through the circuit to zero, we obtain

$$iR_2 + \frac{1}{C}\int_0^t i\,dt = 0. \qquad (9.29)$$

It is easily shown† that for R = R_2,

$$i = \frac{V}{R} e^{-\frac{t}{RC}} \tag{9.30}$$

is the proper solution to the integral equation, Eq. 9.29. One need only substitute Eq. 9.30 into Eq. 9.29 for a proof.

In the atmosphere V/R is the conduction current of 1800 amperes. The product RC is called the time constant of this electrical network and in a time equal to one time constant the instantaneous current i will decay to 1800/e amperes, which is 36.5 per cent of the initial current of 1800 amperes. It takes only 4.6 time constants for the current to decay to 1 per cent of the initial current.

Therefore if we consider the atmosphere to be effectively discharged at the 1 per cent level, we calculate the time necessary to reach this level by setting

$$t = 4.6\mathrm{RC} = 4.6 \frac{\mathrm{R}(z)}{4\pi R^2} \frac{R^2}{z} = \frac{4.6}{4\pi} \frac{\mathrm{R}(z)}{z} \cdot ‡ \tag{9.31}$$

R is the resistance of the atmosphere and C is its capacitance. R(z) is the columnar resistance in esu units and z is the height of the atmospheric layer through which the columnar resistance is measured. R is the radius of the earth. Average values of the quantities at Kew Observatory are $\mathrm{R}(z) = 4 \times 10^{21} \frac{\text{ohms}}{\text{cm}^2} = 4.45 \times 10^9 \frac{\text{esu}}{\text{cm}^2}$ and $z = 18$ kilometers $= 1.8 \times 10^6$ cm. The corresponding resistance for the entire atmosphere at this value of columnar resistance is 730 ohms. The atmosphere as a condenser has a capacity of about 0.25 farad. These values are high for the entire earth, as the resistance is estimated to average 200 ohms. There is some question as to how high the equipotential surface§ forming the upper plate of the condenser should be

† See A. E. Fitzgerald, *Basic Electrical Engineering*, McGraw-Hill Book Co., New York (1945), for an example.

‡ The capacitance of a sphere between two spherical surfaces is equal to

$$C = \frac{1}{\frac{1}{R_1} - \frac{1}{R_2}} \cong \frac{R^2}{z} \text{ esu} = \frac{R^2}{9 \times 10^{11} z} \text{ farads}$$

when R_1 and R_2 are the radius of the surface of the earth and the effective radius of the conducting layer in centimeters. $z = R_2 - R_1$ is the effective height of the conducting layer, determined in this case from the condition that at z

$$\frac{d\mathrm{R}(z)}{dz} \to 0.$$

§ This equipotential surface will be called the conducting layer to distinguish it from the ionosphere proper, an ionized region important at radio frequencies.

taken in the atmosphere, but it seems to be at a considerable distance below the ionosphere as measured at radio frequencies. Measurements have indicated that the columnar resistance is not increased by over 10 per cent in the distance between 18 kilometers and the ionosphere proper at 60 to 100 kilometers. Thus, 18 kilometers rather than the 60 to 100 kilometers that separate the earth from the ionosphere will be taken as the height of the conducting layer. This result means that on the average there is little or no vertical transport of ions above 18 kilometers, the bottom of the stratosphere. Putting in the values indicated for Kew Observatory gives a time constant of about 183 seconds. Taking 4.6 time constants for effectively complete discharge (1 per cent level) of the atmosphere indicates the time for discharge as 14 minutes. Another means for making the same estimate has been given by Scrase.

Scrase[(B1)] estimated the time required for the flow of positive ions from the atmosphere to the earth to neutralize completely the bound negative charge at the earth's surface. At the earth's surface

$$\sigma = \mathrm{j}t = \frac{E_z}{4\pi} \tag{9.32}$$

when average values are used. Solving

$$t = \frac{E_z}{4\pi \mathrm{j}}. \tag{9.33}$$

Average values at Kew are $E_z = 3.65 \text{ volts/cm} = 1.22 \times 10^{-2} \frac{\text{esu}}{\text{cm}}$ and $\mathrm{j} = 1.12 \times 10^{-16} \frac{\text{amp}}{\text{cm}^2} = 3.36 \times 10^{-7} \frac{\text{esu}}{\text{cm}^2}$. The corresponding time taken to neutralize the bound charge, which alternately is to discharge the atmosphere, is 2880 seconds, which is 48 minutes. A similar computation for values over the sea gave 6 minutes. Averaging for the ratio of sea to land areas (3:1) gives a value of t for the entire earth of 17 minutes. Considering the approximate nature of the calculations, the agreement is good between the two methods for the time taken to discharge the atmosphere.

As the atmosphere remains in a charged rather than a discharged state, the inescapable conclusion is that the atmosphere is being charged at the rate given by the conduction current,

$$\mathrm{i} = \mathrm{j}4\pi R^2 = 1800 \text{ amp}, \tag{9.34}$$

where j is the average conduction current density over the earth, a quantity equal to $3.5 \times 10^{-16} \frac{\text{amp}}{\text{cm}^2}$. R is the radius of the earth.

Estimates indicate that only lightning discharges from thunderstorms are capable of creating a supply current of the magnitude required.

It has been estimated that there are between 1800 to 6000 thunderstorms in progress over the world at the same time. For a supply current of 1800 amperes, this requires that each thunderstorm supply between 0.3 and 1.0 ampere of current. Surveys of typical thunderstorms by Gish and Wait[3] indicate that each thunderstorm gave a current ranging from 0 to 6.5 amperes, with an average of 0.6 ampere for 24 storms. The flow of the current was in the proper direction, i.e., upward from the ground. These experimental facts check the hypothesis that the conduction and supply currents are equal and that the site of the supply current is the thunderstorm.

The measurement of thunderstorm electricity

Study of the details of the supply current requires understanding of the electrical nature of thunderstorms; the mechanism of charge formation, the distribution of charge in the thunderstorm, and the mechanism of the lightning discharge. We shall begin our study by defining space charge. If Eq. 9.14 be differentiated in respect to the radius, for a spherically symmetrical system

$$\frac{\partial^2 V}{\partial r^2} = \frac{\partial E_r}{\partial r} = 4\pi \frac{\partial \sigma}{\partial r} = 4\pi q. \tag{9.35}$$

This equation is a specialized case of a more general equation in three dimensions; namely,

$$\nabla \cdot \mathbf{E} = \nabla^2 V = 4\pi q. \tag{9.36}$$

Equation 9.36 is called Poisson's equation and relates q, the electrical charge per unit volume of space, to the divergence of the electric field $\nabla \cdot \mathbf{E}$. q is often called the space charge. In atmospheric electricity, we shall use the less general Eq. 9.35. Use of Eq. 9.35 does not signify that the charge in clouds must be enclosed by a spherical volume only, but rather that the charge can be enclosed by a solid figure, all of whose dimensions are of the same order of magnitude. Electrical effects measured at a distance from the space charge will be almost independent of the details of the charge distribution, so that the charge may be thought of as concentrated in a sphere of radius r, or at a point in space if the distance is very great.

Measurements of the charge distribution in clouds have been made by both direct and indirect methods. The direct method consists of sending instruments into thunderclouds to make an electrical sounding of the cloud. Extensive measurements of this nature were made by Simpson and Scrase[20] in the middle and late 1930's. They used balloons to carry aloft into active thunderclouds an instrument which they called the antielectrograph. The antielectrograph consists of a long wire trailing below and insulated from the balloon. The wire terminates in a series of fine points at one end. In theory, if this wire is in an electric field, there will be a separation of charge along the conductor, the amount and sign depending on the strength of the electric field (see Eq. 9.10). The pointed end of the wire so deforms the electrical field (increases σ at the point) that point discharge occurs. In this process, a wealth of ions is formed as the few ions initially present are accelerated by the electric field and collide with neutral molecules. The corona discharge that occurs can be detected through using paper impregnated with chemicals that will react with ions of only one electrical sign. It was found not only that the paper indicated the direction of the field, but also that the width of the trace could be correlated with the strength of the electric field. Through Eq. 9.35, the latter can be correlated with the space charge. Of course there is a drastic deformation of the undisturbed electric field by the wire, and the results are only semiquantitative. Another obvious difficulty lies in the lack of control on the flight path of the balloon, making a cloud sounding a hit-or-miss affair. Results from these experiments will be quoted after the indirect method of charge observations has been cited.

The indirect method of space-charge measurement requires a knowledge of the effect of the space charge distribution on the electric field at some distance away from the charge itself. Mathematically, this statement is equivalent to saying that a knowledge of the boundary conditions of a system will specify many of the conditions inside the system when the proper physical laws are used in the interpretation.

We shall consider a thundercloud as our system and attempt to deduce the distribution of charge in a thunderstorm from the variation of the vertical component of the earth's electric field with the approach and recession of a thundercloud. Although numerous types of variations occur, it has been substantiated by the many workers in atmospheric electricity in England,[20] South Africa,[13] and other places[17] in the world that by far the greatest number of electric field variations are of the type shown by Fig. 9.6. For the locality, and in the absence of thunderstorms a fair-weather electric field of a characteristic value is established, of the order of 100 volts per meter positive in respect to the earth. As the thunderstorm approaches, the value of the field decreases and may even go negative, the magnitude of the negative value increasing to a maximum at the point of closest approach of the storm. As the storm recedes, the electric field gradually recovers to its fair-weather value.

The typical variation of the electric field on the approach of a thunderstorm can be explained in broad detail through considering the thunderstorm as a giant electric dipole, with the center of positive charge oriented vertically above a negatively charged region. The

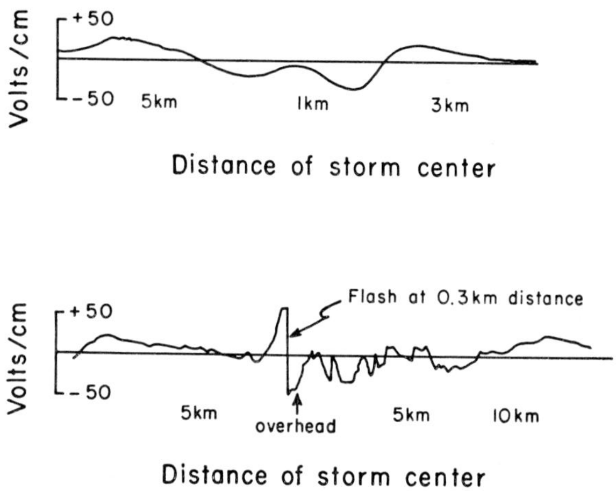

Fig. 9.6 (*After Simpson and Scrase.*) Potential gradient traces at a fixed point on the ground as a function of distance from thunderstorm centers. These two cases were selected as typical, with the upper portion of the cloud positively and the bottom of the cloud negatively charged. The upper trace was obtained from a thunderstorm labeled "small intensity," whereas the bottom figure was from a thunderstorm of "violent intensity."

vertical distance from the earth to the lowest center of charge will be taken as z_1, from earth to upper charge center as z_2. Thus the problem to be solved is to find the magnitude of the vertical component of the electric field measured at O (see Fig. 9.7) from two charge centers of magnitude Q_1 and Q_2 situated a distance z_1 and z_2, respectively, above an equipotential surface, the earth. The distance along the equipotential surface from O to the projection of charge centers Q_1 and Q_2 on this surface will be called x. The diameter of the equivalent spheres containing charges Q_1 and Q_2 will be of the order of the spacing between Q_1 and Q_2, i.e., $z_2 - z_1$.

The solution of the problem indicated in the last paragraph, the geometry of which is shown by Fig. 9.7, is conveniently solved by use of the method of electrical images. Figure 9.8, drawn to one-half the scale of Fig. 9.7, indicates that the shape of the electric field will be unchanged when an equipotential is inserted along the plane of symmetry between the poles of an electric dipole. The earth is such a plane of symmetry (equipotential) for a thunderstorm. In order

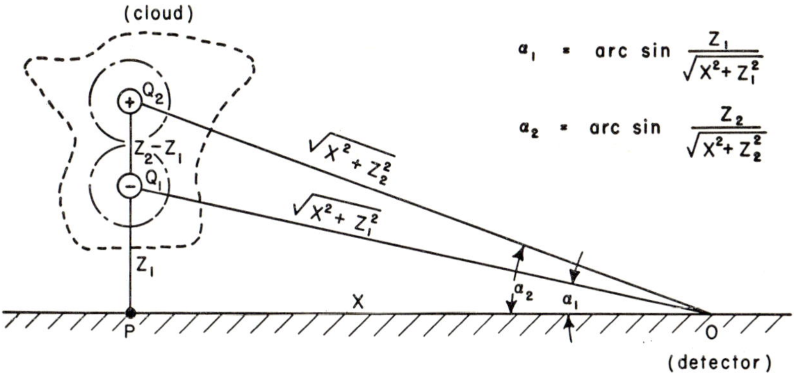

Fig. 9.7 A diagram illustrating the geometry attending the measurement of the vertical component of the electric field at O from a dipolar cloud above P.

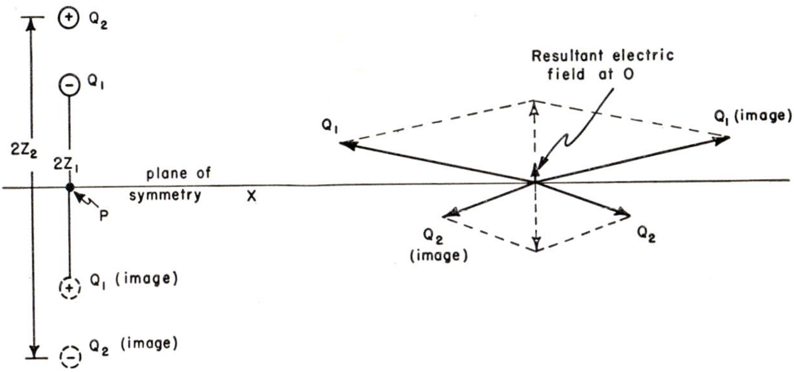

Fig. 9.8 A diagram illustrating the vector electric fields at point O from charges Q_1 and Q_2 and image charges $Q_{1(\text{image})}$ and $Q_{2(\text{image})}$ located respectively above and below point P in the plane of symmetry OP. The solid arrows are the vector field, with the small vertical arrow the vector sum of these fields. The dashed lines and arrows are intermediate steps in the computation of the resultant electric field.

to compute the electric field at point O, the complete dipole must be used; i.e., both the charge center and its image charge below the earth.

Coulomb's law of electrical flux states that the electrical field, a flux density concept, varies as the square of the distance from a point source. Using Eqs. 9.14 and 9.15, with A interpreted as the surface area of a sphere of radius r, we find that

$$E_r = 4\pi\sigma = \frac{4\pi Q}{4\pi r^2} = \frac{Q}{r^2}. \tag{9.37}$$

Equation 9.37 is a statement of Coulomb's law when the electric field is measured in air. It is identical with Eq. 9.1.

The vertical component of the electric field at O from each dipole is twice the value of the field from one pole because both the charge and its image contribute to the field. Thus, for any electrical charge center situated above the earth, the contribution to the earth's electric field is from Fig. 9.7

$$E_z = 2E_r \sin\alpha = \frac{2Q}{r^2}\sin\alpha. \tag{9.38}$$

As $r^2 = x^2 + z^2$ and $\sin\alpha = z/(x^2 + z^2)^{1/2}$, the electric field of the earth from the lower charge center is

$$E_1 = \frac{-2Q_1 z_1}{(x^2 + z_1^2)^{3/2}} \tag{9.39}$$

and from the upper charge center is

$$E_2 = \frac{+2Q_2 z_2}{(x^2 + z_2^2)^{3/2}}. \tag{9.40}$$

The signs of Eqs. 9.39 and 9.40 are opposite, because Q_1 is of the same order of magnitude as Q_2 but opposite in charge. Q_1, the lower charge, is usually negative in sign and Q_2, the upper charge, is usually positive in sign. The total electric field of the earth is the sum of all the electric fields from the dipoles. From Eqs. 9.39 and 9.40, the earth's electric field becomes

$$E = E_1 + E_2 = \frac{2Q_2 z_2}{(x^2 + z_2^2)^{3/2}} - \frac{2Q_1 z_1}{(x^2 + z_1^2)^{3/2}}. \tag{9.41}$$

Over the range of x in which this analysis is valid, there is no fair-weather field because the electrical state of the atmosphere is dominated by the thunderstorm, which in itself gives a very strong field. However, at large values of x, Eq. 9.41 fails because the thunderstorm is no longer the dominant factor in producing the earth's electric field. The limit approached for a thunderstorm at large distances is the earth's fair-weather field, not zero. Within the limitations stated, Eq. 9.41 is a useful tool in determining the large-scale charge distribution in clouds. By the proper choice of the parameters Q_1, Q_2, z_1, and z_2, the general features of an electric field trace in the presence of a

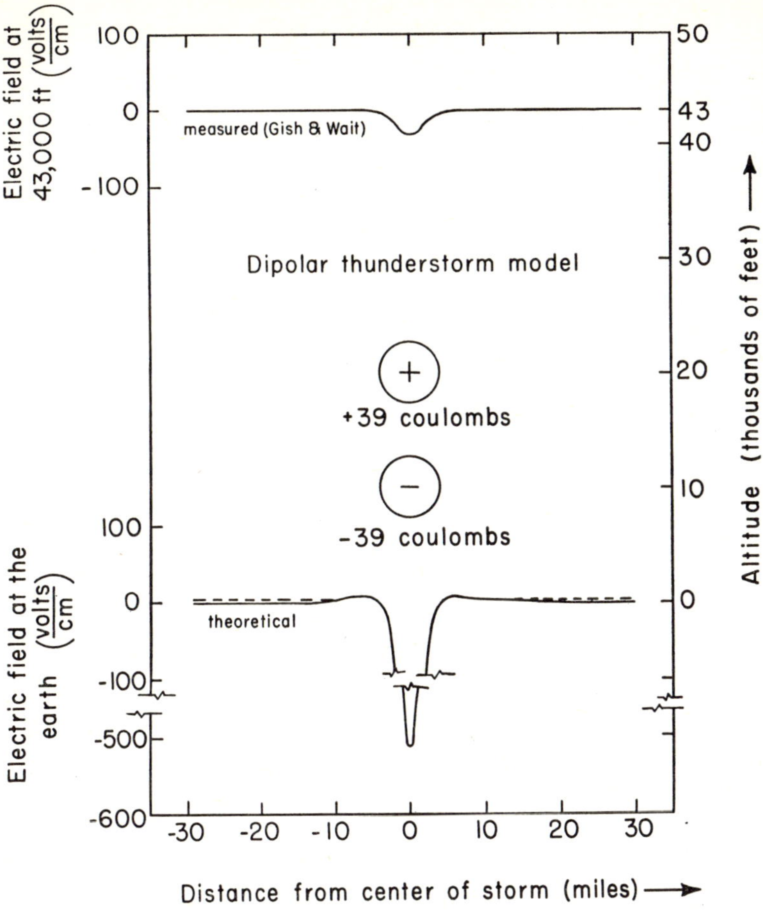

Fig. 9.9 The electric-field variation of a thunderstorm at the ground and at 43,000 feet. The data at 43,000 feet were reported and analyzed by O. H. Gish and G. R. Wait. The data fitted to a high degree of precision the dipole model of a thunderstorm pictured in the insert. Using Eq. 9.41, the corresponding electric-field variation at the ground was computed and is graphed in this figure. At large distances from the storm, the actual boundary condition is the fair-weather electric field (the dashed line) rather than the zero value predicted by the dipole model (solid line). At 43,000 feet, one concludes from actual measurement that the fair-weather field is substantially zero.

The high average value of the field at the ground beneath the storm is close to the breakdown potential of air. Thus, one would expect cloud-to-cloud and cloud-to-ground lightning strokes from the model, but not cloud-to-aircraft.

thunderstorm can be duplicated. For example, in the upper trace of Fig. 9.6, the value of the electric field for the approaching storm is approximately zero at between 2 and 4 kilometers, rising to a maximum value of 25 volts per centimeter near 3 to 6 kilometers. Thus, by setting $Q_1 = Q_2$ and $z_2 - z_1 = 1$ kilometer ($1 \text{ km} < z_2 - z_1 < 4 \text{ km}$, on the average) the salient features of the trace in this region can be obtained by setting $E = 0$ when $x = 3$ kilometers, and $E = 25$ volts per centimeter when $x = 5$ kilometers. Solving Eq. 9.41 under these conditions, we find that $z_1 = 1.7$ and $z_2 = 2.7$ kilometers. The charge $Q_1 = Q_2 = 43$ coulombs. In this example, values of the field at greater than 8 kilometers' distance are dominated by the fair-weather

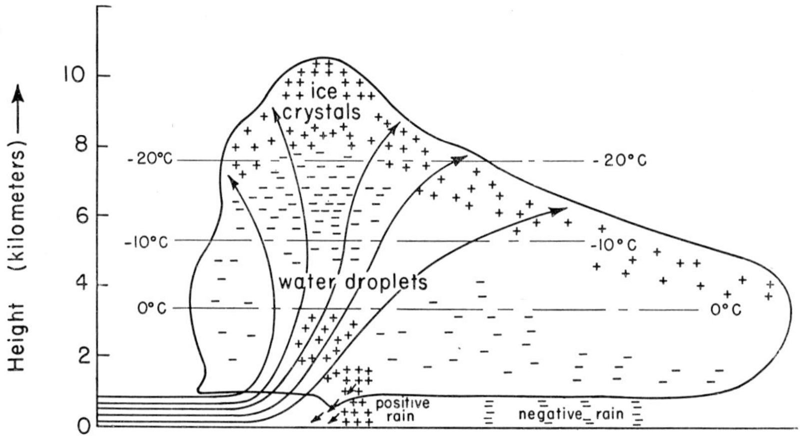

Fig. 9.10 (*After Simpson and Scrase.*) Generalized diagram showing air currents and distribution of electricity in a typical air-mass thunderstorm cloud.

field and for values less than about 2.5 kilometers, the dipole analysis becomes too crude to give correct results. Figure 9.9 shows the results of the field changes obtained from a properly instrumented airplane flying over a thunderstorm and the magnitude of the electric-charge centers in the cloud as deduced from these data.

Data that have been accumulated on the charge distribution in clouds indicate that in a cloud of large vertical development there are usually two distinct regions of charge. In nearly every case, the cloud is negatively charged (a net negative charge) at temperatures warmer than the melting point of ice. A positive center of charge (net positive charge) does not appear as a usual rule until there is evidence that the top of the cloud contains ice crystals. It is in this region of ice crystals that the cloud is positively charged. The region of supercooled water in between the 0° C isotherm and the center of positive charge is predominantly negatively charged. The evidence therefore favors the conclusion that on the average, net negative charge exists in regions of clouds that are composed of liquid water droplets irrespective of temperature and that the net positively charged parts of clouds are in regions where ice crystals predominate (see Fig. 9.10). It is to be understood that this arrangement is only typical, and that individual clouds show complex arrangements of charge distribution. In fact, many clouds have a region of positive charge near the base of the cloud in the region of heavy rain. Moreover, Gunn[(5)] has found that cellular regions perhaps 2 kilometers across exist in active thunderclouds where all the raindrops carry electrical charge of the same sign. Below freezing levels he found high values of electrification per drop and a mixture of positive and negatively charged drops in the same volume of air. Table 9.5 presents a summary, after Gunn, of the electrical charge carried by precipitation particles inside an active thundercloud. It must be emphasized that his instrument measured the charge on raindrops, graupel, or snow, and not the charge on the much smaller cloud particles.

Table 9.5

Electrical Charge Carried by Precipitation Particles inside Active Thundercloud of July 24, 1945

(After R. Gunn)[(5)]

Altitude, feet	Temperature °C	Positive Droplets		Negative Droplets		Particle Space-Charge Density		Total Droplet Density, droplets/cm³
		Number	Charge, esu/drop	Number	Charge, esu/drop	Positive esu/cm³	Negative, esu/cm³	
5,000	+14.7	89	+0.081	171	−0.063	5.61×10^{-5}	3.14×10^{-5}	4.8×10^{-4}
7,500	+10.3	43	+0.280	47	−0.267	15.5×10^{-5}	8.98×10^{-5}	3.2×10^{-4}
10,000	+6.75	71	+0.148	133	−0.112	4.9×10^{-5}	2.88×10^{-5}	2.5×10^{-4}
12,500	+2.4	0	...	196	−0.135	0	7.0×10^{-5}	4.42×10^{-4}
15,000	−0.71	130	+0.123	150	−0.0766	8.5×10^{-5}	4.8×10^{-5}	5.7×10^{-4}
17,500	−5.5	45	+0.036	79	−0.041	1.54×10^{-5}	1.63×10^{-5}	4.1×10^{-4}
20,000	−9.9	76	+0.052	74	−0.062	1.09×10^{-5}	3.6×10^{-5}	3.43×10^{-4}

Some estimate of the electric-field intensity within this particular thundercloud can be obtained by assuming that the particle space-charge density is an approximation to the actual space-charge density in the cloud. From both experiment and the theory of electrostatic fields we find that it is impossible to determine whether a particular set of equipotential surfaces arise from a surface distribution of charge or from a volume distribution of charge provided that the measurements are made at or away from the external surface of the volume where the charge is concentrated. We find then that we may represent the total electric charge Q within a volume V of surface area A as

$$\mathrm{Q} = \int_0^A \sigma \, dA = \int_0^V \mathrm{q} \, dV. \tag{9.42}$$

When the surface is a sphere of radius r and contains either a uniform space charge of density q or surface charge of density σ, then

$$\sigma \int_0^{4\pi r^2} d(4\pi r^2) = \mathrm{q} \int_0^{\frac{4}{3}\pi r^3} d(\tfrac{4}{3}\pi r^3). \tag{9.43}$$

Solving Eq. 9.43 for σ shows

$$\sigma = \frac{\frac{4}{3}\pi r^3 \mathrm{q}}{4\pi r^2} = \frac{r}{3}\,\mathrm{q}. \tag{9.44}$$

On substituting Eq. 9.44 into Eq. 9.14 we find that the electric-field intensity at the surface of the sphere is given by

$$\mathrm{E}_r = \frac{4\pi}{3} r\mathrm{q}. \tag{9.45}$$

Assuming that r for the thunderstorm reported in Table 9.5 is 1 kilometer and that q is the net droplet-space charge density from the same table averaged between 5000 and 12,000 feet, we find that the electric-field intensity has a magnitude

$$\mathrm{E}_r = \tfrac{4}{3}\pi \times 10^5 \times 3.7 \times 10^{-5} = 15.5 \text{ statvolts/cm},$$

which in practical electrical units is 4650 volts per centimeter. Through the averaging process on Table 9.5, q assumed a value of 3.7×10^{-5} esu per cubic centimeter. From Eq. 9.42, Q becomes

$$\mathrm{Q} = \tfrac{4}{3}\pi r^3 \mathrm{q} = \tfrac{4}{3}\pi (10^5)^3 \times 3.7 \times 10^{-5} = 15.5 \times 10^{10} \text{ statcoulombs}$$
$$= 52 \text{ coulombs.}$$

52 coulombs of charge is of the same order of magnitude as the charge inferred from electric field measurements made outside a thundercloud (see Fig. 9.9 where Q = 39 coulombs). The results of the computation suggest that there are neutralizing charges (probably from many small cloud particles) that were not able to be measured, reducing Q to less than 52 coulombs.

Using the experimental and theoretical evidence indicated in this chapter, it is possible to present a unified picture of atmospheric electricity through the device of an equivalent electrical circuit. The schematic flow of electric current in the atmosphere between the conducting layer and earth, the former being maintained at a large positive potential in respect to the latter, is given by Fig. 9.11. The diagram is nearly self-explanatory. The thundercloud, which on a world-wide basis occupies a vertical column less than 1 per cent of the earth's

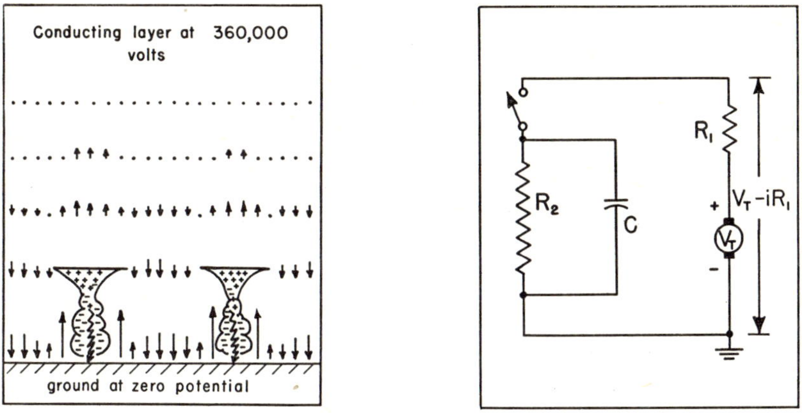

Fig. 9.11 Schematic diagram and equivalent circuit for atmospheric electricity calculations. The thunderstorms all over the earth build up charge to cause a potential difference V_T between the earth and the top of the clouds. A further potential drop of iR_1 occurs between the top of the cloud and the conducting layer to make over the earth $V_T - iR_1 = 360{,}000$ volts. The current i is 1800 amperes. A resistance-capacitance circuit with $R_2 = 200$ ohms and $C \cong 0.25$ farad is the fair-weather field resistance and capacitance. In the atmosphere, the switch is always closed, making C an open circuit for d-c currents. The arrows represent the electric field vectors.

surface, is the generator of electricity and has a constant electrical polarity. There is a voltage drop between the top of the thundercloud and the upper conducting layer through a resistance R_1. All resistances making up R_1 would be measured as columnar resistances and then summed over the appropriate area of the thundercloud. A voltage drop through resistance R_2 occurs between the conducting layer and earth. The value of R_2 for the entire atmosphere is taken to be 200 ohms and is a characteristic of the fair-weather electric field of the earth. The resistance-capacitance circuit represents the aperiodic surge of current when charge is transferred between cloud and ground. Since the current in any series electrical circuit is constant, the average current transferred by lightning throughout the world must equal the conduction current of the fair-weather field. Experiment has shown this requirement to be satisfied because the supply and conduction currents are the same, approximately 1800 amperes for the atmosphere.

Two topics remain to be covered in this elementary discussion of atmospheric electricity. The first topic will investigate the mechanism of the lightning discharge and the second topic will discuss the problem of charge separation in clouds. Toward an explanation of each topic, various theories have been advanced to explain the observed facts and these theories will be discussed. However, it is well for the student to note carefully which statements follow from observation and which statements stem from hypothesis, because the former are not likely to be substantially changed over the years and the latter may not fare so well in the test of time.

The Lightning Discharge. Neglecting for the moment the question of how the high concentration of space charge is generated, the pertinent observations of lightning discharges can be discussed. The information[13] on lightning discharge has been obtained through use of the Boys camera, by electric-field changes, and by radar-scope interpretation.

Historically, the Boys camera has given the main body of information on the visible part of the lightning stroke. In principle, the camera consists of photographic film rotating at high speed past two diametrically opposed lenses. Because of the high rotational rate, any vertical lightning flash having a finite speed of travel will appear on the film with the lower end displaced from the upper end by an amount proportional to the rotational rate of the film. By simple geometric considerations, the speed of travel of the lightning stroke can be calculated. Should multiple strokes exist, the Boys camera separates the strokes so that the number of strokes can be counted. No quantitative estimate of the amount of electricity transferred in the lightning flash is obtained by this means. The results of research with this instrument indicate that lightning flashes are sequenced in the following manner.

1. A dark *leader stroke* is present, visible to the Boys camera, but because of the overwhelming brightness of the main strokes is not observed visually. The leader stroke may be either a *stepped leader* or a *dart leader*. The stepped leader stroke, the more common, proceeds in a series of short paths, separated by a finite interval of time, each path being an extension of the last path. The process continues until the leader stroke makes its way from one charge center to another of opposite sign, either from cloud to ground or from cloud to cloud. It has been found that each step of the leader stroke is of the order of 50 meters, traversing this distance with a velocity equal to 5×10^9

centimeters per second (one-sixth the speed of light). A pause of about 100 microseconds (10^{-4} second) occurs between steps. The velocity of the stepped leader averaged over both the step travel and the pause is of the order of 5×10^7 centimeters per second. The stepped leaders show at least all of the branching that occurs in the main stroke when the latter is of the forked-lightning type. When particularly heavy branching occurs in the upper part of the flash, the stepped leader is called a "type-β leader" in the literature. "Type-α leader" is unbranched in the upper part.

The dart leader does not show steps, but is a continuous stroke from, for example, cloud to ground. The speed of the dart leader is of the order of 2×10^8 centimeters per second, and the leader itself shows no branching. A dart leader precedes each successive stroke of a multiple lightning flash, irrespective of the type of leader stroke preceding the initial flash.

2. The *main stroke* is that which is popularly known as the lightning stroke. It is a flash of such brilliance that it can so oversaturate the retina of the eye as to cause the sensation of light to persist for the order of a second. The camera, however, may show only one main stroke of great brilliance traveling the same path initiated by the leader stroke at a typical velocity of 3.5×10^9 centimeters per second. The main stroke may be and usually is followed by other strokes in the same path, but with each of the multiple strokes showing successively less brilliance and less branching than the preceding stroke. A typical visual lightning flash is a multiple of 3 or 4 strokes, with 10 to 12 not uncommon. The average time between strokes has been measured to be of the order of 0.07 second and the average duration of the entire flash to be 0.25 second. A large number of multiple strokes will give a longer apparent flash.

Coincident with the lightning strokes as measured by the Boys camera are characteristic electrical field changes at the ground. From the measurements of the magnitude of the electric field, quantitative estimates of the amount of charge transferred by the lightning strokes can be made. It has been estimated that a lightning flash (the sum total of the lightning strokes) represents during each stroke an average current of 20,000 amperes of electricity with peak currents of the order of 200,000 amperes. In general, about 6 microseconds are required to reach peak current value and about 24 microseconds to fall to one-half the peak value. Still using average values, we find that about 30 coulombs of electricity are affected by the lightning flash, of which 10 coulombs are utilized in neutralizing space charge while the other 20 coulombs are transferred from one region to the other.

These figures enable us to make some estimates of the supply current needed to complete the electrical circuit of Fig. 9.11. It has been estimated that 2000 to 6000 thunderstorms are in progress over the world at every instant, representing a worldwide total of lightning flashes of the order of 100 to 300 discharges per second. At 20 coulombs per discharge, calculation shows a total current of 2000 to 6000 amperes from these storms. Only about 1800 amperes of this current is needed as the supply current from ground to cloud. The remaining current over 1800 amperes represents the charge transferred between sections of clouds in cloud-to-cloud lightning. One deduces that there is more cloud-to-cloud than cloud-to-ground lightning, perhaps in a ratio as great as 2 or 3 to 1.

The detailed changes in the electric field can be made to yield an estimate of the height at which each lightning stroke in a multiple flash originates.(15) Theory indicates that electrostatic field changes may vary inversely as the cube, the square, or as the first power of the distance from the site of the lightning discharge, the type of variation being a function of the electrical property causing the field change. The electric moment of the lightning discharge causes the electric field to vary as $1/x^3$; the electromagnetic induction field arising from a changing current of electricity in the lightning stroke varies as $1/x^2$, while the acceleration of charge causes a radiation field that varies as $1/x$. Thus, at short distances the electric dipole moment of the cloud and subsequent lightning flash is the principal contributor to the electric field. Equation 9.41 is an example of dipole variation causing an inverse cubic variation in electric field and Fig. 9.12 illustrates the electric-field change from this cause during a lightning flash at 3 kilometers distance. At large distances, the radiation field predominates in the leader strokes (see Fig. 9.12 for the electric field at 15 kilometers) but the influence of the radiation field is absent from the main and successive multiple strokes.

In Fig. 9.12, three types of lightning discharge are illustrated and are labelled as Cases I, II, and III. In each case, the top figure illustrates the visual lightning strokes as registered by a Boys camera. The middle figure illustrates the time variation of the electric field close on to the flash (taken at 3 kilometers); the bottom figure illustrates the changes in the far electric field (at 15 kilometers).

In Case I, the leader stroke is of the stepped leader type. Its path is deduced as being the end points of the leader in the camera representation and as the wiggles in the curve of the far field representation (all figures in a vertical array have the same time scale). Absence of the wiggles from the near-field representation shows that the electrical effects of the stepped leader are unimportant at short distances. However, at short distances the field becomes more negative in a smooth fashion, suggesting that the dipole moment of the leader stroke is changing at a uniform rate. From this evidence, a *pilot streamer* not visible to the camera but electrically significant is postulated. The speed of the pilot streamer corresponds to the average speed of the stepped leader plus delay times and is the smooth envelope of the ends of the stepped leader strokes. When the main and multiple strokes occur (the heavy lines in upper figure) there is

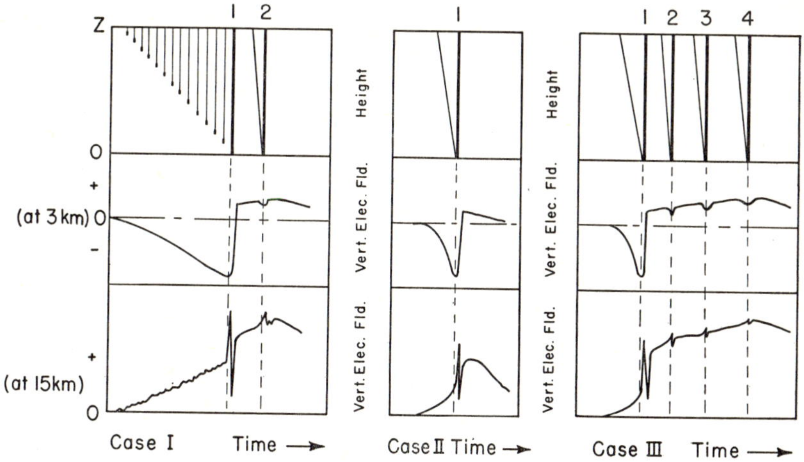

Fig. 9.12 (*After Malan and Schonland.*) The lightning flashes as recorded by a Boys camera, with the corresponding near and far field changes in the earth's electric field. Three typical cases are presented.

a characteristic change in both the near and far electric fields. These characteristic changes are shown in the figure at the position of the dashed lines and are the electromagnetic induction field effects.

Cases II and III are similar, but a dart rather than a stepped leader stroke is shown. In these cases there is no pilot streamer. The strokes are numbered in each case, with number 1 being the main stroke. The correlation of the stroke number with the time that the stroke influences the electric field leads to Fig. 9.13. From Fig. 9.12, Case III, careful scrutiny will show that in the near-field representation the influence on the field of each successive stroke becomes longer than the effect of the previous stroke. This effect is characteristic. Under the assumption that the speed of each stroke is a constant, it follows that the height at which each stroke originated must be higher than that of the stroke it succeeds. Theory provides the numerical values for which Fig. 9.13 is drawn. This figure is not a plot of the three cases in the preceding figure.

The origin of the second stroke (the first measurable one) is in the neighborhood of — 5° C on the average and no strokes come from regions lower in temperature than about — 40° C. This result is in agreement with investigations by radar[23] and suggests that the phase state and the change in phase state (from undercooled water to ice) may have a bearing on the charge-generating mechanism of the thundercloud. This point will be further discussed.

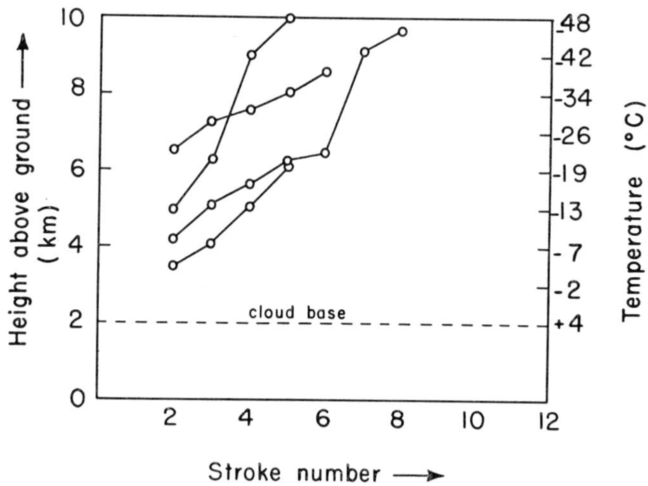

Fig. 9.13 (*After Malan and Schonland.*) The measured increase of the height at which the successive strokes of a lightning flash originate. The ground elevation was 1.8 kilometers above mean sea level (British South Africa).

The lightning discharge is visualized to proceed in the following manner.

By the buildup of charge in a thundercloud, high electric fields are created. Local projections and space-charge configurations can cause high local values of the electric field, much higher than those that would be measured conventionally. The high field that exists will accelerate the ions that are ever present in the atmosphere. In the course of the acceleration more ions will collide with other neutral air molecules, forming new ions, the process as it continues becoming an avalanche effect. In general, the positive ions will be heavier than the negative ions (free electrons) and therefore will have lower mobilities. As a result, a positive space charge will be created that will continue to grow as more air molecules are ionized by the fast-moving negative ions. Finally, a conducting path is formed between two charged centers or a charged center and ground. The bulk of the charge in the charged center is neutralized, with an at-

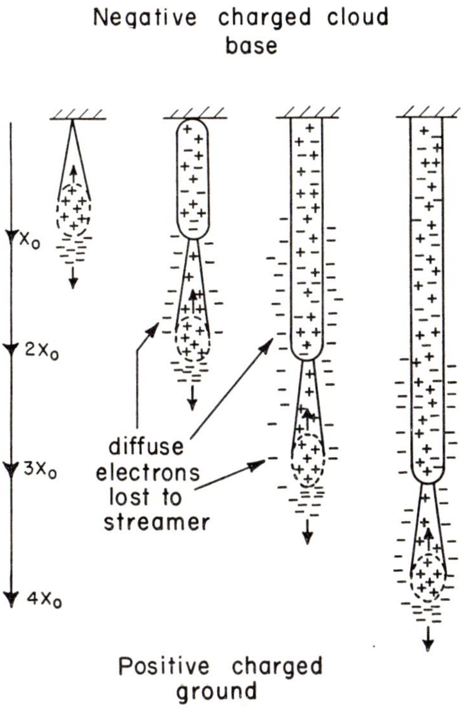

Fig. 9.14 (*After Loeb.*) Schematic diagram illustrating a streamer mechanism from a negative cloud base. The electron avalanche precedes the positive space charge into the gap between the negative cloud base and positive earth, advancing respectively distances x_0, $2x_0$, $3x_0$, $4x_0$, . . . nx_0. A streamer of positive charge moves backward along the conducting plasma (the ionized channel) counter to the motion of the advancing electron avalanche. When the electron avalanche approaches sufficiently close to the ground (the anode) a positive streamer may move from the anode, completing the conducting channel from cloud to earth. A discharge of the cloud, the lightning flash, follows immediately.

tendant flash of light. The ionization in the path after the first stroke may still persist. Since the centers are not completely neutralized in the first discharge, subsequent discharges may take place. The process described, involving the leader stroke, the main stroke, and additional multiples, is analogous to the breakdown of the dielectric of a condenser when too great a field is put across the condenser. An arcing condenser is known as a spark gap.

In air, fields as low as 200 volts per centimeter will cause a corona, or glow discharge. This is especially likely to occur at the masts of ships (St. Elmo's fire), at radio antennas, at wing tips of aircraft, or at points of any type. The high local field accelerates the few ions, causing collision and the creation of new ions. Thus, a corona discharge is a source of ions under relatively low potential gradients. The light or glow that is noticed is a result of positive ions becoming neutralized by electrons. There is a release of energy which can be shown by quantum theory to be at the frequency of visible light. Such a process will account for the visible light in the leader stroke. Similarly, the lightning flash observed visually involves the same process but on a grand scale. It has been found that electric sparks do not occur in dry air at sea-level pressure and temperature until fields of 30,000 volts per centimeter, the breakdown potential, have been attained. As the breakdown potential of air decreases with a decrease in pressure, an increase in moisture, and in the presence of large particles that may elongate under electrical tension (clouddrop, raindrop, or ice needles) it is assumed that lightning does not occur until fields near 10,000 volts per centimeter are created. As mentioned before, fields of this magnitude are not measured at the ground, where measurements seldom exceed 300 volts per centimeter. This is not surprising, because the path taken by a lightning bolt is not large in area and is directed toward projecting points on the ground. On the other hand, electric field meters are usually set up in regions where the field is expected to have minimum distortion. Figure 9.14 illustrates the successive steps in the creation of an ionized channel.

Theories for the separation of charge in thunderstorms

There are several theories for the separation of charge in thunderstorms, and it is conceivable that all may enter to some degree in the charging process. These theories will not be discussed in historical order, but in an order which seems to the author to be in the sequence that might occur in a thundercloud.† In this somewhat subjective treatment, the first theory of the separation of charge to be discussed will be that attributable to C. T. R. Wilson of cloud-chamber fame.

In the Wilson[22] theory, it is assumed that an electric field, ions, and cloud droplets of various sizes coexist. All these assumptions are justified in a young cumulus cloud that is growing vigorously. A large

† Some very serious objections to these theories are presented by Professor P. A. Sheppard.[18] In view of the lack of an affirmative substitute for these mechanisms, admittedly inadequate, a conventional discussion is given.

cloud drop, falling in respect to the rising air, is initially electrically neutral, but it is polarized with the positive charge concentrated on the side of the drop facing the negative side of the field, the negative charge facing the positive side of the field (see Fig. 9.15). In falling

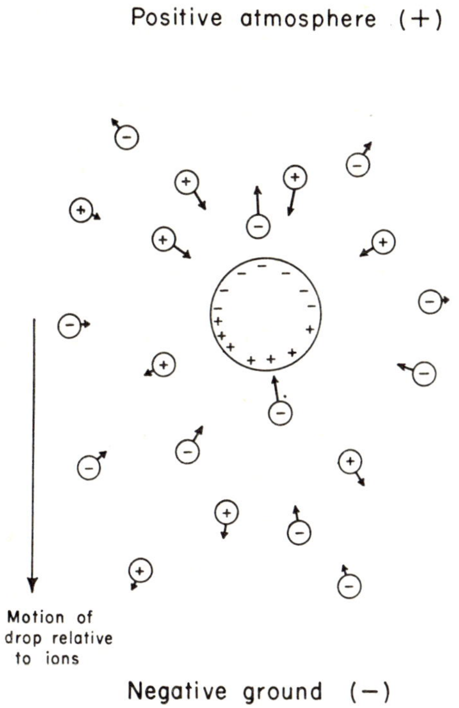

Fig. 9.15 Schematic drawing of the Wilson mechanism for charging drops. A drop originally neutral is polarized by the action of the earth's electric field. When a high relative fall velocity between drop and ions exists, only those ions in the path of the drop will tend to be captured. Because of electrical attraction, more negative than positive ions will be captured.

through air that contains both negative and positive ions, the collection efficiency of the drop will be greater for the ions opposite in sign to the concentration of charge on the leading or windward surface of the drop than to the ions of the same sign. The collection efficiency of the trailing or leeward surface of the drop is poor for both classes of ions because the drop is moving away from the ions and because of an adverse flow pattern of the air. However, ions with sign opposite to that of the trailing surface of the drop will be slightly favored for capture. Consideration of these factors leads one to the conclusion that a falling drop initially neutral should become charged. The amount of charge gained cannot exceed a value determined by the size of the drop and its dielectric constant, but this value is probably seldom attained. It would be expected from a consideration of this process that all drops would not have the same sign at any level in the cloud but that there would be a tendency for the larger drops to become more negative (because of the normal fair-weather field of the earth when the process started) and concentrate in the lowest section of the cloud. The smaller drops will tend to be lifted to higher levels in the cloud and in the process collect the excess of positive ions left by the larger droplets. It is important always to remember that we are talking about net charges and that all degrees of ionization and deionization by the proper sequence of ion capture are possible. The fact that there has been some theoretical difficulty in obtaining high enough rates of charge separation to explain the values of charge distribution observed in mature thunderclouds indicates that this process probably is not the important process in the mature thundercloud when lightning strokes are observed.

The second theory, advocated by Simpson,[(19)] makes use of the Lenard effect. It was found by Lenard that drops on being shattered were left with a predominantly positive charge. The possibility always exists that the smaller drops which also form on impact were negatively charged but that they evaporated or floated away in the air before their charge could be measured. In support of the breaking drop theory, it is known experimentally that for drop sizes much in excess of 5 to 7 millimeters surface tension will not hold a drop together under oscillations set up when the drop falls through the air. Raindrops are never observed with sizes greater than this value; 1 to 3 millimeters is a common value. It has also been observed that rain hitting the ground is predominantly positively charged as is the base of many clouds in the rain area. Simpson's theory for the production of charge by breaking raindrops appears to be reasonable explanation for the charge on rain and at the base of the cloud.

Both mechanisms mentioned have been subject to criticism.[(18)] The Wilson theory, although explaining the sign of the observed charge distribution, fails to meet quantitatively the space-charge density of electric charge observed in clouds. On the other hand, the Simpson theory, although showing quantitative agreement with observation, leads to the wrong sign of the charge distribution in clouds and as a result has been largely discarded as a primary charge-separating mechanism.

Experimentation[24] indicates that the freezing of supercooled water can lead to large charge separations. This experimental fact is receiving considerable attention because it has been observed that the centers of positive and negative charge in a thundercloud are located in the ice and liquid-water sections of a cloud, respectively.

The evidence at the moment is conflicting. It is generally conceded that ice tends to assume a positive charge when charge has been separated by friction, i.e., the rubbing action of blowing snow. Also, experiments as far back as the early nineteenth century showed that water on freezing assumed a positive charge. However, recent experiments by Workman and Reynolds on the freezing of very weak solutions of salt (nearly pure water) indicate that such water assumed a negative charge on freezing to ice. In all cases, it is agreed that the separation of charge is great enough to explain quantitatively the amount of charge required in a thunderstorm. A theory of the mechanism for producing the observed charge distribution in thunderclouds will be contingent on the electric charge that ice assumes when undercooled water freezes. Further work is needed to determine this point.

Miscellaneous Effects of Thunderstorms. Thunder and lightning have terrified man and beast for ages. This terror of lightning has resulted in much superstition and symbolism. For example, Zeus, one of the primary Greek and Roman gods, is often shown as hurling thunderbolts as a symbol of his power. The burning bush that is not consumed by the fire in the Bible and St. Elmo's fire at the mastheads of ships are now taken to be evidences of corona discharge. Even today such manifestations can be awe-inspiring spectacles; in less sophisticated ages they were inevitably the object of superstition. Furthermore, explanations can be given for the following natural manifestations so puzzling to the ancients:

Lightning Forms. The lightning strokes that we have discussed are true of all lightning. Other forms of lightning are sheet or heat lightning, wherein the entire sky appears to light up. The explanation for the general illumination of cloud and sky lies in the lightning flash from a distant storm being obscured from direct observation by intervening clouds, mountains, or horizon. Usually, the flashes are so far away that no thunder is heard.

Ball lightning has been reliably reported and is a luminous ball of light about the size of a basketball that races over the earth or over the floor indoors. It is believed that ball lightning is a highly ionized pocket of air resulting from the high values of electric field accompanying lightning strokes. Little danger has been reported from contact with ball lightning.

Thunder. It has been estimated that temperatures in the return stroke of a lightning flash momentarily reach 10,000 to 18,000° K. The energy released from deionization causes the adiabatic expansion of the air in the channel of the stroke. Initially, the ionized channel has a diameter estimated as of the order of $\frac{1}{2}$ to 1 centimeter. During the adiabatic expansion, the channel radius increases at a rate of 10^5 centimeters per second, with the same process occurring for every subsequent discharge in the same lightning flash. Since there may be as many as 10 or 20 strokes in a flash separated by intervals of 50 milliseconds, an adiabatic pressure wave is set up that vibrates at audiofrequencies. A 50-millisecond period for the pressure wave represents an audiofrequency of 20 cycles per second. This frequency is in the audible frequency range of the human ear. Of course, no thunderclap is a pure tone, but is a combination of many low frequencies and overtones. Because of echoes and reverberations from local terrain, thunder is most usually heard as a low-pitched, rumbling noise.

The speed of sound is roughly 1000 feet per second, so that the rule that the lightning flash originated at a mile for every 5 seconds of time that elapsed between flash and sound is scientifically correct. However, air is a relatively poor conductor of sound and the absorption of the sound waves is so great that thunder is seldom heard when the observer is over 3 to 5 miles from the flash.

Sferics.[C7] We noted earlier that the lightning stroke radiates pulses of electromagnetic energy because of the strong and rapid changes in the electric field accompanying the lightning discharge. For example, the stepped leader has a period between successive advances of the streamer of about 10^{-4} second. This period corresponds to a frequency of 10 kilocycles per second (kc). The presence of higher harmonics and the roughness of the approximations further indicate that lightning is a cause of sharp bursts of static even at broadcast frequencies (10^3 kc). Other forms of radio static can be caused by snow, rain, and variations in the ionosphere. Precipitation static from snow or rain is caused by direct contact of the precipitation with the antenna and the subsequent transfer of charge from precipitation to antenna. A rapid distortion of the electric field results. Ionospheric variations are often extraterrestrial in origin and will be discussed under the general topic of the upper atmosphere.

In the favorable region of 5 to 30 kilocycles per second, the signal (radio static) emitted by a lightning flash can be detected for long distances by a radio tuned to a frequency in this range. By means of a directional antenna, the azimuth of the lightning flash can be ascertained. When two or more stations separated by a known distance fix the direction from their station of lightning flashes from the same storm, calculation of the geographical position of the storm becomes a simple problem in geometry. Storm detection by these means is known as *sferics*. As the range of storm detection by sferics is of the order of hundreds of miles (much greater than radar), the method has potential advantages for tracking thunderstorms (presumably frontal) across the vast steppes and oceans inaccessible to or uninhabited by mankind.

References

1. Bandel, H. W., "Corona from Ice Points," *J. Appl. Phys.*, *22*, 984–985 (1951).
2. Byers, H. R., "The Thunderstorm," Final Report of the Thunderstorm Project, Washington, D.C. (1949). U.S. Government Printing Office.
3. Gish, O. H., and Wait, G. R., "Thunderstorms and the Earth's General Electrification," *J. Geophys. Research*, *55*, 473–484 (1950).
4. Golde, R. H., "Frequency of occurrence of lightning flashes to earth," *Quart. J. Roy. Meteorol. Soc.*, *71*, 307–308 (1945).
5. Gunn, R., "Free Electrical Charge on Precipitation Inside an Active Thunderstorm," *J. Geophys. Research*, *55*, 171–178 (1950).
6. Gunn, R., "The electrification of cloud droplets in non-precipitating cumuli," *J. Meteorol.*, *9*, 397–402 (1952).
7. Hogg, A. R., "Air-Earth Current Observations in Various Localities," *Archiv für Meteorol., Geophys., Bioklimatol.*, *3*, No. 1–2, 40 (1950).
8. Hutchinson, W. C. A., "Point-discharge currents and the earth's electric field," *Quart. J. Roy. Meteorol. Soc.*, *77*, 627–632 (1951).
9. Hutchinson, W. C. A., and Chalmers, J. A., "The electric charges and masses of single raindrops," *Quart. J. Roy. Meteorol. Soc.*, *77*, 85–95 (1951).
10. Israël, H., "The diurnal variation of atmospheric electricity as a meteorological-aerological phenomenon," *J. Meteorol.*, *9*, 328–332 (1952).
11. Kuettner, J., "The electrical and meteorological conditions inside thunderclouds," *J. Meteorol.*, *7*, 322–332 (1950).
12. Loeb, L. B., "The Mechanism of Lightning Discharge," *J. Franklin Inst.*, *246*, 123–147 (1948).
13. Malan, D. J., and Schonland, B. F. J., "Progressive Lightning IV," *Proc. Roy. Soc.*, (A), *191*, 485–503 (1947).
14. Malan, D. J., and Schonland, B. F. J., "An Electrostatic Fluxmeter of Short Response-time for Use in Studies of Transient Field-changes," *Proc. Phys. Soc.*, (B), *63*, 402–408 (1950).
15. Malan, D. J., and Schonland, B. F. J., "The Distribution of Lightning in Thunderclouds," *Proc. Roy. Soc.*, (A), *209*, 158–177 (1951).
16. Nolan, J. J., "The Control of the Electrical Conductivity of the Lower Atmosphere," *Archiv für Meteorol., Geophys., Bioklimatol.*, *3*, No. 1–2, 17 (1950).
17. Reynolds, S. E., and Harris, C. W., "Thunderstorm Electricity," Final Report under Contract W36-039-sc-38196, April 10, 1951. New Mexico School of Mines (unclassified).

18. Sheppard, P. A., "The cumulonimbus and thunderstorm: Part II: Electrical phenomena," *Science Progr., 38,* 488–496 (1950).
19. Simpson, G. C., "The Mechanism of a Thunderstorm," *Proc. Roy. Soc.,* (A), *114,* 376–401 (1927).
20. Simpson, G., and Scrase, F. J., "The Distribution of Electricity in Thunderclouds 'I'," *Proc. Roy. Soc.,* (A), *161,* 309–352 (1937).
21. Simpson, G., and Robinson, G. D., "The Distribution of Electricity in Thunderclouds 'II'," *Proc. Roy. Soc.,* (A), *177,* 281–329 (1941).
22. Wilson, C. T. R., "Some Thundercloud Problems," *J. Franklin Inst., 208,* 1–12 (1929).
23. Workman, E. J., and Reynolds, S. E., "Electrical Activity as Related to Thunderstorm Cell Growth," *Bull. Am. Meteorol. Soc., 30,* 142–144 (1949).
24. Workman, E. J., and Reynolds, S. E., "Electrical Phenomena Occurring during the Freezing of Dilute Aqueous Solutions and their Possible Relationship to Thunderstorm Electricity," *Phys. Rev., 78,* 254–259 (1950).

Source Books

B1. Chalmers, J. A., *Atmospheric Electricity,* Oxford, Clarendon Press (1949).
B2. Fleming, J. A., editor, *Physics of the Earth—VIII, Terrestrial Magnetism and Electricity,* McGraw-Hill Book Co., New York (1939).

Articles from the *Compendium of Meteorology,* American Meteorological Society, Boston (1951). T. F. Malone, editor.

C1. Gish, O. H., "Universal Aspects of Atmospheric Electricity," 101–119.
C2. Gunn, R., "Precipitation Electricity," 128–135.
C3. Hagenguth, J. H., "The Lightning Discharge," 136–143.
C4. Israël, H., "Instruments and Methods for the Measurement of Atmospheric Electricity," 144–154.
C5. Israël, H., "Radioactivity of the Atmosphere," 155–161.
C6. Wait, G. R., and Parkinson, W. D., "Ions in the Atmosphere," 120–127.
C7. Wanta, R. C., "Sferics," 1297–1300.

Problems

9.1 The average fair-weather electric field of the earth has been found to vary according to the empirical relationship

$$E = 90e^{-3.5\times10^{-3}z} + 40e^{-0.23\times10^{-3}z}.$$

E is in volts per meter and z is in meters. (*a*) Find the potential difference between the 10-kilometer level and ground. (*b*) What is the maximum potential difference between the atmosphere and ground as predicted by this equation? (*c*) Other methods have estimated the maximum potential difference between atmosphere and ground as 360,000 volts. Is this answer consistent with the direct measurements of the potential gradient given above? What factors might account for differences?

9.2 The average current density in the atmosphere near the ground has been estimated as 3.5×10^{-16} amp/cm^2. (*a*) Assuming the current density throughout the atmosphere is equal to its average value at sea level, find the columnar resistance of the atmosphere as a function of height. Use the expression given in Prob. 9.1 for the variation of the electric field. Plot the results. (*b*) Find the average resistance of the atmosphere as a function of height. (*c*) What is the maximum value of the columnar resistance? Using this value, find the resistance of the atmosphere.

9.3 Using the variation of resistance with height found from Prob. 9.2*b*, find the time constant of the atmosphere as a function of height. Plot the results and note that the time constant converges to a constant independent of height near the surface of the earth. Comment on the magnitude of the time constant relative to values given in the text.

9.4 An ice particle and a water droplet of the same mass coexist at the same electric potential near the top of a large cumulus cloud. The mass of each cloud element is 4×10^{-9} gram. The ice particle is assumed to have the form of a disk 10 microns thick and has a density of 0.8 g/cm^3. (*a*) What is the ratio of the maximum charge that can be carried by the ice particle to that of the cloud droplet? (*b*) Suppose that there are enough of these particles in a cloud to form 1 gram of condensed water per cubic meter of air. Each raindrop has a maximum negative charge of 1×10^{-16} coulomb. What is the net space charge if all the ice particles assume their maximum positive charge and all the water droplets assume their maximum negative charge when the ice particles are 9 times as abundant as the water droplets? When the ice particles and water droplets are in equal numbers? (*c*) Find the magnitude of the minimum electric field required to charge the water droplet. In electrostatic units, the field is related to the maximum charge that can be carried by a droplet of radius r through

$$Q = 3Er^2.$$

9.5 An airplane and a ground station are symmetrically located in respect to the dipolar cloud of Fig. 9.7. The airplane is located $z_2 + z_1$ units directly above the ground station at O. For the problem, assume that $Q_1 = Q_2 = 20$ coulombs and that $z_1 = z_2 - z_1 = 1$ kilometer. (*a*) Is there any difference in the magnitude of the undistorted electric field at the site of the airplane from that at O? (*b*) How large is the electric field at O? At the airplane?

9.6 When a corona discharge occurs from the tip of an insulated conductor a current flows. This fact is utilized to measure rapid changes in the atmosphere's electric field at Kew Observatory. An empirical expression

$$i = a(E^2 - 4)$$

was found that relates the corona current in microamperes to the electric field in volts per centimeter. For positive currents $a = 9 \times 10^{-4}$ and for negative currents $a = 5 \times 10^{-4}$. (*a*) Corona currents as high as 4 microamperes have been measured and are found to follow this law. What is the magnitude of the electric field under this extreme? (*b*) What is the average value of the space charge density in the neighborhood of the discharge point if a positive current of 4 microamperes is observed? Consider the ions that contribute to the corona current to be small ions whose mobility is 1 cm/sec. The collecting point is a spherical surface 1 mm in radius. State the assumptions used in solving the problem.

CHAPTER 10

The Ionosphere and the Ozonosphere

The ionosphere. Introduction

Mention of the role of the ionosphere in the electrical state of the atmosphere was first made in the chapter on atmospheric electricity, where the extreme lower boundary of the ionosphere, like the earth, was considered as a surface of constant potential for direct currents of electricity. Under this assumption, the earth-atmosphere electrical state was likened to a simple d-c electric circuit, with the earth and ionosphere acting as resistanceless wires, or, in the reciprocal sense, surfaces with infinite conductivity.

This point of view has many implications. The most significant implication for our purposes is that a direct current of electromagnetic energy will not penetrate the ionosphere but rather that a current of radiation of such low frequency as to approach for a short interval of time a direct current is perfectly reflected from its lower surface. Using conventional terminology, we may say that for d-c energy, which may be considered as propagated by a wave of infinite wavelength and zero frequency, the reflectivity of the ionosphere $\mathcal{R}$ is unity, the transmissivity τ is zero, the conductivity Λ approaches infinity, and the absorptivity $\hbar$ also approaches infinity.

On the other hand, we know that visible light waves penetrate the ionospheric regions with little absorption and reflection. If this were not so, it would not be possible to observe from the earth's surface planets and stars outside our atmosphere. Light, a limited section of the electromagnetic spectrum, has a wavelength of the order of 6×10^{-5} centimeter and frequency of the order of 5×10^{8} megacycles per second (Mc). For radiation of this wavelength, the ionosphere represents a region the reflectivity of which approaches zero, the transmissivity approaches unity, and the conductivity and absorptivity approach zero.

Since the two examples cited are at different extremes of the electromagnetic spectrum, with the ionosphere being a very important consideration at the low-frequency end and nearly negligible at the high-frequency end, it seems reasonable to expect a zone of transition at some intermediate frequency. With the advent of radio transmission over long distances, experiment and theory showed that the frequency range 1 to 15 megacycles per second (20 to 300 meters wavelength) marked the division between considering and neglecting the influence of the ionosphere on radiant energy. Thus, it follows that studies of light and microwave transmission in the atmosphere can, in the main, neglect the ionized state of the atmosphere, but that short-wave and standard broadcast frequency radio can not.

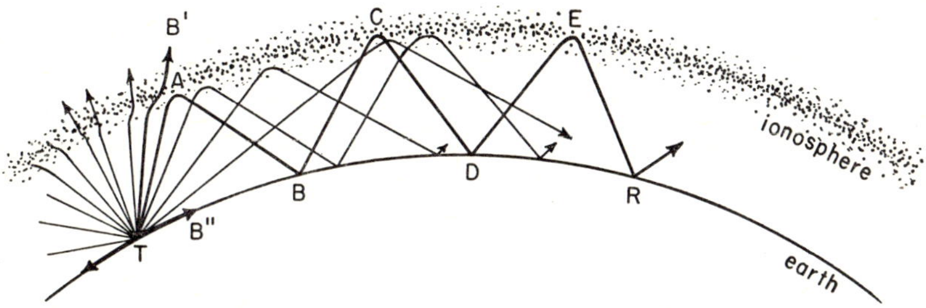

Fig. 10.1 The reflection of radio waves from the ionosphere. Ray *TB''* is the ground wave; ray *TABCDER* is the sky wave. The region between *B''* and *B* receives neither the sky nor the ground wave and is therefore called a skip zone.

For long radio waves, greater than 300 meters, long-distance radio transmission is possible because the energy emitted by an antenna can be reflected with little loss from both the ionosphere and the earth, so that the ray path followed by a radio wave can be path *TABCDER* (see Fig. 10.1) as well as the direct path *TB''BDR*. If the transmitter is located at *T* and the receiver is at *R*, the former path is that taken by the *sky wave* and the latter by the *ground wave*. Because of high absorption losses as energy is absorbed by the earth from the section of the wave front propagated in the atmosphere next to the earth, the ground wave is quickly attenuated and may never be detected at *R*. The sky wave however may reach *R* as a relatively strong signal. In the critical range of 1 to 15 megacycles per second, a particular frequency may penetrate the ionosphere at a small angle of incidence yet be reflected at a larger angle of incidence, schematically shown in Fig. 10.1. Because of this theoretically implied and experimentally justified fact, there are regions such as the one indicated in the figure where no reception is possible by either the ground or sky wave. These regions are called *skip zones.* It is the purpose of the following paragraphs to indicate the elementary theory, methods of observation, and structure of the ionosphere. The study of the ionosphere dates from 1902, when Kennelly and Heaviside[B5] independently postulated the existence of an ionized layer at high elevations in the earth's atmosphere. The basic theory of the ionosphere was adapted from existing theories of the conductivity of ionized gases by Eccles[11] in 1912, but direct experimental confirmation was lacking until 1925, when Appleton and Barnett[B5] proved the existence of the sky wave by analyzing the interference pattern caused by the difference in phase between the sky and ground wave at a point many miles away from the transmitter.

Basic theory

In an elementary discussion of the ionosphere, we consider the change in velocity of a polarized ray of energy at radio frequencies as the ray passes from a nonionized, nonmagnetic, and nonabsorbing atmosphere with characteristics essentially that of free space into a region where a high concentration of free electrons exists (of the order of 10^4 to 10^6 free electrons per cubic centimeter). This region with its high concentration of electrons is called an ionospheric layer (a misnomer in that the effects to be discussed are produced mainly by the free electrons and not the heavier ions also existing in the region). This region will also be considered primarily as nonmagnetic and nonabsorbing. In order to consider the fine details of ionospheric propagation, the magnetic field of the earth must then be considered and the direction and plane of polarization of the radio wave (determined by the orientation of the emitting antenna, whether vertically or horizontally positioned) must be known.

Absorption of radio-frequency (RF) energy by the ionosphere occurs, but in general these absorption losses are slight and can be neglected. It is only in the lowest ionospheric region, the *D* layer, where absorption is so strong that this layer is considered primarily an absorbing rather than a reflecting layer.

Scattering of radio-frequency energy by the ionosphere also occurs, but the attenuation produced is usually not of practical importance. However, scattering can produce important effects not explainable by ray geometry. One example of its importance is indicated by the fact that radio signals are often received in regions of the earth that by ray geometry are skip zones. Signals are received in these zones because part of a radio signal reflected from the *F* layer of the ionosphere is scattered out of its expected ray path as it penetrates the E layer. This latter layer is nearer the earth than the F layer and it is the scattered signal from the E layer that is detected in a skip zone.

In the absence of a magnetic field, an electromagnetic wave will be propagated through a cloud of free electrons at a phase velocity v. The ratio of c to v defines the index of refraction for the ionized layer and it is found to be

$$m = \frac{c}{v} = \sqrt{1 - \frac{e^2}{\pi M}\frac{n}{\nu^2}}. \tag{10.1}$$

c is the speed of light, v the phase velocity of the wave in the ionized gas, e the charge on and M the mass of an electron, n the number of free electrons in a unit volume of atmosphere, and ν the frequency of the polarized radio wave incident on the layer of free electrons. When the effect of the earth's magnetic field is considered, the index of refraction of the medium assumes two values for the propagation of a single wave frequency of which Eq. 10.1 is one of the values.† The effect of a single frequency being propagated with two different velocities through a single medium is analogous in effect to the optical case of birefringence exhibited when polarized light is passed through certain crystals. The two components of velocity can either be combined to cause elliptical polarization through the phase retardation of one mode of propagation in respect to the other, or the two components may be separated by refraction as in the case where a birefringent crystal is sliced on an angle and recemented together so that a second medium exists between the two pieces. The ray that passes undeviated through the birefringent crystal is called the ordinary ray and the deviated ray is called the extraordinary ray. Similarly, because Eq. 10.1 gives the index of refraction of one mode of propagation both in the presence of and in the absence of a magnetic field, the ray following this law is called the ordinary ray. The second ray, which in the case of ionospheric reflection can be separated in time at the critical frequency (to be defined) from the ordinary ray rather than in space as is done when light passes through a crystal, is called the extraordinary ray. The splitting of a polarized radio wave by the earth's magnetic field is called the magneto-optic effect.

We may use Eq. 10.1 to indicate how a radio wave will be refracted by an ionospheric layer where the electron concentration first increases

† In the special case where the direction of propagation is exactly parallel to the earth's magnetic field (longitudinal propagation) this statement is not true. In this case, m from Eq. 10.1 lies between the two indices.

with height to a maximum of n_0 and then decreases with a further increase in height. Equation 10.1 indicates that an increase in n will cause an increase in the phase velocity v. The phase velocity v will always be greater than c, the speed of light, and this point needs a word of explanation.

Because the index of refraction of the ionospheric layer varies with the frequency of the incident radiation, the ionospheric layer consisting of n free electrons per unit volume is called a dispersive medium. In a dispersive medium, a differentiation must be made between phase and group velocity, a distinction that does not exist in a nondispersive medium. Because energy can be propagated and detected only over a frequency interval centered at a dominant frequency, a dispersive medium propagates this small band of frequencies at different phase velocities. A beat frequency is produced and it is this frequency that is propagated with a velocity called the group velocity. The group velocity u is related to the phase velocity v by†

$$u = c\,\frac{d\nu}{d(m\nu)}\cdot \tag{10.1a}$$

From an inspection of Eq. 10.1a, one can see that since m is constant in a nondispersive medium, $u = c/m$, and the group and phase velocities are equal. If the range of frequencies is small enough so that a meaningful average index of refraction can be taken, we see that $u \cong v$. However, for larger frequency intervals $u \neq v$, and a distinction must be made. With ionospheric propagation, we have the case where $v > c > u$, and the electromagnetic energy is propagated with a speed less than the speed of light.

It turns out that we may use Eq. 10.1,‡ which in reality refers to a phase wave, to predict the *path* through the ionosphere taken by a *group* wave whose dominant frequency is ν. However, the velocity of the wave packet along this path must be the group velocity, and this velocity will be less than the speed of light.

If then, a segment of a plane wave front is propagated along the ray path TA in Fig. 10.2, no deviation from this path will occur as long as $n = 0$. When this wave enters a region where n begins to increase in the upward direction (assuming that ν is a constant) m

† The phase velocity v is included in the definition of the index of refraction m (see Eq. 10.1).

‡ Equation 10.1 also has imaginary values, but these will not be considered, since we will use the equation to denote the path of the group ray. Physically, an imaginary value for the index of refraction means that a phase wave will undergo total reflection.

becomes less than 1, the upper portion of the wave front travels at a greater speed than the lower, and the ray is caused to bend more and more parallel to the earth. The refractive process continues until one of two conditions occurs. The first possibility is that the ray becomes parallel to the earth at a height where dn/dz is still positive.

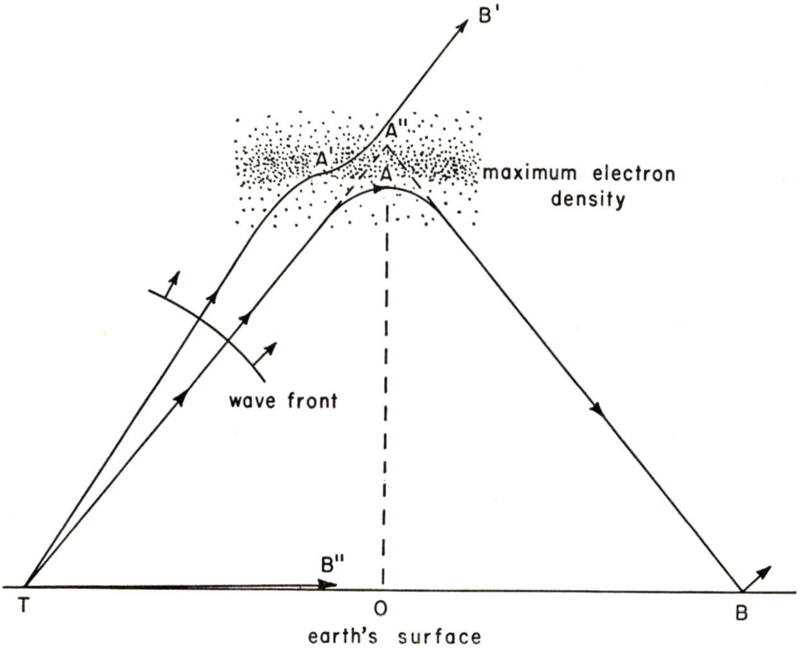

Fig. 10.2 An illustration of the paths followed by two rays of the same wave front incident on the ionosphere at different angles. A'' is the point at which reflection appears to occur, and A is the highest point reached by the ray. Ray TB'' is the ground wave. B'' may or may not extend to point B. If it does not, a skip distance $B''B$ exists. If it does, a receiver at B will detect both a ground and a sky wave.

Under this condition the ray continues to curve past the earth-parallel position and becomes directed more and more toward the earth until it passes out of the ionized layer. This ray AB will return to the earth to be detected or reflected at B. The second possibility is that the ion concentration will begin to decrease with height or in the limit remain zero; i.e., $dn/dz < 0$, before the ray becomes parallel to the earth. Under all but the limiting condition the ray will reverse its curvature, with the lower portion of the wave front moving faster than the upper end. As a consequence, the ray will curve in an arc away from the earth and pass through the ionospheric layer. The path $TA'B'$ illustrates this possibility. This behavior has been observed and has led to a simple picture of the distribution of electrons in an ionospheric layer in which it is assumed that the ions follow

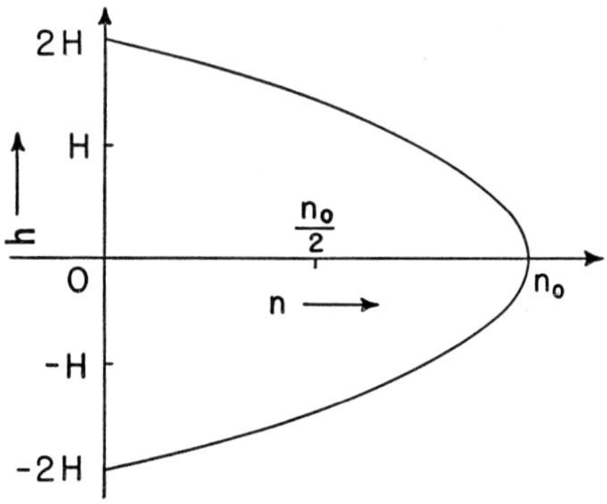

Fig. 10.3 The geometry of the simple Chapman region.

a parabolic distribution of electron concentration with the thickness of the layer. Consulting Fig. 10.3, if h is the thickness of a simple Chapman region, as this ionospheric layer is called, and h is zero where $n = n_0$, the level of maximum electron concentration, then we may represent n as

$$n = n_0\left(1 - \frac{h^2}{4H^2}\right). \tag{10.2}$$

H is called the scale height. From theory†

$$H = \frac{kT}{Mg}. \tag{10.3}$$

M is the mean molecular mass of the layer of scale height H, k is the Boltzmann constant, g the acceleration of gravity and T the absolute temperature. $2H$ is called the semithickness of the layer and represents the distance in which the concentration of electrons goes from essentially zero to the maximum concentration n_0. Values

† The integration of $\left(\frac{\partial p}{\partial z}\right)_{x,y,t} = -\rho g$ for an isothermal atmosphere yields

$$p = p_0 e^{-\frac{Mg}{kT}z} = p_0 e^{-\frac{z}{H}}$$

which gives the identity, Eq. 10.3. p is the pressure at height z and p_0 is the pressure at $z = 0$.

of the semithickness are tabulated in Table 10.2 on page 333. Equation 10.2 is a quantitatively accurate picture to about ± 5 per cent and is the type of model that aids in explaining the refraction of radio waves in the ionosphere. This particular picture of electron distribution arises quite naturally when one considers that the formation of free electrons in the atmosphere is a result of absorption of the ionizing radiations in the ultraviolet spectrum of the sun. Since both air and radiation are required to ionize the air, one would expect no ions but plentiful radiation in the absence of air, and an increase in ionization as the air density increases, until a point is reached where so much radiation has been absorbed by layers above that the increase in air density is no longer important. Beyond this point the ionization decreases to zero as the ionizing radiations are absorbed in the layers of air above the layer under consideration. Such a consideration leads to a region of maximum ionization with a minimum both above and below, as is given by the simple Chapman region representation.

Methods of measuring the equivalent height of the ionosphere

The height of the ionosphere is conventionally measured by reflecting radio waves at either vertical or oblique incidence. The vertical-incidence case will be discussed first, since the theory is the more simple.

When an electromagnetic wave is normally incident on a surface of different index of refraction from the media in which it is initially traveling, reflection occurs. The reflectivity, the ratio of the reflected to incident energy, is a function both of the relative index of refraction between the two media and the absorption coefficients of the media. If we consider the ionosphere as a region where the index of refraction of the atmosphere changes abruptly but without absorption of the wave, Fresnel's reflectivity formula

$$\mathcal{R} = \left(\frac{m-1}{m+1}\right)^2 \tag{10.4}$$

may be used. Complete reflection occurs when $m = 0$ or ∞. Equation 10.4 is valid only for $m \geq 0$, but a complete treatment of the reflection of electromagnetic waves shows that total reflection occurs when m is a pure imaginary number, the case when $m < 0$. Table 10.1 lists some typical reflectivities for land and sea at various frequencies as a comparison to the reflectivity of the ionosphere.

From the condition for complete reflection, we may define a critical frequency ν_c, obtained by setting m in Eq. 10.1 equal to zero. The

Table 10.1

Reflection Characteristics (*at Normal Incidence*) of the Earth's Surface (Dry Land and Sea Water) for Electromagnetic Waves

The numbers are only typical values to indicate trend. All units not labeled are dimensionless. *e* and *s* denote dry earth and sea water, respectively.

Region of Electromagnetic Spectrum	Frequency ν	Wavelength λ		Index of Refraction m	Absorption Coefficient mk	Dielectric Constant ϵ	Conductivity, sec^{-1} Λ	Reflectivity $\mathcal{R}$
Radio	<12 Mc	>25 meters	*e*	2	0.03	4	1×10^7	11%
			s	54	53	81	4.3×10^{11}	96%
Microwave	1900 Mc	16 cm	*e*	—	—	—	—	—
			s	8.7	1.0	75	2.1×10^{11}	63%
Infrared	3.3×10^7 Mc	9 microns	*e*	1.01	0.01	1.02	4.2×10^{12}	0%
			s	1.27	0.05	1.61	2.6×10^{13}	2%
Light	5×10^8 Mc	0.6 micron	*e*	1.01	0.01	1.02	6.3×10^{13}	0%
			s	1.33	0	1.78	$\sim 3 \times 10^{10}$	2%

At small grazing angles all values increase toward 100% reflectivity.

critical frequency is reached when $m = 0$, and $n = n_0$, giving

$$\nu_c^2 = \frac{e^2}{\pi M} n_0. \tag{10.5}$$

As a result, Eq. 10.1 may be expressed as

$$m = \sqrt{1 - \frac{\nu_c^2}{\nu^2}}. \tag{10.6}$$

Equation 10.6 is capable of experimental measurement and guides our interpretation of results.

A radio transmitter of variable frequency and a tuned receiver are set up to emit and detect the vertically reflected waves. The energy is emitted in pulses so that it is possible to measure the delay time t, measured from the instant the pulse leaves the transmitter to the instant the detector receives the pulse (see Fig. 10.4).

Assuming no retardation of the group wave when it is reflected from the ionosphere, one obtains the height of the reflecting layer from the definition of velocity, realizing that the path traveled at the speed of light c, before detection, is twice the height of the ionosphere. We deduce that

$$z' = \frac{ct}{2}. \tag{10.7}$$

z' is called the virtual height of the ionosphere. The pulse of energy on entering the ionosphere slows down from the speed of light to zero, reverses direction, and attains the velocity c again as it leaves the ionosphere for the earth. z' is the maximum height from which the energy could have been reflected, and is obtained through neglecting the slowing down of the wave on reflection. In general, a correction term must be computed to convert virtual height to

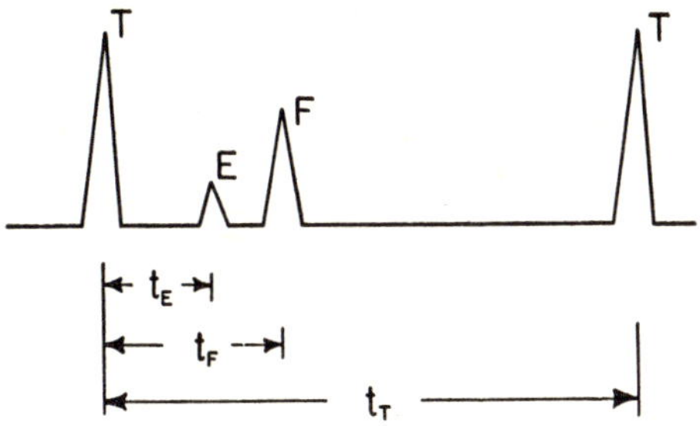

Fig. 10.4 A schematic illustration of a pulse of energy (T) emitted by a transmitter. The time t_T is the pulse repetition rate. The pulse T is detected t_E microseconds later after having been reflected from the E layer, and detected after t_F microseconds after reflection from the F layer.

geometric height z. Theory shows that the level of maximum ionization occurs at a geometric height z_0, equal to the virtual height, provided that the virtual height is measured at a point where

$$\frac{\nu}{\nu_c} = 0.834. \tag{10.8}$$

The geometric height of the bottom of the ionospheric layer is found through measuring the virtual height at a point where for the layer there is a minimum increase in virtual height with increased frequency. That is where

$$\frac{dz'}{d\nu} \rightarrow 0. \tag{10.9}$$

At this point the virtual and geometric heights are again equal. The semithickness $2H$, is then the difference between the virtual heights measured under the conditions of Eqs. 10.8 and 10.9.

Let us then consider Fig. 10.5, which we assume is experimental data derived from measurements of the delay times of pulses sent to and reflected from the ionosphere at normal incidence.

As the frequency is increased, the delay time correspondingly increases, with extremely high delay times at certain critical frequencies. The region giving the shortest delay time we call the E region, the next highest the F_1, then the F_2, up to a G region near 400 kilometers, about which little is known. Each region is separated by a critical frequency of penetration for the region. A D layer, lower than the E region, exists during the daytime, but this layer is an absorbing rather than reflecting layer. The virtual height for each

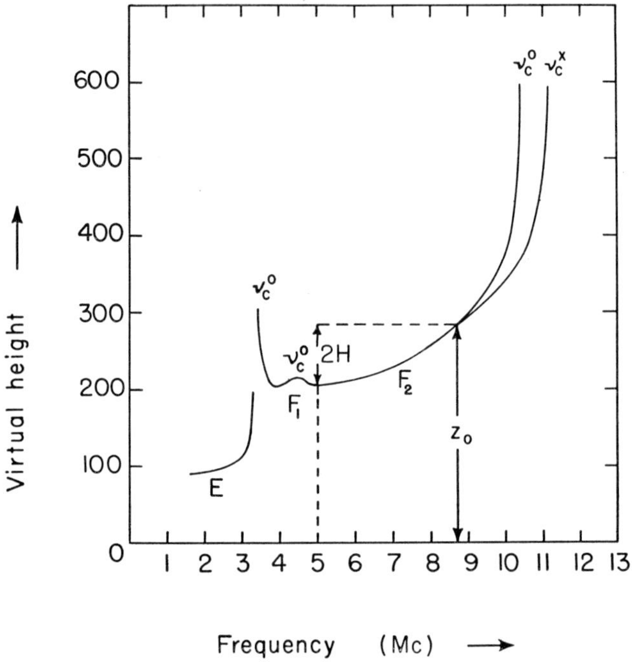

Fig. 10.5 A plot of the virtual height of the ionosphere versus the frequency of a probing radio wave. The critical frequencies for the various layers are labeled as $\nu_c{}^0$ for the ordinary ray, $\nu_c{}^x$ for the extraordinary ray. z_0 is the true height and $2H$ the semi-thickness of the F_2 layer.

layer beginning with the E is computed from Eq. 10.7. The critical frequency for a layer is that point where the virtual height tends toward infinity, and identified as the cusp in the frequency versus virtual height curve. We can see from Fig. 10.5, which may be taken as typical of a radio sounding of the ionosphere, that three critical frequencies exist: one each for the E and F_1 layers and a split critical frequency for the F_2 layer. The critical frequencies are 3.4, 4.5, 10.4 and 11.2 megacycles per second, respectively. Only the 10.4 megacycles per second critical frequency will be considered, as it is to this ray that the conditions given by Eqs. 10.8 and 10.9 apply. The superscripts 0 and x represent the ordinary and extraordinary rays, respectively.

From Eq. 10.8 we find that the true height of the F_2 layer would be measured at 8.7 megacycles per second. The height from Fig. 10.5 is 285 kilometers. The semithickness of the layer is labeled $2H$ on Fig. 10.5 and is found to be 80 kilometers. Equation 10.9 was used. No account of the retarding effect on the radio waves by the E and F_1 layers was considered as the waves passed through. The effect though small would have to be considered in a more detailed examination of the problem.

In a similar fashion, the heights of the F_1 and E layers are respectively 205 and 105 kilometers. The semithickness of the first is not determinable from the figure and for the second is 15 kilometers. Figures 10.6 and 10.7 are plots of the average concentration, height, and semithickness of the various ionospheric regions for day and night, and Table 10.2 lists some additional data. It is common for the F_1 and F_2 regions to merge into a single F region at sunset and persist as a single layer near the height of the F_2 region until dawn. The photochemical action of the sun is responsible for the additional ionization during the day to form the F_1 region. Occasionally, two E layers are noted, an E_1 and a higher E_2.

For the F_2 region, two critical frequencies are noted. After the language of optics, the lower frequency is called the ordinary ray of critical frequency $\nu_c{}^0$ and the higher frequency has a critical frequency $\nu_c{}^x$ and is called the extraordinary ray. The cause for the splitting lies in a factor, the earth's magnetic field, which was discussed earlier. Both the ordinary and extraordinary rays are elliptically polarized both before entering and after leaving the ionosphere.

The second method of measuring the virtual height of the ionosphere follows from a re-examination of Fig. 10.2. Should a transmitter be located at T and the receiver at B, there exists the possibility that a signal emitted from T will travel to B, both as the ground wave $TB''B$ and the sky wave TAB. Because of the shorter distance, the section of the wave traveling as the ground wave will reach B before the section of the wave traveling as the sky wave. The difference in time can be measured accurately. In one method, the signal is emitted in pulses whose duration is short compared to the delay times to be measured between ground and sky waves. In the second method, the more common, the transmitter sends out a wave of slowly varying frequency. The two components of the same frequency, the ground and the sky wave, are recombined at the receiver. On recombination the two components will, in general, be at different phases

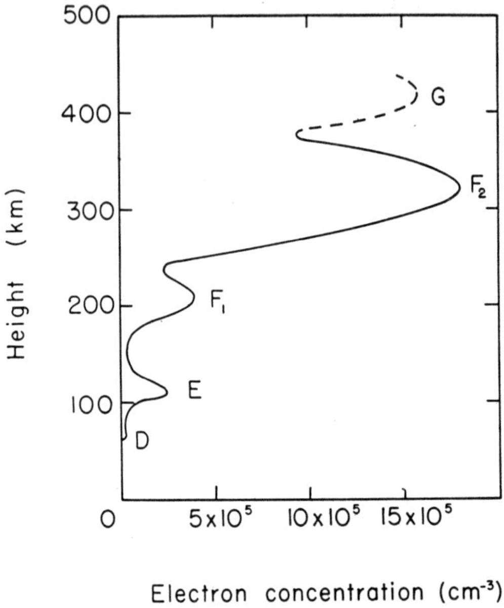

Fig. 10.6 A typical height versus maximum electron concentration profile of the various ionospheric layers by day.

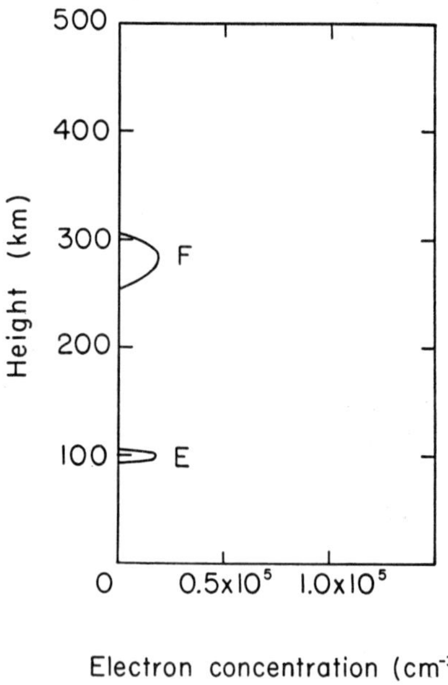

Fig. 10.7 A typical height versus maximum electron concentration profile of the various ionospheric layers by night. Note the concentration scale is increased by a factor of 10 over the scale of Fig. 10.6.

because of the vastly different lengths of path traveled. As a result, we expect that if the difference in path remains nearly constant, and the frequency is varied slowly, the signal will vary alternately from strong to weak as the phases become alternately coincident and totally anticoincident. The principle is the same as the optical interferometer technique of fringe counting. From this information the delay time can be computed. This method is called the frequency-change method.

Table 10.2

Characteristics of the Various Ionospheric Layers

The heights and semithicknesses of the various regions are stated as absolute geometric heights above mean sea level. To obtain approximately the virtual height values of the layers and the virtual semithicknesses, multiply E layer values by the factor 1.1, the F_1 layer by the factor 1.25, and the F_2 layer by the factor 1.4.

Layer	Height (km) z_0		Semi-thickness (km) $2H$	Maximum Electron Concentration (cm^{-3}) n_0		Electron-Positive Ion Recombination Coefficient ($cm^3 \ sec^{-1}$) Γ
	Day	Night		Day $\times 10^5$	Night $\times 10^5$	$\times 10^{-11}$
D	$\sim$60	Absent	Thin	0.15 to 2.0	0	5,000,000†
E	90 to 130	90 to 100	5 to 20	1.5 to 3	0.2	100 to 1000
F_1	160 to 280	Absent	10 to 30	2.5 to 4	0	100
F_2	300 to 350	280 to 300	30 to 100	10 to 30	0.2	1 to 10
G	400 to 500	..	..	..	..	..

† Complete recombination of electrons with positive ions occurs in a few seconds in the D layer.

The information gained by this method of determining virtual heights is substantially the same as that gained by vertical soundings. The actual path traveled in the atmosphere will be TAB of Fig. 10.2 instead of $TA''B$, although both paths are possible in the same time. Path $TA''B$, however, is ruled out because of index of refraction considerations. OA'' is the virtual height and is obtained through considering the ray as traveling at constant velocity c, the speed of light in free space, and reflected from a surface located at A''. It is evident from the geometry that $OA'' > OA$.

In general, radio waves reflected from the ionosphere from other than vertical incidence show higher critical frequencies for penetration, the critical frequency increasing with increasing angle of incidence. The reflected ray (strictly, "refracted") exhibits the elliptical polarization phenomenon of the vertically incident radiation.

Much amateur short-wave radio is on frequency bands reflected by the *E* layer. Communication between certain widely separated stations via the sky wave had been noticed to be highly variable in signal strength and not explainable in terms of the diurnal variation of the *E* layer. Examination of reports from many amateurs, carefully analyzed with respect to position and time, has indicated that the *E* layer has a highly variable horizontal concentration, with regions of high concentration (high reflectivity) separated by regions of low electron concentration (high transparency to radio waves). Further analysis has shown that these regions move as an identifiable entity with speeds of the order of 200 kilometers per hour. If the observations are indicative of the mass motion of the electrons and not an indication of ion formation on the leading edge and dissipation on the trailing edge of the electron cloud, studies of the motion of the "sporadic *E*" layer can yield wind velocities at the 100-kilometer level.

Electron production in the ionosphere

The net rate of production of electrons† in a unit volume of the ionosphere, dn/dt, is equal to the rate of production of electrons less the rate of destruction of the electrons by recombination. By letting

$$\frac{dn}{dt} = Q - \Gamma n^2, \tag{10.10}$$

we interpret Q as the rate of production and Γn^2 as the rate of destruction of the electrons. n is the total number of free electrons per unit volume and Γ is called the recombination coefficient.

Ion pairs are produced principally by the absorption of ultraviolet radiation from the sun by the gases of the atmosphere. For example, atomic oxygen absorbs radiation of wavelengths below 0.0910 micron, molecular oxygen those below 0.175 micron, and molecular nitrogen those below 0.795 micron. With the absorption of ultraviolet radiation and consequent ionization of the gas as it makes the transition to the ionized state, it is assumed that the *D* layer is formed by the ionizing of O_2 gas to $O_2^+ + e$, the *E* layer by the ionizing of

† A positive ion is produced every time an electron is ejected from a molecule, forming in the process an ion pair. In the theory, except for the *D* layer, it is the electrons that contribute to the major effects on radio waves entering the ionosphere.

O_2 gas to $O_2^+ + e$ at a higher ionization potential than in the D region, the F_1 layer by the ionizing of N_2 to $N_2^+ + e$, and the F_2 region by the ionizing of O to $O^+ + e$. The ejection of an electron to form the ion is symbolized by e.

Theory shows that Q is dependent on the zenith angle of the sun measured at the ionospheric layer itself. Therefore,

$$Q = Q' \cos \zeta \tag{10.11}$$

if Q' is the rate of ion production at normal incidence; i.e., when $\zeta = 0$.

At equilibrium, there is no net electron production, making

$$\frac{dn}{dt} = 0 = Q - \Gamma n^2. \tag{10.12}$$

Substitution of Eq. 10.11 into Eq. 10.12 and solving yields

$$n = \left(\frac{Q'}{\Gamma}\right)^{1/2} \cos^{1/2} \zeta = n' \cos^{1/2} \zeta, \tag{10.13}$$

showing that the equilibrium concentration of electrons at any latitude is a function of the zenith angle of the sun. During the daylight hours the E layer follows this equilibrium condition quite closely, because for the E layer dn/dt is small compared to Γn^2 or Q. n' is the maximum number of electrons per unit volume formed when the sun is directly overhead.

The nonequilibrium case occurs when dn/dt does not equal zero. This condition is true during the daylight hours for the F_2 layer especially and for all layers at night. The nighttime variation is especially interesting because once the sun has set and no longer illuminates the ionospheric region under consideration, then $Q = 0$ and Eq. 10.10 reduces to

$$\frac{dn}{dt} = -\Gamma n^2. \tag{10.14}$$

Integration of Eq. 10.14 with Γ constant gives a means of experimentally evaluating Γ, the recombination coefficient. With n_1 the value of n at $t = 0$, and n_2 the value of n at a time t units later, and with both measurements made after sunset, integration of Eq. 10.14 shows that

$$\int_{n_1}^{n_2} \frac{dn}{n^2} = -\Gamma \int_0^t dt, \tag{10.15}$$

which yields

$$\Gamma = \left[\frac{1}{n_2} - \frac{1}{n_1}\right] \frac{1}{t}. \tag{10.16}$$

We see that we now have the elements of a simple theory for predicting and measuring the variation of critical frequency or electron concentration with the time of day. Mathematically, the latter is most conveniently expressed as the zenith angle of the sun. We have Eq. 10.5 that expresses the variation of the maximum concentration of electrons in an ionospheric layer with the critical frequency. By combining this equation with Eq. 10.13 we obtain†

$$\nu_c{}^2 = \frac{e^2}{\pi M}\left(\frac{Q_0'}{\Gamma}\right)^{1/2} \cos^{1/2} \zeta = \text{constant} \times \cos^{1/2} \zeta. \tag{10.17}$$

Equation 10.17 is surprisingly well followed during the period from just after sunrise‡ to just before sunset for the E and the F_1 layer. A plot of Eq. 10.17 would show a minimum critical frequency at sunrise, rising to a maximum critical frequency at local noon (ζ a minimum value), and falling to the same low value at sunset as occurred at sunrise. Throughout the night, the value of ν_c remains relatively constant at the low value given at sunset. The nighttime value is not given by this theory. We also see that the variation of frequency during the typical summer day should be much greater than the variation of the critical frequency during the typical winter day because the sun is at a higher solar altitude (smaller zenith angle) during any given hour during the summer than during the winter. In the same way, we expect greater concentrations of electrons in any layer during the summer day than during a corresponding time of a winter day. Considerable deviation from this typical picture may be expected for any particular set of measurements of the state of the ionosphere. We shall mention a few.

The behavior of the E region most closely follows this simple theory. However, important variations exist. The "sporadic E" layer has been discussed. During meteoric showers, the ionization of the E layer is increased in the vicinity of the meteors. The meteors entering the earth's atmosphere with speeds in the vicinity of 40 km/sec form a high-pressure, high-temperature gas cap at their leading edge. Within this region ionization by collision between the air molecules occurs. This ionization is most marked at the height of the E layer, causing sharp increases or "bursts" in reflected radio signals as the meteor leaves a trail of highly ionized particles in its path.

Three types of ionospheric disturbances are intimately connected with solar disturbances. These disturbances are called the *sudden*

† Q_0' expresses the value of Q at the level of maximum electron concentration for a zenith sun.

‡ $\zeta \leq 85°$.

ionospheric disturbance (S.I.D.), which greatly increases the absorptivity of the D layer and causes radio *fadeout* for periods of an hour or two; *the ionospheric storm,* which hampers radio communication by decreasing the ionization and hence the reflectivity of the F layer; and lastly the *magnetic storm,* which causes violent fluctuations of the geomagnetic field of the earth. All these disturbances appear during sunspot maxima, and are a result of the solar-flare activity that accompanies sunspot activity.

The solar flares are extremely hot masses of gas that leave the chromosphere of the sun in an arching path and return to the surface. It appears that not all the material in the flare returns to the sun, but that some is directed into space as corpuscular emanations. These corpuscles, which are highly ionized, appear to travel at speeds of about 1.4×10^3 kilometers per second. They travel in straight lines until, if they are properly directed initially, they reach the earth.

Simultaneous with the emission of corpuscular radiation from the intensely hot solar flares, there is released an unusually large amount of ultraviolet radiation from the same flares. The ultraviolet radiation, being electromagnetic in character, travels at 3×10^5 kilometers per second, traversing the distance between sun and earth in about 9 minutes. It is this ultraviolet radiation that causes the S.I.D. of the D layer and its effects are noticeable only in the sunlit hemisphere of the earth almost immediately after the appearance of solar flares. The increase in ionization is more marked at low than at high latitudes.

Because of the slower speeds of the corpuscular radiation, the ionospheric and magnetic storms do not begin until about 30 hours after the appearance of solar flares, whose orientation could direct corpuscular radiation toward the earth. Those disturbances that arise from corpuscular bombardment of the earth's upper atmosphere are not confined to the earth's sunlit hemisphere. Rather, they are mainly confined to the polar regions of the earth. The storms may last for days. The concentration of the disturbances to polar regions is marked, and arises from the charged nature of the particles. The particles, being highly ionized and traveling at great speed, constitute an electric current. It is well known that an electric current has an accompanying magnetic field. It is with this magnetic field of the corpuscles that the earth's magnetic field interacts, causing the corpuscles to be deflected toward the magnetic poles of the earth as the magnetic field accompanying the corpuscles tries to line up with the earth's magnetic field. Collisions frequently occur between the corpuscles and the molecules of the upper atmosphere, ionizing the latter.

The effect is concentrated in the polar regions, the region of the earth's magnetic poles, and may extend to as low as 80 kilometers in the earth's atmosphere. This corpuscular radiation is responsible not only for ionospheric and magnetic storms, but also for the Polar Aurora which is the visible manifestation of the magnetic storm and occurs at the same time.

Aurorae. The polar aurora, the *aurora borealis* of the northern hemispheric polar regions and the higher-temperate latitudes in winter and the corresponding *aurora australis* of the southern polar regions, arise because of the bombardment and subsequent ionization of the molecules of the upper atmosphere by corpuscular emanations from solar flares. The ionized air molecules on recapturing electrons and readjusting to their normal energy states emit quanta of energy, some of which is in the green and red of the visible spectrum. This effect leads, in general, to a bright-green luminosity of the northern sky which we call the aurora. It has been definitely proved that it is the green line of atomic oxygen (0.5577 micron) and the red doublet line of oxygen (0.6363 and 0.6300 micron) that are responsible for the color of the aurora. As might be expected, many lines and bands of nitrogen have also been identified, but except for the red band of nitrogen at 0.6500 micron they are much weaker than the two oxygen lines. Because the aurorae occur at heights of from 80 to 300 kilometers (although some heights as great as 800–1000 kilometers are reported) where the pressure ranges from 10^{-3} to 10^{-6} millimeter of mercury, and because the atomic transformations are carried out without the wall effects common to laboratory spectroscopy, the study of the spectrum of the aurora (and of the night sky) yields valuable information on the structure of matter.

The aurorae can assume many characteristic patterns, and an international classification of the forms exists.(C5) The two great divisions are based on whether or not the aurorae show ray structure. Those aurorae having ray structure have such descriptive names as rays, draperies, and corona, while those aurorae without ray structure may be known as homogeneous quiet arc, pulsating arc, homogeneous bands, pulsating surfaces, or feeble glow. The classifications are based mainly on the descriptions and photographs of Störmer,(23) whose name is virtually synonymous with aurora.

The light of the night sky

In the absence of the moon and artificial light, the sky possesses a measurable luminosity which is called the light of the night sky. The sources of this light may be radiations from atmospheric gases at high elevations, polar aurora, zodiacal light (origin not definitely known), cosmic dust particles (an average of 1 per cubic centimeter), and stars and nebulae. A study of the spectrum of the night-sky light is extremely valuable for at least two reasons: first, the spectrum yields information on the composition of the high atmospheric and interstellar constituents, and second, the spectrum serves as a cosmic laboratory for measuring energy transformations in the atom and molecule, some of which are not duplicated easily or at all in terrestrial laboratories.

Study of the night-sky radiations is not so intimately connected with the meteorology of the troposphere as is the study of atmospheric ozone. Ozone is an important constituent of the atmosphere, because it absorbs solar ultraviolet radiation and therefore may be a significant heat source for driving the circulation of the lower stratosphere and of the troposphere.

Ozone in the atmosphere

The solar spectrum is unobserved at sea level for wavelengths less than 0.29 micron. Between 0.29 and 0.34 micron, the amount of solar energy reaching the earth's surface varies markedly with the zenith angle of the sun, the wavelength of cutoff of the solar spectrum being observed to move toward longer wavelengths with increasing zenith angle of the sun. The behavior of the energy spectrum of the sun can be explained by postulating an absorbing layer of gas in the amosphere whose absorption coefficient increases with decreasing wavelength in the region $0.29 < \lambda < 0.34$ micron. It was shown by Fabry and Buisson[8] in 1921 that the absorption spectrum coincided with that of ozone, a result confirmed by subsequent investigators of whom Dobson[8,9] in England and Götz[8,14] in Switzerland deserve particular mention. Figure 10.8 shows the principal absorption regions of ozone and oxygen, which are for ozone the Hartley region (0.20 to 0.30 micron); the Huggins bands (0.30 to 0.35 micron); the Chappuis bands in the visible (0.40 to 0.80 micron); and a band in the infrared centered at 9.6 microns. Molecular oxygen absorbs very strongly in the Schumann region (0.13 to 0.20 micron), with peak absorption near 0.14 micron.

The first investigators[9] attempted to measure the mean height of the ozone layer and the total amount of ozone in the earth's atmosphere. As it is reasonable to expect a distribution of ozone concentration with height, the term ozone layer implies a region where a large concentration of ozone exists when compared to a region a few kilometers above or below it. We shall consider that this layer has an optical thickness (viewed at the zenith) of u_0 and that the optical thickness when viewed at a zenith angle ζ measured *at the base of the ozone layer* is $u_0 \sec \zeta$. u_0 may be expressed as $(\bar{\rho}/\rho_0)(z_2 - z_1)$. It is understood that u_0 is the thickness of the ozone layer expressed in terms of the height to which a unit volume of ozone of density ρ_0 would reach if compressed

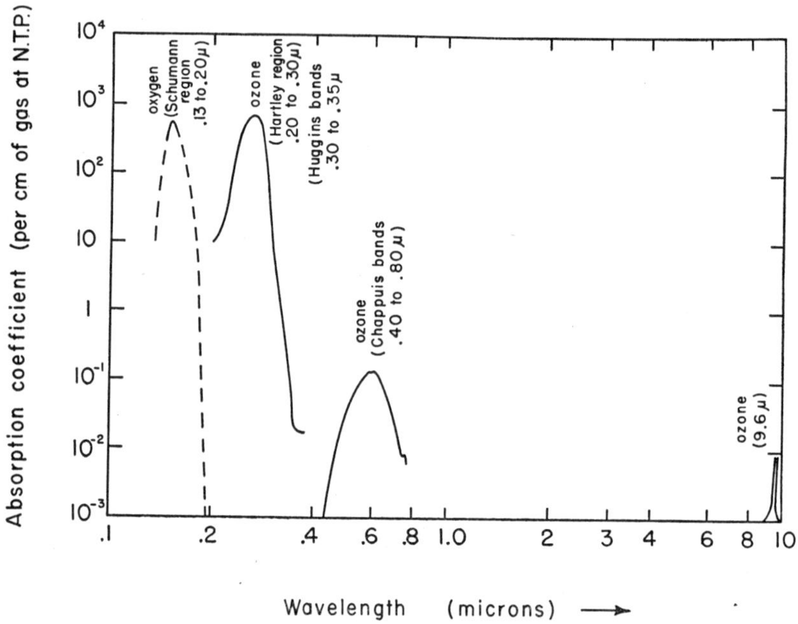

Fig. 10.8 The absorption spectra of ozone (full lines) and oxygen (dashed line). Only the principal regions of absorption are indicated.

to 1 atmosphere at a temperature of 0° C. $\bar{\rho}$, the mean atmospheric density of ozone, is small and is a function of z. The standard condition is called the normal temperature and pressure state (N.T.P.). σ_λ is the extinction coefficient for ozone at N.T.P. for the wavelengths used. In order to prevent complete absorption these wavelengths must lie between 0.29 and 0.34 micron. We shall assume that σ_λ is made up of an absorption coefficient k_λ for ozone, a scattering coefficient of the form $\beta\lambda^{-4}$ (see section on Rayleigh scattering), and a scattering coefficient δ independent of wavelength, to account for the scattering by particles larger than molecular size.

At any wavelength where ozone absorbs, we may write for the flux density

$$\ln \frac{E_0}{E} = k_\lambda u_0 \sec \zeta + \beta\lambda^{-4} z_0 \sec \theta + \delta\, z_0 \sec \theta. \qquad (10.18)$$

This equation is a statement of the Bouguer-Lambert law, Eq. 3.11. We understand k_λ† to be the absorption coefficient per unit length of optical path through the ozone. The sun is viewed from the height at which the ozone layer exists, making an angle ζ with the zenith. θ is the zenith angle of the sun when viewed from the ground with the entire atmosphere as the scattering medium. The term $\beta\lambda^{-4}z_0 \sec\theta$ can be evaluated by the standard atmosphere technique, with the methods described on page 49 of Chapter 2. $\sec\zeta$ is related to $\sec\theta$ through

$$\sin\zeta = \frac{\sin\theta}{1 + (z_0/R)}. \tag{10.19}$$

The geometry‡ is depicted by Fig. 10.9 and the answer is deduced from

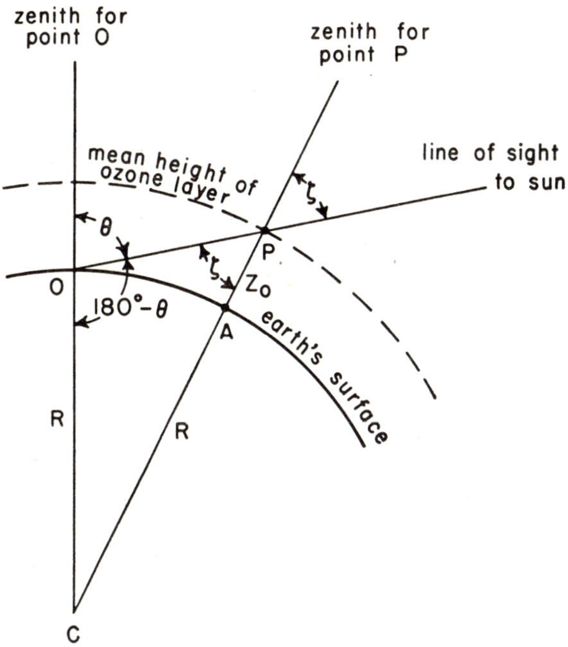

Fig. 10.9 Geometry for the relationship between the sun's zenith angle θ measured from point O at the surface of the earth to the required zenith angle ζ when the sun is viewed from point P in the ozone layer at a height z_0 above the surface of the earth. R is the radius of the earth. The law of sines requires that

$$\frac{\sin(180° - \theta)}{R + z_0} = \frac{\sin\zeta}{R}.$$

Atmospheric refraction is neglected.

† k_λ is identical with k_a in other sections. Here it is desired to stress the wavelength dependency.

‡ It is only at zero zenith angle that ζ and θ are equal.

the law of sines. z_0 is the mean height of the ozone layer and R is the radius of the earth. When $\sec^2 \theta \gg z_0/R$, the usual case, $\sec \zeta$ reduces to this approximation formula,

$$\sec \zeta = \sec \theta \left[1 - \frac{z_0}{R} (\sec^2 \theta - 1) \right]. \tag{10.20}$$

The effect of scattering by large particles may be eliminated by measuring the energy at two wavelengths λ_1 and λ_2 in the absorbing region of ozone. In general these two wavelengths though close together have widely different absorption coefficients. Two wavelengths often used are $\lambda_1 = 0.311$ micron and $\lambda_2 = 0.329$ micron. λ_1 is very strongly absorbed and λ_2 is very weakly absorbed by ozone. By writing Eq. 10.18 for both λ_1 and λ_2, subtracting the two equations, and simplifying, we obtain

$$\ln \frac{E_2}{E_1} = (k_1 - k_2) u_0 \sec \zeta + (\lambda_1^{-4} - \lambda_2^{-4}) \beta z_0 \sec \theta + \ln \frac{E_{02}}{E_{01}}. \tag{10.21}$$

We assume that $\ln (E_{02}/E_{01})$, the ratio of the solar flux density at these two wavelengths incident on the top of the atmosphere, is constant and known and that the term in λ^{-4} can be evaluated for any zenith angle θ. $\ln (E_2/E_1)$ is measured at the ground, as is θ. Thus, from Eqs. 10.19 and 10.21 we see that we have two equations and three unknowns; u_0, z_0, and $\sec \zeta$.

A solution can be arrived at in the following manner. If u_0 and z_0 remain constant over the period of measurement (an assumption with no *a priori* evidence either for or against it) Eq. 10.21 must be the equation of a straight line when $\ln (E_2/E_1)$ is plotted against $\sec \zeta$. The slope of this line is $(k_1 - k_2)u_0$.

From our set of observations of θ we deduce a series of values of ζ by assuming for each set of θ a constant z_0, using Eq. 10.19 to relate the quantities. Then each set of $\sec \zeta$'s is plotted against the observed $\ln (E_2/E_1)$, yielding a family of curves where each member of the family is characterized by the parameter z_0. It is found that during days when the meteorological conditions are relatively steady, there will be one member of the family which to within the limits of experimental error of the observations is a straight line. The z_0 value of this curve is the height of the ozone layer and the optical thickness of the layer is the slope divided by $(k_1 - k_2)$. The method yields a height indicative of a weighted mean height of the ozone and this height is usually taken to be that of the center of mass of the ozone. By this method, the height of the ozone layer usually is measured to be 20 to 25 kilometers with an error of less than $\pm$ 5 kilometers. The optical path at N.T.P. varies between 0.2 to 0.4 centimeter in thickness with a probable error of measurement of about 2 per cent of the total amount.

A second method for determining the ozone concentration and distribution in the atmosphere was introduced by Götz,[14,18] and called by him the "Umkehreffect"† because of the characteristic shape of the curves when the logarithm of the ratio of the intensity of the zenith light at two wavelengths (usually 0.311 and 0.329 micron) is plotted versus the fourth power of the zenith angle of the sun. In order to obtain the umkehr curves, the zenith-sky light is observed from about 3 hours before to $\frac{1}{2}$ hour after sunset or from $\frac{1}{2}$ hour before to 3 hours after sunrise so that the sun's zenith angle is between 70 and 95°.

The direct rays of the sun are shielded from the instrument, so that the light reaching the instrument has been scattered at least once (once is assumed by the theory with empirical corrections added for secondary scattering) from the zenith sky.

The amount of light that reaches the instrument at a given wavelength from the zenith sky depends on two factors, the first being dependent on the amount of air at a height favorable for scattering, and the second depending on the absorption of the light before and after scattering has occurred. It turns out that when the sun is low the scattering of light can be assumed to come from a relatively small layer of air high in the atmosphere, the height depending mainly on the difference in the magnitude of the absorption coefficients when two wavelengths close together are chosen and the sun is at a fixed zenith angle. The wavelength having the greater absorption coefficient (0.311 micron in the example) is effectively scattered from the greater height (see Fig. 10.10), a height above most of the ozone layer, because the combination of direct plus scattered light has a minimum path length through the ozone. It is implied therefore that the greatest amount of energy reaches the instrument on the ground from this height. On the other hand, a wavelength (such as 0.329 micron) having a low absorption coefficient for ozone has an effective scattering height much lower down in the atmosphere than the wavelength having the higher absorption. The air is more dense at this level but the smaller ozone absorption per unit length allows transmission of the direct plus the subsequently scattered light over a longer optical path.

With this information, it is possible to predict how the intensities at these two wavelengths will behave as the sun goes toward the horizon. For each individual wavelength, the intensity of the zenith sky will

† *Umkehr* in German means reversal. The umkehr curves show a characteristic minimum point, a reversal of slope.

decrease as the zenith angle increases. However, the rates of decrease are so different that the ratio of the more strongly absorbing to the less strongly absorbing wavelength, $I_{0.311}/I_{0.329}$, will show the following characteristics. As the 0.311-micron wavelength scatters from above most of the ozone in the atmosphere and this height increases as the zenith angle increases, the scattered light passes through an unchanging amount of ozone during the course of the umkehr measurements and the intensity shows a gradual decrease as the sun sets. At the

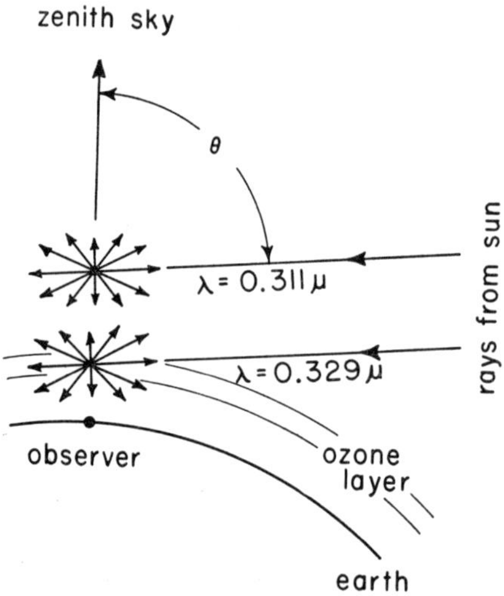

Fig. 10.10 Schematic drawing of the geometry of the umkehr effect. Only the rays scattered downward along the zenith angle are detected at the observer's position on the surface of the earth. Note that the 0.329-micron wavelength has the longer path through the ozone layer from sun to observer. Only the principal wavelengths that are scattered and detected from the optimum height are shown.

same time, the 0.329-micron wavelength that initially is scattering from lower down in the atmosphere is having its effective height of scattering also increased as the sun sets, but at such a rate that the intensity of the 0.329-micron line falls off less rapidly than the 0.311-micron line. The difference in rates causes the ratio $I_{0.311}/I_{0.329}$ to decrease at first. Finally, when the sun is near the horizon, the rate at which $I_{0.329}$ falls off becomes equal to and finally exceeds that at which $I_{0.311}$ changes, causing the ratio to reach a minimum value and then increase again until darkness makes further measurements impossible. Three umkehr curves are shown in Fig. 10.11.

With a distribution of ozone in the atmosphere assumed, an umkehr curve can be computed by theoretical means. The distribution is so modified until the theoretical curves agree in essential detail with the umkehr curve measured. At this point, it is assumed that the amount and distribution of ozone required by theory are representative of the ozone distribution in the atmosphere. A typical ozone distribution measured by the umkehr effect is shown in Fig. 10.12, together with

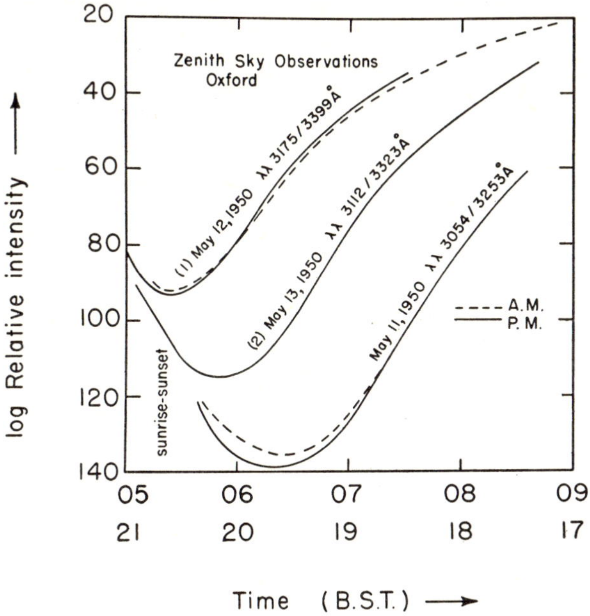

Fig. 10.11 (*After Dobson.*) Measurements of zenith skylight near sunrise and sunset. The points give the measured rates of the intensity of two adjacent wavelengths for three such pairs of wavelengths. The small scatter of the points (not shown) from a smooth curve is an indication of the small random error of the observations. The shape of such curves depends on the vertical distribution of ozone in the atmosphere. The fact that the curves reverse their slopes near sunrise and sunset is due to the fact that at these times much of the light of the shorter wavelengths is scattered by the air at a great height, which is above most of the ozone.

a couple of direct measurements of ozone concentrations by spectrographs carried by balloons or rockets. There are at least one and possibly two heights where the ozone concentration is a maximum, at 20 to 25 kilometers, although ozone is found at sea level and as high as 75 kilometers.

A third method of measuring the mean height of the ozone layer has been brought forth by Strong.[24] He has found that the absorption of energy by the 9.6-micron band of ozone shows a one-fourth power dependence on total atmospheric pressure. The absorptivity by ozone in the Hartley band is relatively independent of pressure, so that a measurement in this region indicates the total concentration of ozone in the atmosphere. A measurement of the absorptivity in the ϕ band of water vapor indicates the total amount of water vapor in the atmos-

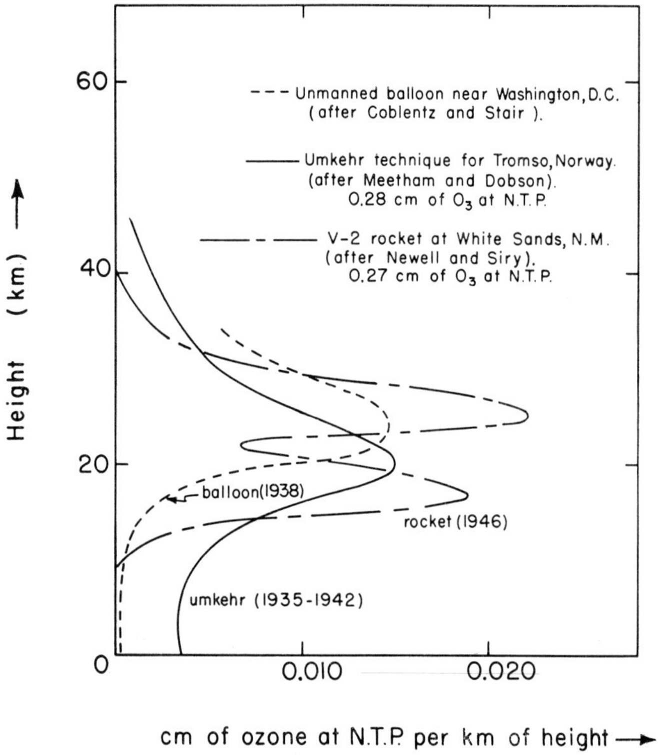

Fig. 10.12 Three ozone concentrations as a function of altitude. Note the agreement among the curves. Detail as shown by the rocket (double maximum) is not attainable by the indirect umkehr method.

phere below the ozone layer. With these data, the energy measured at the ground in the 9.6-micron band can be corrected for the effect of water vapor to give the total absorption due to ozone. From a laboratory calibration it is known that the combination of concentration and absorptivity which has been measured must arise from a particular mean pressure of the layer where most of the ozone is concentrated. Atmospheric soundings plus extrapolation will give the density of the atmosphere so that the height of the pressure level where the ozone is concentrated may be determined.

Although direct measurements of the ozone layer are possible by rocket- or balloon-borne apparatus, the expense is so great that it is reasonable to expect that these indirect methods will continue to serve for some time to come.

Observations of Ozone Content in the Atmosphere. Ozone observations have been conducted at various isolated sites in the world since 1925. Much of our information on ozone distribution comes from a series of measurements made by Dobson(9) and co-workers and veri-

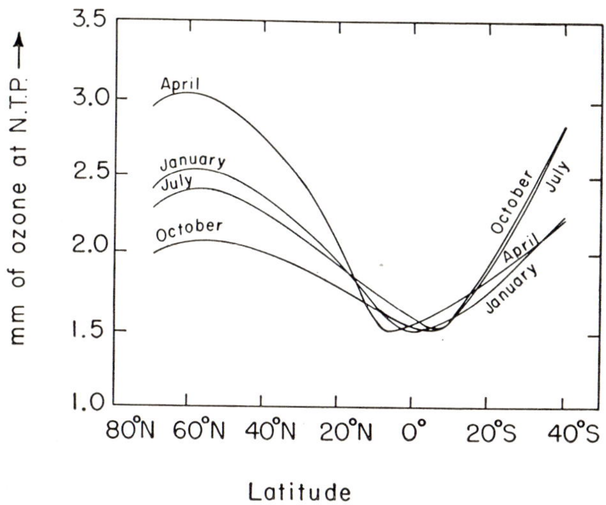

Fig. 10.13 (*From data collected by Craig from various sources.*) The latitude distribution of monthly mean ozone concentrations in the atmosphere.

fied by other workers throughout the world. Since no synoptic network of ozone observing stations has been attempted, the results to be presented are mean conditions of the ozone layer.

It has been found that in a given month of the year the total ozone concentration in the atmosphere is a minimum in equatorial regions, increasing to a maximum at latitudes near 70° from the equator and decreasing again poleward. There is, however, a seasonal shift of the ozone pattern causing (from 15° to 60° N) a maximum of ozone to be concentrated above a northern hemisphere station in the spring months of March and April when compared to the minimum values found during the fall months of October and November. Figure 10.13 shows the mean concentration of ozone for selected months.

Measurements of the ozone concentration have been made on a time scale small enough to obtain a picture of the ozone distribution about typical cyclones and anticyclones. The picture has been found to be consistent with those upper-level circulation patterns, associated with surface pressure centers, that are found at a height within but near the base of the ozone layer; i.e., 10 to 15 kilometers. These heights are near the standard 200-millibar pressure surface. Air, on entering

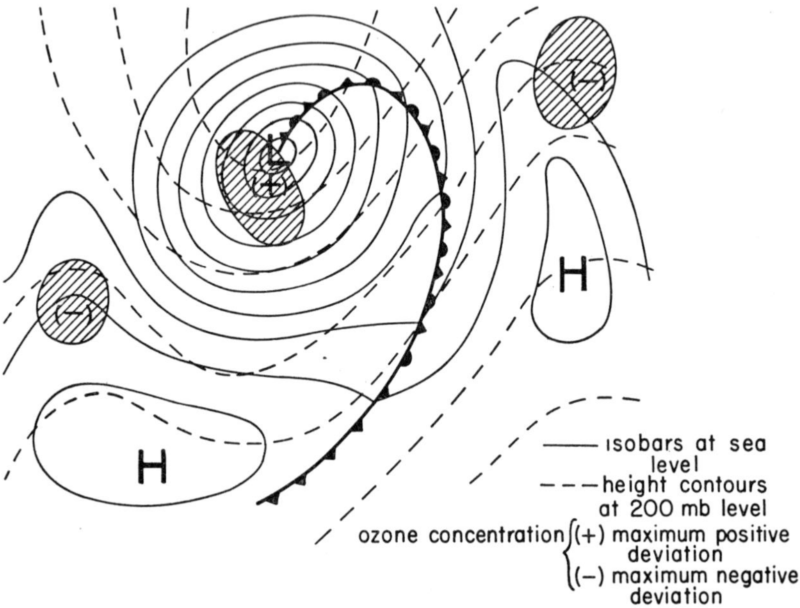

Fig. 10.14 The distribution of the center of the maximum deviations from the normal amounts of ozone (both positive and negative deviations) about a typical low- and high-pressure center.

these upper level troughs, suffers convergence and subsidence. These dynamic effects are reflected in a lowered tropopause and an increased amount of ozone. Upper-level troughs are nearly always to the west of the surface low and the maximum effect is generally southwest of the surface low. The converse effect in front of the surface low, a region at these upper levels of ascending and diverging air motion, is less well marked. Figure 10.14 shows a typical surface low-pressure center, with corresponding upper trough and ozone concentration. The explanation for these observed changes in ozone concentration that correlate so well with the dynamics of the atmospheric circulation lies in a consideration of the equilibrium conditions attending the formation and dissipation of ozone.

The photochemical equilibrium of ozone

The absorption of sunlight in the ultraviolet leads both to the condition that causes ozone formation and ozone destruction. Earlier in this chapter we saw that oxygen gas absorbs strongly in the Schumann region ($\lambda < 0.200$ micron). Oxygen also absorbs in the weak Herzberg bands centered near 0.240 micron. Molecular oxygen will dissociate completely into atomic oxygen when it absorbs a quantum of energy whose energy corresponds to a free-space wavelength of less than 0.240 micron, according to the chemical equation

$$O_2 + h\nu\ (\lambda < 0.240\ \mu) \rightarrow 2O. \tag{10.22}$$

$h\nu$ is the energy of a photon of light and is equal to hc/λ ergs when $h = 6.55 \times 10^{-27}$ erg second, c, the speed of light, $= 3 \times 10^{10}$ centimeters per second, and λ is a wavelength in centimeters less than 0.240 micron $= 2.40 \times 10^{-5}$ centimeter.

The collision of one molecule of diatomic oxygen with one atom of monatomic oxygen in the presence of a third neutral molecule (either oxygen or nitrogen) forms ozone by the following reaction

$$O_2 + O + M \rightarrow O_3 + M. \tag{10.23}$$

M is the third molecule required because considerations of the conservation of energy and momentum require a three-body collision. O_3 is the chemical formula for ozone.

Concurrently with ozone formation there is ozone destruction caused by the absorption of energy by ozone at wavelengths less than 1.100 micron, in either the Chappuis, Huggins, or Hartley region. The low energy hc/λ necessary for dissociation indicates that ozone is very unstable in the presence of sunlight. We have, therefore, ozone destroyed according to the following reaction

$$O_3 + h\nu\ (\lambda < 1.100\ \mu) \rightarrow O_2 + O \tag{10.24}$$

and then

$$O_3 + O \rightarrow 2O_2. \tag{10.25}$$

The four reactions shown are believed to be the important reactions in the formation and destruction of ozone at the concentrations and temperatures found in the atmosphere.

The process of formation and destruction is not separate but is carried out in the same volume of space and at the same time. In order to see how the process works, let us assume that n_1, n_2, n_3, and n_M

are the number of molecules of O, O_2, O_3, and M respectively present in a unit volume of air. Let q_2 be the number of quanta of energy impinging on the unit volume of air in a unit time, whose wavelength is favorable for being absorbed by O_2. Similarly q_3 is the number of quanta incident on the same unit of volume in the same time that can be absorbed by O_3. A fraction α_2 of the incident quanta q_2 will be absorbed by the oxygen and another fraction α_3 of the incident quanta q_3 will be absorbed by the ozone. The total number of oxygen molecules dissociated is $n_2\alpha_2 q_2$ and the total number of ozone molecules dissociated is $n_3\alpha_3 q_3$. On dissociation, as each molecule of O_2 forms 2O and each molecule of O_3 forms simply O, the number of molecules of O formed in a unit time in this volume is $2n_2\alpha_2 q_2 + n_3\alpha_3 q_3$.

At the same time, the probability that a molecule of monatomic oxygen will combine with its neighbors and be destroyed is proportional to $k_{12}n_1 n_2 n_M$ when the reaction given by Eq. 10.23 is considered and $k_{13}n_1 n_3$ when Eq. 10.25 is considered. The joint probability of collision is the product of the numbers of molecules that might enter the collision, and k_{12} and k_{13} are the fraction of the total possible collisions that are successful. k_{12} is the fraction when an O_2 molecule collides with an O atom, and k_{13} is used for an O_3 molecule colliding with an O atom. The number of atoms of O destroyed in a unit time in the volume under consideration is $k_{12}n_1 n_2 n_M + k_{13}n_1 n_3$.

At equilibrium, the rate of formation of O atoms, dn_1/dt, must equal the rate of destruction so that

$$\frac{dn_1}{dt} = 0 = 2n_2\alpha_2 q_2 + n_3\alpha_3 q_3 - k_{12}n_1 n_2 n_M - k_{13}n_1 n_3. \qquad (10.26)$$

By similar reasoning the equilibrium rate of formation of ozone molecules must be

$$\frac{dn_3}{dt} = 0 = k_{12}n_1 n_2 n_M - k_{13}n_1 n_3 - n_3\alpha_3 q_3. \qquad (10.27)$$

If n_1 is eliminated from these two equations through solving them simultaneously, we find that n_3 will be expressed as

$$n_3 = \frac{k_{12}}{k_{13}} n_2 n_M \frac{n_2\alpha_2 q_2}{n_3\alpha_3 q_3 + n_2\alpha_2 q_2}\cdot \qquad (10.28)$$

Equation 10.28 may be simplified into the following expression

$$n_3 = n_3{}^* \frac{Q_2}{Q_3 + Q_2} \qquad (10.29)$$

if we let

$$n_3{}^* = \frac{k_{12}}{k_{13}} n_2 n_M = n_3{}^*(z), \tag{10.30}$$

$$Q_2 = n_2\alpha_2 q_2 = n_2(z) \sum_\lambda \alpha_2(\lambda) q_2(\lambda, z), \tag{10.31}$$

$$Q_3 = n_3\alpha_3 q_3 = n_3(z) \sum_\lambda \alpha_3(\lambda) q_3(\lambda, z). \tag{10.32}$$

Examination of Eq. 10.29 shows that this equation is quadratic in n_3 because Q_3 is a function of n_3. In the evaluation, n_3, $n_3{}^*$, n_2, n_M, q_2, and q_3 are functions of height, so that the evaluation must be carried out for selected height surfaces. Once adopted, $n_3{}^*$ is considered known because the temperature and pressure conditions can be measured for the heights in the atmosphere chosen. The ratio k_{12}/k_{13} is known from laboratory measurements. The number of quanta q_2 and q_3 entering a layer is dependent on the number of quanta reaching the top of the atmosphere decreased by the number absorbed and back scattered from layers above. Recent evidence indicates that the sun radiates much less energy than a 6000° K black body in the regions where oxygen and ozone absorption are most important for photochemical reactions. Figure 10.15 illustrates the measurements of the solar spectrum made from a rocket at an altitude where the ozone concentration is so small as to be essentially negligible.

The terms $\sum_\lambda \alpha_i q_i$ are evaluated through subdividing the spectrum into narrow wavelength intervals and selecting the appropriate mean value of α_i and the proper number of q_i for the interval and summing the results.

As might be expected, the actual computations, first completed by Wulf and Deming,[28,29] are tedious, and the reader is referred to a monograph by Craig[8] for details. It is possible, once the equilibrium case has been computed, to make estimates of the rate at which equilibrium is approached. Only the results of these computations will be presented. The vertical and latitudinal distribution of ozone required by the photochemical theory is in major details consistent with observational data.

All investigators of the photochemical equilibrium of ozone agree that the ozone at altitudes in excess of 30 kilometers can be considered to be in a state of photochemical equilibrium while at lower levels calculations show that weeks and even years are required for equilibrium to be established. The distinction is important because the nonequilibrium state below 30 kilometers suggests a means for explaining day by day changes in the ozone content of the air, which may be as large in magnitude as the mean semiannual changes shown in Fig. 10.13.

Studies of air motion in the stratosphere show that air entering upper-level troughs associated with surface cyclones suffers both de-

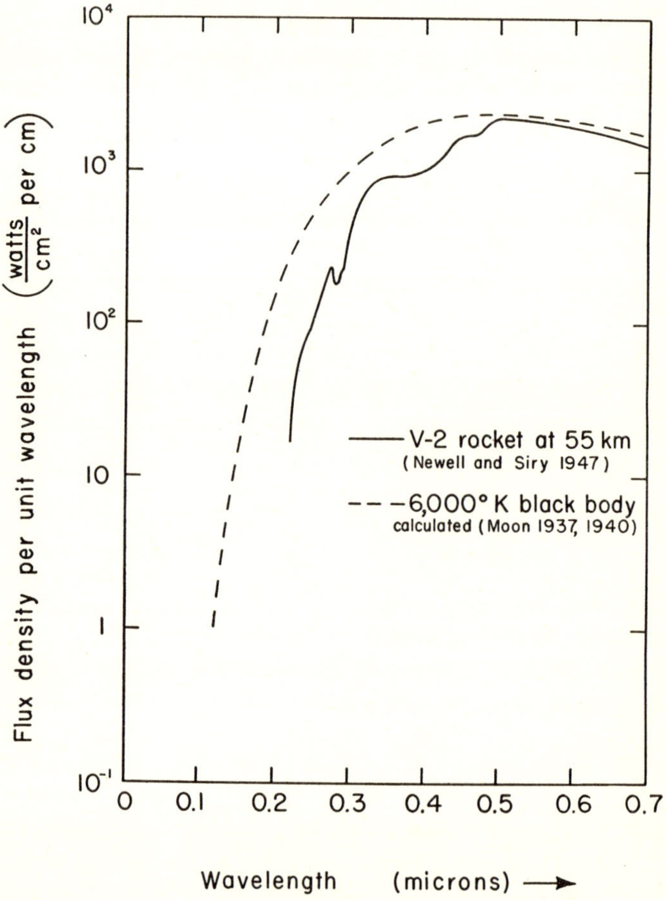

Fig. 10.15 (*After Craig.*) The solar spectrum in the visible and near ultraviolet as measured directly by V-2 rocket and calculated on the assumption of a 6000° K black body. Note that this one measurement gives much less ultraviolet radiation than a 6000° K black body.

scending and converging motion. The ozone present in the air moves therefore with the general air motion to lower elevations. Because of the relatively large gradient of equilibrium time, ozone that sinks to levels lower than 30 kilometers maintains the same relative density when compared to the surrounding air, being destroyed so slowly that this effect can be neglected. Meanwhile, the air above 30 kilometers has lost ozone to lower levels, but the equilibrium amount of ozone is quickly restored in this region by photochemical action. The subsidence process therefore leads to an increase in the total amount of ozone in a vertical column. By the same analysis, upward motion causes all ozone in excess of the equilibrium amount to be rapidly destroyed in passing to levels above 30 kilometers, yet creating no new ozone at levels below 30 kilometers. The effect of upward motion in the high stratosphere is to decrease the total amount of ozone in a vertical column.

Concurrent with vertical velocities in the atmosphere is also horizontal motion which can likewise transport ozone from regions of high ozone content to regions of low ozone content, and the reverse. Below 30 kilometers such transport should conserve the total amount of ozone leaving one region and entering another, a condition implying that below 60° N latitude northerly winds should tend to increase the total amount of ozone above a station to the south, and southerly winds decrease the total amount. However, as in most meteorological studies, it is difficult to determine in many situations which process, vertical motion or horizontal advection, plays the more important role in changing the total ozone content of the air above a given locality. At present, it seems that the day-to-day and season-to-season changes in the ozone content of the atmosphere are correlated less with solar influences than with the day-to-day and season-to-season changes in atmospheric circulation.

The composition of the atmosphere

Extensive surveys have been made of the oxygen, carbon dioxide, argon, and helium contents of tropospheric air. The results indicate that these gases and others (including nitrogen) maintain constant ratios. The surveys have been extended into the stratosphere and to date no definite variation in the relative per cent composition of the principal constituents of the air up to 70 kilometers in the atmosphere has been found. In all these surveys, one excludes water vapor which is in highly variable amounts (up to 3 per cent by mass) in the troposphere. Water vapor in the stratosphere will be discussed in the next chapter. The constancy of the composition of the air indicates turbulence and lack of radiative equilibrium in the stratosphere.

Lack of a quiescent stratosphere prevents stratification below 70 kilometers of the atmospheric gases, precluding the lighter gases from increasing in relative abundance as the height increases. The lack of an increase percentagewise in hydrogen and helium with altitude may be taken as evidence against diffusive equilibrium.

Table 10.3 is presented to show the composition of the dry air of the atmosphere.

Table 10.3

The Atmospheric Constituents of Dry Air are Given in Per Cent by Volume or for the Rare Gases in Parts per Million (ppm) by Volume

(After Glueckauf, E., "The Composition of Atmospheric Air," *Compendium of Meteorology,* American Meteorological Society, Boston, 1951.)

Conventional chemical symbols are used. Many of the variable components are industrial in origin.

Constituent	Content, per cent	Content, ppm
Nonvariable Components of Atmospheric Air		
N_2	78.084 ± 0.004	
O_2	20.946 ± 0.002	
CO_2	0.033 ± 0.001	
A	0.934 ± 0.001	
Ne		18.18 ± 0.04
He		5.24 ± 0.004
Kr		1.14 ± 0.01
Xe		0.087 ± 0.001
H_2		0.5
CH_4		2
N_2O		0.5 ± 0.1
Variable Components of Dry Air (Earth's Surface)		
O_3		0 to 0.07
SO_2		0 to 1
NO_2		0 to 0.02
CH_2O		Uncertain
I_2		up to 10^{-4} g/m^3
NaCl (solid)		10^{-4} g/m^3
NH_3		0 to trace
CO		0 to trace

References

1. Appleton, E. V., "The Bakerian Lecture: Regularities and Irregularities in the Ionosphere," *Proc. Roy. Soc.*, (A), *162*, 415–479 (1937).
2. Appleton, E., "Magnetic and Ionospheric Storms," *Archiv für Meteorologie, Geophysik, und Bioklimatologie, 3,* 113 (1952).
3. Bailey, D. K., et al., "A New Kind of Radio Propagation at Very High Frequencies Observable over Long Distances," *Phys. Rev., 86,* 141–145 (1952).
4. Berning, W. W., "Charge densities in the ionosphere from radio doppler data," *J. Meteorol., 8,* 175–181 (1951).
5. Chackett, K. F., Paneth, F. A., and Wilson, E. J., "Chemical Analysis of Stratospheric Samples from 50 to 70 Km. Height," *J. Atmos. Terrest. Phys., 1,* 49–55 (1950).

6. Chackett, K. F., Paneth, F. A., Reasbeck, P., and Wiborg, B. S., "Variations in the Chemical Composition of Stratosphere Air," *Nature, 168,* 358–360 (1951).
7. Clark, K. C., "Ionospheric Absorption by N_2 and O_2 of Certain Extreme Ultraviolet Solar Wavelengths," *Phys. Rev., 87,* 271–276 (1952).
8. Craig, R. A., "The Observations and Photochemistry of Atmospheric Ozone and Their Meteorological Significance," *Meteorol. Monographs, I,* (1950). American Meteorological Society, Boston.
9. Dobson, G. M. B., "Observations of the amount of ozone in the earth's atmosphere and its relation to other geophysical conditions, part IV," *Proc. Roy. Soc.,* (A), *129,* 411–433 (1930).
10. Dobson, G. M. B., "Recent Work in the Stratosphere," *Quart. J. Roy. Meteorol. Soc., 77,* 488–492 (1951).
11. Eccles, W. H., "On the Diurnal Variations of the Electric Waves Occurring in Nature, and on the Propagation of Electric Waves Round the Bend of the Earth," *Proc. Roy. Soc.,* (A), *87,* 79–99 (1912).
12. Elvey, C. T., and Farnsworth, A. H., "Spectrophotometric Observations of the Light of the Night Sky," *Astrophys. J., 96,* 451–467 (1942).
13. Gerson, N. C., "Abnormal *E* Region Ionization," *Can. J. Phys., 29,* 251–261 (1951).
14. Götz, F. W. P., Meetham, A. R., and Dobson, G. M. B., "The Vertical Distribution of Ozone in the Atmosphere," *Proc. Roy. Soc.,* (A), *145,* 416–446 (1934).
15. Herman, R., Herman, L., and Gauzit, J., "Infra-red Spectrum of the Night Sky," *Nature, 156,* 114–115 (1945).
16. Jones, M. W., and Jones, J. G., "Tidal effects in the ionospheric F-layer," *J. Meteorol., 7,* 14–20 (1950).
17. Penndorf, R., "Absorption of solar energy by oxygen molecules in the E-layer," *J. Meteorol., 7,* 243–244 (1950).
18. Ramanathan, K. R., Moorthy, Bh. V. R., and Kulkarni, R. N., "The effect of secondary scattering in the calculation of the vertical distribution of atmospheric ozone from the Götz inversion—effect," *Quart. J. Roy. Meteorol. Soc., 78,* 625–626 (1952).
19. Reed, R. J., "The Role of Vertical Motions in Ozone—Weather Relationships," *J. Meteorol., 7,* 263–267 (1950).
20. Reed, R. J., and Julius, A. L., "A qualitative analysis of two proposed mechanisms for vertical ozone transport in the lower stratosphere," *J. Meteorol., 8,* 321–325 (1951).
21. Roach, F. E., and Barbier, D., "The Height of the Emission Layers in the Upper Atmosphere," *Trans. Am. Geophys. Union, 31,* 7–12 (1950).
22. Seaton, S. L., "Temperature and Recombination Coefficient in the Ionosphere," *J. Meteorol., 4,* 197–200 (1947).
23. Störmer, C., "Résultats des mesures photogrammétriques des aurores boréales observées dans la Norvège méridionale de 1911 à 1922," *Geofys. Publikasjoner, 4,* No. 7 (1926).
24. Strong, J., "On a New Method of Measuring the Mean Height of the Ozone in the Atmosphere," *J. Franklin Inst., 231,* 121–155 (1941).
25. Watson, G. N., "Diffraction of Electric Waves by the Earth," *Proc. Roy. Soc.* (A), *95,* 83–99 (1919).
26. Watson, G. N., "Transmission of Electric Waves round the Earth," *Proc. Roy. Soc.* (A), *95,* 546–563 (1919).
27. White, F. W. G., and Geddes, M., "The Antarctic Zone of Maximum Auroral Frequency," *Terrestrial Magnetism and Atm. Elec., 44,* 367–377 (1939).

28. Wulf, O. R., and Deming, L. S., "The Theoretical Calculation of the Distribution of Photochemically-Formed Ozone in the Atmosphere," *Terrestrial Magnetism and Atm. Elec., 41,* 299–310 (1936).
29. Wulf, O. R., and Deming, L. S., "The Distribution of Atmospheric Ozone in Equilibrium with Solar Radiation and the Rate of Maintenance of the Distribution," *Terrestrial Magnetism and Atm. Elec., 42,* 195–202 (1937).
30. Yerg, D. G., "Ionospheric wind systems and electron concentrations of the *F* layer," *J. Meteorol., 8,* 244–250 (1951).

Source Books

B1. Bennington, T. W., *Radio Waves and the Ionosphere, Wireless World,* Ififfe and Sons, London (1943).
B2. Fleming, J. A., *Terrestrial Magnetism and Electricity,* McGraw-Hill Book Co., New York (1939).
B3. Kuiper, G. P., *The Atmospheres of the Earth and Planets,* University of Chicago Press, Chicago (1949).
B4. Mitra, S. K., *The Upper Atmosphere,* Royal Asiatic Society of Bengal, Calcutta (1948)
B5. White, F. W. G., *Electromagnetic Waves,* fourth edition, Methuen's Monographs on Physical Subjects, John Wiley & Sons, New York (1950).

From the *Compendium of Meteorology,* T. F. Malone, editor, American Meteorological Society, Boston (1951).

C1. Chapman, S., "Photochemical Processes in the Upper Atmosphere and Resultant Composition," 262–274.
C2. Craig, R. A., "Radiative Temperature Changes in the Ozone Layer," 292–302.
C3. Glueckauf, E., "The Composition of Atmospheric Air," 3–10.
C4. Götz, F. W. P., "Ozone in the Atmosphere," 275–291.
C5. Harang, L., "Aurorae and Magnetic Storms," 347–355.
C6. Hulburt, E. O., "Night-Sky Radiations from the Upper Atmosphere," 341–346.
C7. Lettau, H., "Diffusion in the Upper Atmosphere," 320–333.
C8. Mitra, S. K., "General Aspects of Upper Atmospheric Physics," 245–261.
C9. Seaton, S. L., "The Ionosphere," 334–340.

Problems

10.1 Measurements of the critical frequency of the *E* layer gave the following data.

Time, local sun time	Hour Angle of Sun, degrees	Critical Frequency, megacycles
12 noon	0°	4.05
3 P.M.	45°	3.70
5 P.M.	75°	3.30

The measurements were made on June 21, the summer solstice, when the sun's declination is 23.3° north. The station from which the observations were made is located at 33° north latitude. Find the electron-positive ion recombination coefficient for the *E* layer under the assumption that this coefficient remains constant over the 5-hour period. The data also indicated that the level of maximum electron concentration remained at 116 km throughout. (See Prob. 2.4 for formula to compute the zenith angle of the sun.)

10.2 Using whatever data from Prob. 10.1 is necessary, compute the mean temperature of the E layer if its semithickness at noon was measured to be 22 km. On the assumption that the mean temperature indicates the temperature at the level of maximum electron concentration and the lapse rate is −4° C/km, find the temperature at 82 km. Is this temperature reasonable?

10.3 The temperature in an ionospheric layer is found to be 400° K. Find the recombination coefficient Γ for one atom of ionized oxygen O^- and for one atom of nitrogen N^-. The relationship between the recombination coefficient and the absolute temperature is(B2)

$$\Gamma = 1.02 \times 10^{-11} \frac{Z^2}{T^{1/2}} \cdot \quad (\mathrm{cm}^3/\mathrm{sec})$$

Z is the charge on the nucleus: i.e., the atomic number.

What is the average recombination coefficient of the layer if the number of ionized nitrogen atoms is 4 times the number of ionized oxygen atoms?

10.4 Outside the earth's atmosphere, the ratio of the solar intensity at $\lambda = 0.329$ micron to the intensity at $\lambda = 0.311$ micron is 1.55. The absorption coefficients for ozone at these two wavelengths are respectively 0.3 cm^{-1} and 3.2 cm^{-1}. The unit of absorption is measured in pure ozone at a pressure of 1 atmosphere and a temperature of 0° C.

Measurements at sea level of the relative intensities (or relative flux densities) of the solar energy at these two wavelengths show the following variation with zenith angle.

Zenith angle	60°	70°	75°	80°	85°
$\dfrac{E_{0.329}}{E_{0.311}}$	12.5	31.3	76.0	385.	1150.

How much ozone exists in the atmosphere and what is its mean height?

10.5 In the wavelength interval contained between $\lambda = 181$ millimicrons and $\lambda = 182$ millimicrons, the solar energy at normal incidence on the top of the atmosphere is 5×10^{-7} watt/cm². The average absorption coefficient of oxygen in this region is 0.7 cm^{-1} at standard temperature and pressure. (*a*) Show that this energy represents about 4.6×10^{11} quanta per cm² and the absorption coefficient becomes 2.6×10^{-20} per molecule. This latter value corresponds to α_2 of the text. (*b*) Show that the average number of oxygen molecules between 70 and 80 km for each centimeter² of horizontal cross section is 3.7×10^{14}. The average pressure and temperature of the layer are respectively 5.7×10^{-2} mb and −12° C, and 23 per cent of the molecules by mass are molecular oxygen. (*c*) What is the number of quanta penetrating to the 70-km level, calculated on the assumption that the exponential law of decay is valid? How many quanta are absorbed?

CHAPTER 11

The Temperature, Density, Pressure, and Humidity of the Upper Atmosphere

Introduction

In the troposphere and lower stratosphere below about 20 to 30 kilometers in height, temperature measurements are made with the thermometers, thermistors, and thermocouples common to the laboratory. The air is dense enough so that conduction of heat from the air to the thermometer causes the thermometer to indicate reliably the air temperature. The air is dense enough for the equations of motion of classical fluid dynamics to be valid, yet of low enough pressure so that the equation of state for an ideal gas

$$p = \rho R_G T \dagger \tag{11.1}$$

holds. Equation 11.1 relates the pressure p to the density ρ and absolute temperature T of the air. R_G is the gas constant per unit mass of air. Although Eq. 11.1 can be assumed to hold to the upper limit of the atmosphere, the temperature and pressure at these low densities cannot be measured by instruments used at sea-level pressures.

On ascending to levels above 25 to 50 kilometers, material bodies in the atmosphere are in air of such low ambient density that they tend toward radiative equilibrium with the sun, earth, and outer space. Usually such bodies (either man made or extraterrestrial in origin) are traveling at such high speeds relative to the air as they enter or leave the earth's atmosphere that compression of the air at the leading edge of the missile rather than radiative equilibrium determines the missile's temperature. Radiational equilibrium between the missile and its surroundings determines the missile's temperature outside the earth's atmosphere and is independent of the missile's speed in this region.

Measurement of the ambient temperature and density of the air above 25 kilometers, therefore, is usually made by careful measure-

† For low pressures, $p = nkT$ is preferred as the equation of state. n is the number of molecules per unit volume and k is the Boltzmann constant.

ment of some physical phenomenon having a variation that is temperature dependent. Several of these phenomena will be mentioned in this introduction and further described in later sections.

The theory of sound predicts that the velocity of sound in a gas depends on the gas temperature. This principle is utilized to determine the temperature of the atmosphere from 20 to 60 kilometers. Above 60 kilometers the air density is so slight that all but the lowest frequencies (below the audible) are attenuated strongly, and detectable sound energy is not propagated over long distances.

A second class of temperature measurement derives from spectroscopic measurements of the intensity of absorption bands in a gas such as ozone or nitrous oxide. These temperatures as well as those derived from acoustical measurements are the same thermodynamic temperatures that would fit Eq. 11.1. At very low pressures (heights in excess of 100 kilometers), where temperature must assume a statistical nature, the line broadening of emission spectra from the ionized molecules and atoms causing the aurora and some of the light of the night sky determine a temperature.

A similar philosophy governs temperature measurements from exploration of the ionosphere by radio waves. It has been found that the equilibrium between the formation and destruction of ions is temperature dependent. The same statement may be made for the equilibrium amounts of ozone; namely, that the production and destruction of ozone molecules are, at equilibrium, a temperature-dependent process.

Finally, at heights greater than 100 kilometers, the spacing between molecules is so great that collisions between individual molecules occur only after the molecules have traveled, on the average, distances reckoned in centimeters and meters rather than the angstrom units characteristic of sea-level densities. This average distance between molecular collisions is called the *mean free path* of a molecule, and is a function of the pressure of the air. Because at these heights any temperature-indicating device will have dimensions comparable to or smaller than a mean free path, a temperature can be inferred only from Maxwell-Boltzmann statistics. These statistics relate a thermodynamic temperature to the average momentum transfer from molecules to instrument.

Measurements of the luminosity of meteors and many measurements from rockets are in reality primarily density measurements. On the assumption of an isothermal atmosphere at the heights where the density is measured, a characteristic temperature may be inferred. The following sections will elaborate on the temperature and density measurements mentioned in this introduction. Radiative equilibrium and its temperature dependence have been discussed in Chapters 4 and 5 and will not be discussed further.

Temperature from acoustical measurements

Shortly after World War I, Whipple[21,22] in England was able to explain the reason why cannon fire was often heard at sites far removed from the fighting and yet was not heard at intervening sites. These latter were in the straight line between the location where cannon fire was heard and the site where the cannon were fired. The reports were so numerous as to be reliable, and subsequent experiments conducted with a network of competent observers proved that explosions could be heard at long distances, with zones of audibility separated by zones of inaudibility. A typical case is shown schematically in Fig. 11.1. The cause for the zones of inaudibility is analogous to that for the skip zones in radio transmission: namely, a sky wave of sound travels into the upper atmosphere and is refracted enough to be bent back to the earth at some great distance from its source. As the refraction of sound waves in a gas depends on the gradient of the absolute temperature of the gas, whereas the attenuation is a function of the density of the gas, a means exists for determining the temperature of the lower stratosphere by measuring the velocity of the sky wave of sound. These measurements can yield upper air temperatures below the height (about 60 kilometers) where low air density causes strong absorption of the sound waves.

Sound travels as a compressional wave in a physical medium such as a gas, a liquid, or a solid. The molecules of the medium are periodically displaced into more dense and less dense packing than in its normal state by the regularly varying pressure of the sound wave. A thin diaphragm can be set into motion by sound waves, to vibrate at the same frequency as the incident sound wave. Thus, a microphone is a convenient method for detecting sound waves in air. Because of the periodic rarefactions, sound waves will cause a cooling of a wire heated to just below red heat. A hot-wire device whose change in electrical resistance is proportional to the temperature change of the wire can therefore be used to detect sound waves.

In classical physics, the speed of an ordinary sound wave through any medium of density ρ and adiabatic bulk modulus B_s is given by

$$v = \sqrt{\frac{B_s}{\rho}}. \tag{11.2}$$

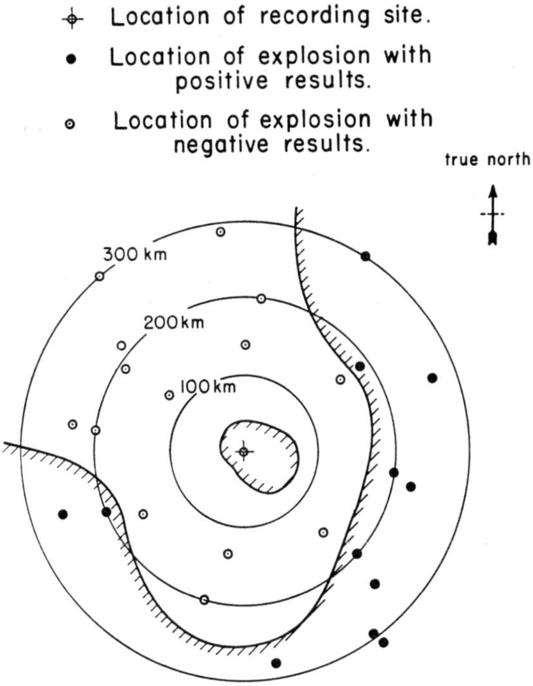

Fig. 11.1 (*After Crary.*) Illustration of zones of audibility and inaudibility from experiments conducted in Alaska in August, 1949. The detecting instruments were set up at the center of the zone and bombs were detonated at the sites marked. The open circles represent sites where the explosions were undetected and the darkened circles the sites where the explosions were heard. This illustration may alternately be considered as one where an explosion set off at the central position would be heard by observers stationed at the darkened circles and not heard by observers stationed at the open circles. Of course, observers close to the site of the explosion would also hear the explosion. The hatching delineates the probable zones of audibility. In Europe, several alternate zones of audibility and inaudibility have been reported.

The thermodynamics of ideal gases indicate that

$$B_s = \frac{c_p}{c_v} p. \tag{11.3}$$

The ratio c_p/c_v is the ratio of the specific heat of the gas at constant pressure to its specific heat at constant volume. The pressure is symbolized by p. On combining Eqs. 11.1, 11.2, and 11.3, the speed of sound through a gas becomes

$$v = \sqrt{\frac{c_p}{c_v} R_G T}. \tag{11.4}$$

For dry air, Eq. 11.4 takes the simple form,

$$v = 20.04T^{1/2} \text{ meters per second.} \tag{11.5}$$

Once a law relating the speed of the wave front to the physical properties of the medium (in this case the temperature) is determined, all the concepts of index of refraction discussed in Chapter 1 hold. For example, differentiating Eq. 11.4 with respect to height shows

$$\frac{\partial v}{\partial z} = \frac{1}{2}\sqrt{\frac{c_p}{c_v}\frac{R_G}{T}}\frac{\partial T}{\partial z} = -\frac{\gamma}{2}\sqrt{\frac{c_p}{c_v}\frac{R_G}{T}}. \tag{11.6}$$

γ is the ambient lapse rate of temperature. Substituting Eq. 11.6 into

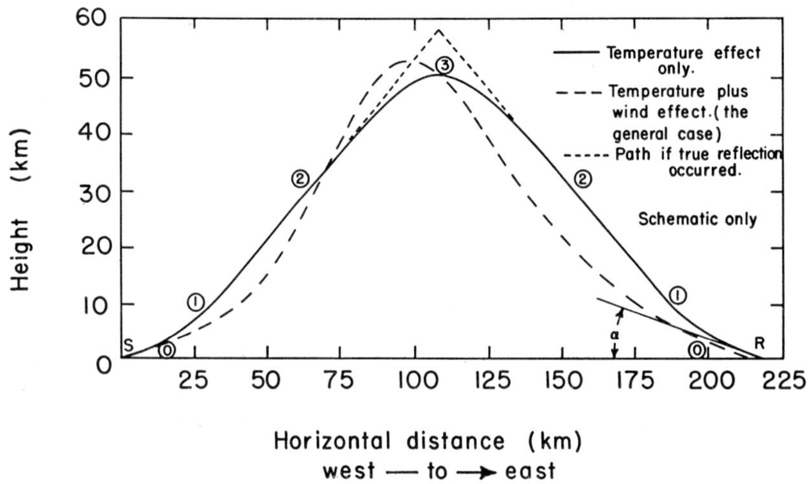

Fig. 11.2 The ray path of a sound wave through the atmosphere for the case of a temperature and wind distribution analogous to the curves for summer, 1949, of Fig. 11.3. The solid curve is the effect of the temperature alone in determining the ray path. The broken curve is the combined effect of temperature and wind in determining the ray path. The drawing is schematic only, for the wind effect is exaggerated.

Eq. 1.5 of Chapter 1 indicates the instantaneous curvature of a plane sound wave as

$$\frac{1}{r} = \frac{1}{v}\frac{\partial v}{\partial z} = -\frac{\gamma}{2T}. \tag{11.7}$$

Suppose that we consider an explosion to occur at point S of Fig. 11.2. By means of an electrical impulse (radio or telephone) we can indicate at R the instant the explosion occurred. In addition, the following quantities can be measured at R.

The time† t is measured from the instant that the explosion occurred until the sound from the explosion is detected at site R; the angle with respect to the horizontal at which the sound ray arrives at R; the temperature and wind velocities in the atmosphere to as high as sounding balloons will go. We may consider that this last measurement gives the temperature and wind profiles of the lower parts of Fig. 11.3*a*, *b*.

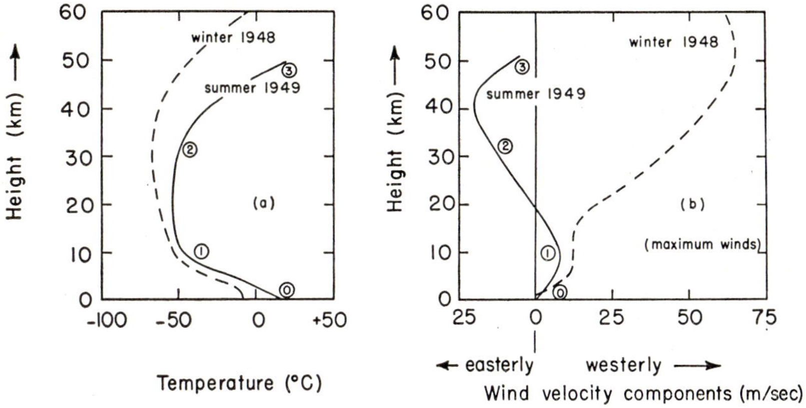

Fig. 11.3 (*After Crary.*) Winds and temperatures in Alaska from acoustical propagation studies. The summer situation is described as the example in the text.

Neglecting for the moment the wind velocity and considering only the temperature effect, we see from Eq. 11.7 that the curvature of a ray of sound should be concave upward from point 0 to 1 of Fig. 11.3*a* and straight from 1 to 2. The latter points represent the isothermal lower stratosphere. Supposing the sounding to be representative of the entire atmosphere between points S and R, the downward path of the ray must be symmetrical with the upward path. This requirement means that the elevation angle α measured at R is the same as the angle at which the sound ray left S. Thus, the path of the ray and, by use of Eq. 11.5, the time required for the wave front to travel from point 0 to point 2 in the atmosphere and back to the earth can be determined. If we call this time which we calculate from theory and upper air soundings t_c, we will find that $t_c < t$ in all cases. The problem, therefore, is to account for a time Δt equal to $t - t_c$ during which the ray reversed direction in the level from 2 to 3.

† It is easy to demonstrate that the time taken for sound to travel between the two points S and R as described is always much longer than is required for sound to travel between the same two points at sea level as a ground wave. Only a sky-wave theory is consistent with experimental data.

If an isothermal atmosphere is assumed between levels 2 and 3, the path would be the dotted line of Fig. 11.2. This path is discarded because the geometry calls for a reflection phenomenon (angle of incidence equal to angle of reflection). More serious still, a wave front takes a longer time than Δt units to travel this path. Therefore, a path must be constructed that can be traversed in exactly Δt units of time and still be consistent with the temperature distribution that is assumed between points 2 and 3. These conditions are, in general, satisfied by postulating an *increase* of air temperature between levels 2 and 3. By Eq. 11.7, the curvature reverses above 2 to concave downward, the path length is shortened over a pure reflection phenomenon and a path such as is shown as the solid line of Fig. 11.2 is constructed. This technique for determining the ray path of a sound wave is another example of a self-consistent solution wherein several parameters must be satisfied simultaneously.

The wind speed, in practice, must always be taken into account. To show the necessity for this, consider that the angle of descent of the sound ray may be as great as 35°. For a stratosphere temperature of −60° C, the horizontal component of the sound velocity in the stratosphere may be as little as 293 cos 35°; i.e. 240 meters per second. Wind speeds of 45 meters per second (100 mph) are not uncommon in the stratosphere. Because the wind may be in the direction of, opposing the direction of, or in general be at an angle to the horizontal component of the sonic velocity in still air, the speed of sound relative to a stationary set of detectors may be measured in the range 270 + 45 to 270 — 45 meters per second in traversing a specified path. For the wind and temperature pattern shown in Fig. 11.3, the ray path of the sound wave through the atmosphere would correspond to the second curve (dashed-line notation) of Fig. 11.2. Although the assumed wind distribution between levels 2 and 3 will influence the speed of travel of the sound wave relative to the ground, there is usually so little wind shear† that an increase of temperature with height is normally required for sound to refract toward the earth.

The sounds that are propagated through the upper stratosphere with slight attenuation are the low-frequency components of the sound wave. The high-frequency components are rapidly damped out and are not detected. For example, Schrodinger‡ gives the transmissivity

† Wind shear is the change of the horizontal wind velocity v_H with height z. Mathematically, horizontal wind shear equals $\frac{\partial v_H}{\partial z}$.

‡ Schrodinger, E., *Physik. Z.*, *18*, 445 (1917).

of sound in air as

$$\tau_\lambda = e^{-30.1\frac{\bar{l}s}{\lambda^2}} \tag{11.8}$$

where $\bar{l}$ is the mean free path of the molecules (a function of height), λ is the wavelength of the sound wave, and s is the path length through the air. Since $\bar{l}s/\lambda^2$ is dimensionless, any consistent length unit may be used. From this equation one finds that a sound wave traveling through 1 kilometer of air at 0° C, 50 kilometers high in the atmosphere, suffers an attenuation of only 1 per cent in its transmitted intensity if the wavelength of the sound is 22 meters (frequency = 15 cps), but it suffers 99.9 per cent attenuation if the wavelength is only 2 meters long (166 cps). Since 20 cps is at the lower limit of audibility for the human ear, sound that is heard at great distances has a low-pitched rumble and has not traveled at elevations greater than 50 to 60 kilometers in the atmosphere. Even instrumental detection of sound waves is frequency-limited to detecting sound waves that have traveled at maximum altitudes not in excess of 80 kilometers in the atmosphere. Physically, Eq. 11.8 indicates that serious attenuation of a sound wave occurs when large numbers of air molecules do not suffer collision with other air molecules within distances comparable to the wavelength of the sound. Because of this motion, large numbers of particles move from a region of high density to a region of lower density, destroying the orderly periodic motion of the air that constitutes the sound wave. This loss in order manifests itself as a decrease in the intensity (periodic pressure effects) of the sound wave. The mean free path is a statistically derived parameter indicating this effect.

Data obtained on the speed of sound in the upper stratosphere indicates an increase in temperature between the 35- and 60-kilometer level in the atmosphere. Data will be summarized in Table 11.1.

The density of the upper atmosphere from meteor observations

The earth in its yearly journey around the sun encounters at certain times of the year large numbers of small particles called meteors. Their average mass is estimated as of the order of 6×10^{-3} gram and their density is comparable to that of iron. As their orbits are not coincident with that of the earth, many of the meteors enter the earth's atmosphere at high relative velocities, of the order of 50 to 200 times the speed of sound at sea level. The resistance of the earth's atmosphere causes them to slow down and become incandescent. Their mass is so small that almost all of these meteors completely volatilize by the time they reach a height as low as 55 kilometers in the atmosphere. Occasionally, a meteor is so large that it penetrates the entire atmosphere and hits the surface of the earth as a meteorite. We shall not be concerned with these. Most of the work on meteors is concerned with the small meteors that first appear at heights as great as 100 to 150 kilometers and disappear no lower than 35 kilometers. The disappearance must be characterized by a smooth burning out rather than by an explosive disintegration for the subsequent theory to be valid.

The equipment for making meteor observations consists of two wide field telescopes situated a known distance apart (of the order of 1000 kilometers) and both monitoring the same section of the heavens. Observations are made on photographic film. The telescope cameras have open shutters about eight-ninths of the time. The other one-ninth of the time, a rotating sector acts as a shutter to cut off any light from the heavens. The shutters rotate at about 20 times per second. Thus, both cameras during the course of an evening are periodically exposed to any light that comes from the heavens. A meteor on entering the atmosphere and becoming incandescent will leave an exposed trail on the photographic film interrupted every one-twentieth of a second by an unexposed mark as the shutter closes. Thus, in the second or so that a meteor is visible, it is possible to measure from the film both the height of appearance and of disappearance of the meteor, the direction that the meteor makes with the zenith (the zenith direction of the apparent radiant), the velocity and deceleration of the meteor at every height, and from the density of the exposed film the intensity of the light from the meteor.

Utilizing these measurable quantities as parameters, Lindemann and Dobson(11) and later Whipple(10,23,24) have given a complete theory for meteor burnout. The only other parameter in the theory is the air density, and in practice it is for this parameter that the equations are solved. The theory follows.

Let us adopt the symbols M for the mass of a meteor whose equivalent spherical radius is r, t for time, ρ for air density, T for air temperature, I for the intensity of the light from the meteor, s the length of atmospheric path traversed by the meteor (given by $z \sec \zeta$ where z is the geometric height in the atmosphere and ζ the apparent zenith angle of the path), p the air pressure, and L the heat required to volatilize a unit mass of the meteor. Factors of proportionality will be denoted by the letter C with subscripts.

A meteor on entering the earth's atmosphere suffers a loss in momentum per unit time of $\frac{d}{dt}(Mv)$. Part of the momentum is lost by deceleration and the rest by a loss of mass, once an incandescent temperature is attained. The air gains the momentum lost by the meteor. Part of this momentum assumes the form of a resisting force and part goes to compress the gas immediately ahead of the meteor, raising in turn both the temperature of the air and the temperature of the meteor. This temperature-raising process continues until the volatilizing temperature of the meteor is reached and the meteor begins to lose mass. At this point, the energy transfer per unit time can be shown to be[10]

$$v \frac{d}{dt}(Mv) = Mv \frac{dv}{dt} + v^2 \frac{dM}{dt} = C_1 \rho v^2 M^{2/3} + C_2 I. \qquad (11.9)$$

The term $C_1 \rho v^2 M^{2/3}$ comes from ballistic theory. This term is just the product of the velocity and the mass of air contained in a column v units long (in each unit of time dt) and cross section πr^2. The meteor is assumed to have an equivalent spherical mass of $\frac{4}{3}\pi r^3$. As M is proportional to r^3, r^2 is proportional to $M^{2/3}$. Figure 7.4 of Chapter 7 may be consulted as a guide in interpretation. The term $C_2 I$ expresses the fact that the high temperatures formed in the compressed gas cap can only be observed as the intensity of light that the glowing meteor gives off.

At the high velocities with which meteors are observed to travel, C_1 and C_2 are not constant but are functions of velocity,

$$C_1 = K_1 v \qquad (11.10)$$

and

$$C_2 = \frac{K_2}{v}. \qquad (11.11)$$

K_1 and K_2 are constants that are empirically observed although theoretical values can be derived.

Equation 11.9 can be identified as the two separate equations

$$Mv \frac{dv}{dt} = C_1 \rho v^2 M^{2/3} \qquad (11.12)$$

and

$$C_2 I = v^2 \frac{dM}{dt}. \qquad (11.13)$$

Symbolizing the deceleration dv/dt by $\dot{v}$ and substituting for C_1 from Eq. 11.10, Eq. 11.12 when solved for the air density yields

$$\rho = \frac{\dot{v}}{K_1 v^2} M^{1/3}. \qquad (11.14)$$

Substituting K_2/v for C_2 in Eq. 11.13 and solving for the change in mass shows

$$dM = K_2 \frac{I}{v^3}\, dt. \tag{11.15}$$

Postulating that the mass of the meteor is zero at t_d, the time of disappearance of the meteor, the mass M at any time t before that is

$$M = K_2 \int_{t_d}^{t} \frac{I}{v^3}\, dt \tag{11.16}$$

provided that the time is reckoned positive in the direction of increasing mass.

By the use of Eq. 11.16, the mass can be eliminated from Eq. 11.14. The air density at any time t is given then by

$$\rho = K \frac{\dot{v}}{v^2} \left[\int_{t_d}^{t} \frac{I}{v^3}\, dt \right]^{1/3}. \tag{11.17}$$

The single constant K has been substituted for the combination $K_2{}^{1/3}/K_1$. All the quantities of Eq. 11.17 can be measured and K is empirically determined by the solution of the equation at low elevations where ρ is known from other independent techniques for measuring the air density.

The techniques whereby Lindemann and Dobson evaluated the air density can be derived from Eq. 11.14, although not precisely in the form in which they expressed their results.

Consider Eq. 11.14 before the exact instant the meteor becomes visible. The air density in outer space is zero. The velocity of the meteor relative to the earth is so high that to be considered infinite causes a small and unimportant error mathematically. Consider then Eq. 11.14 when integrated from the conditions in outer space to the time when the meteor first becomes visible. Denote this instant of time by t_a, the air density by ρ_a, and the velocity of the meteor by v_a. The integration limits in velocity are negative because decelaration occurs with increasing time. Equation 11.14 becomes

$$\int_0^{t_a} \rho\, dt = \int_0^{t_a} \rho(t)\, dt = \frac{M_0{}^{1/3}}{K_1} \int_{-\infty}^{-v_a} \frac{dv}{v^2}. \tag{11.18}$$

The left-hand side of the equation is simply ρ_a, since ρ under the integral sign is a function of time and the value of this integral at time t_a must yield the true instantaneous air density. The right-hand side of the equation is a straightforward integration. The results of the integration show that the density of the air at the time the meteor of mass M_0 is first observed is given by

$$\rho_a = \frac{M_0^{1/3}}{K_1 v_a}\cdot\dagger \tag{11.19}$$

From our equations, we can also deduce the conditions that should be encountered at the point of disappearance of a meteor. We proceed in the following manner.

During the duration of the meteor trail, the heat energy gained by the meteor from the air is reflected in the amount of heat $L\frac{dM}{dt}$ required to volatilize dM units of the meteor in unit time. Since the energy must be supplied by the air, it follows that

$$L\frac{dM}{dt} = C_3\rho v^2 M^{2/3}. \tag{11.20}$$

As before

$$C_3 = K_3 v \tag{11.21}$$

where C_3, K_3, and L are new factors of proportionality that must be empirically determined. Solving Eq. 11.20 for dM after substituting for C_3 from Eq. 11.21 yields

$$dM = \frac{K_3}{L}\rho v^2 M^{2/3} v\, dt. \tag{11.22}$$

Now

$$v\, dt = ds = dz \sec \zeta \tag{11.23}$$

from geometrical considerations of the path the meteor makes with the zenith direction (angle ζ). Equation 11.23 and the hydrostatic equation show

$$dp = -\rho g\, dz = -\rho g v \cos \zeta\, dt, \tag{11.24}$$

and we find thereby that Eq. 11.22, with the help of Eq. 11.24, may be expressed as

$$\int_{M_0}^{0} \frac{dM}{M^{2/3}} = \frac{K_3\overline{v^2}}{L}\int_{t_a}^{t_d} \rho v\, dt \cong -\frac{K_3}{L}\frac{\overline{v^2}}{g}\sec\zeta \int_0^{p_d} dp. \tag{11.25}$$

Integrating Eq. 11.25 and solving for p_d gives

$$p_d = \frac{K_4}{\overline{v^2}} M_0^{1/3} \cos \zeta \tag{11.26}$$

when $3(Lg/K_3)$ is set equal to K_4.

† The case of normal incidence only is discussed. Should the meteor enter the atmosphere at angle ζ with the zenith, this equation should be multiplied by $\cos \zeta$. See also Eq. 11.27 as a guide.

From the equation of state, Eq. 11.1, and Eq. 11.26, we find that the density of the air at the level of disappearance is

$$\rho_d = \frac{K_4}{R_G} \frac{M_0^{1/3}}{\overline{v^2} T_d} \cos \zeta. \tag{11.27}$$

M_0 is the mass of the meteor at the time it enters the atmosphere. T_d, which has to be assumed, is the air temperature at the level of disappearance, and $\overline{v^2}$ is the average value of the square of the velocity over the entire visible trail.

The classical curves of Lindemann and Dobson, showing the disappearance of the meteors, are reproduced as Fig. 11.4. Although these curves have a smooth envelope as drawn, the actual data show considerable scatter. They do show that the majority of meteors appear at heights greater than 100 kilometers and disappear most frequently between 70 and 80 kilometers. A most significant feature, which requires explanation, is the appearance of a minimum in the disappearance curve at about 55 kilometers and a secondary maximum at nearly 48 kilometers.

From equations of the type of Eqs. 11.19 and 11.27, Lindemann and Dobson were able to make the following interpretation of the curves of Fig. 11.4. The conclusions have also been substantiated by Whipple and his co-workers, using Eqs. 11.17 and 11.26.

If one assumes an isothermal atmosphere whose temperature is equal to that of the lower stratosphere (about 220° K), the density can be calculated for any height, provided that a standard density is known. For example, the hydrostatic equation for an ideal gas is

$$dp = -\rho g\, dz = -\frac{p}{R_G T} g\, dz. \tag{11.28}$$

Integrating Eq. 11.28 for an isothermal temperature T_0, letting $z = 0$ at the level where $p = p_0$,

$$\int_{p_0}^{p} \frac{dp}{p} = -\frac{g}{R_G T_0} \int_0^z dz,$$

and

$$p = p_0 e^{-\frac{gz}{R_G T_0}}. \tag{11.29}$$

In an isothermal atmosphere, we have from Eq. 11.1 that

$$\frac{p}{p_0} = \frac{\rho}{\rho_0} \tag{11.30}$$

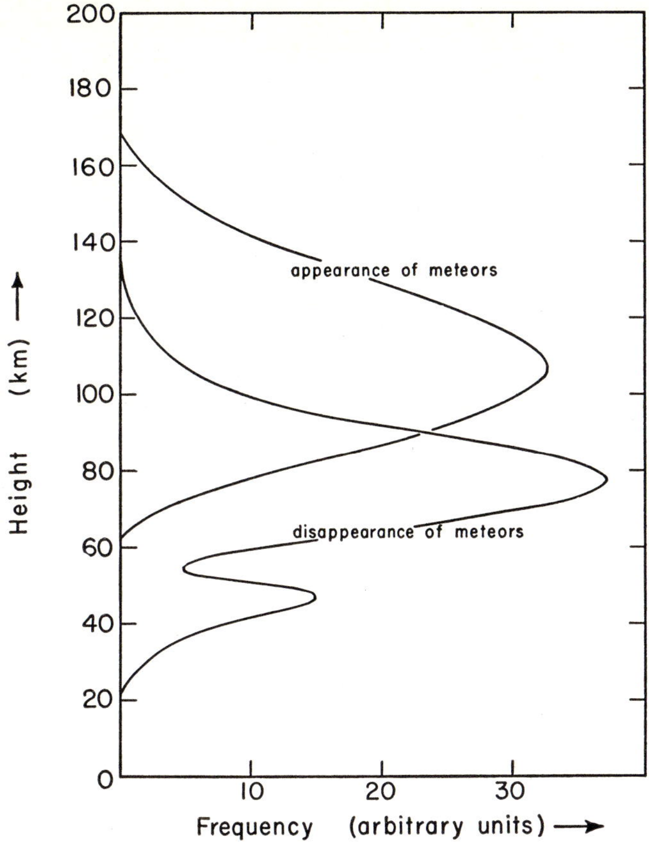

Fig. 11.4 (*After Lindemann and Dobson.*) The relative frequency as a function of the height at which meteors were observed to appear and disappear. The velocities at appearance and disappearance were of the order of 20 to 100 kilometers per second, with a few of higher speed noted at both appearance and disappearance.

so that Eq. 11.29 may be written as

$$\rho = \rho_0 e^{-\frac{gz}{R_G T_0}}. \tag{11.31}$$

Figure 11.5 shows the variation of density with height as computed from Eq. 11.31 when an isothermal atmosphere of 220° K is assumed. The two dashed lines indicate the limits of the density as computed from the heights of appearance and disappearance of meteors, equations of the type of Eqs. 11.19 and 11.27. It is to be noted that the density does not fall off so rapidly with height as is predicted by

Eq. 11.31, but at a rate characteristic of a higher mean temperature. The conclusion is therefore that the temperature begins to increase rapidly with height above 85 kilometers.

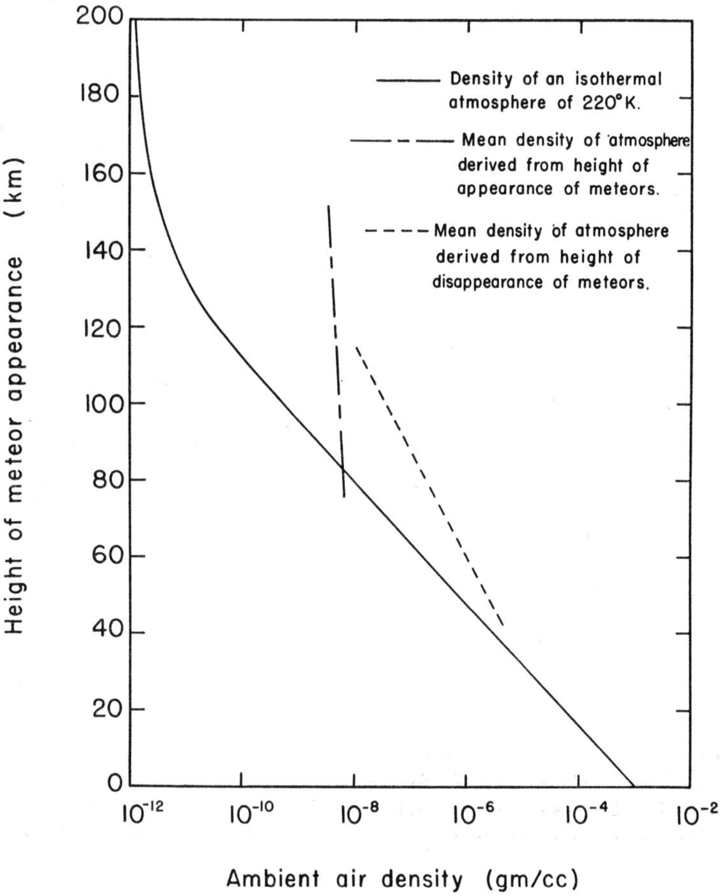

Fig. 11.5 (*After Lindemann and Dobson.*) A comparison of the ambient density of the upper atmosphere from the appearance and disappearance of meteors and a calculation for the density of an isothermal atmosphere of 220° K. The data shows the density of the upper atmosphere to correspond to a much higher temperature than 220° K. Equation 11.31 may be used. The theoretical calculation of density assumed a much greater preponderance of lighter gases above 100 kilometers than below.

The minimum in the height-disappearance frequency curve (Fig. 11.4) at 55 kilometers can be explained if a region of high temperature is assumed to exist just above 55 kilometers. Theory shows that if the density increased uniformly as in an isothermal atmosphere, the height of disappearance of the meteors should be clustered about a single maximum. However, if the meteor passes through a region of cooler air and, thus, by Eq. 11.31, less dense air in the mean for a stated height interval, the rate of evaporation of the meteor will be retarded and burn-out will occur in the next lower region of high temperature and density, i.e., the region beginning at about 60 to 65 kilometers. The burn-out is completed about 15 kilometers below this level, because a finite time is required for complete evaporation, once a high rate of vaporization has been re-established.

Temperature, density, and other properties of the upper atmosphere as measured by rockets

Since the close of World War II, the German V-2 rocket and other smaller unguided missiles propelled on the rocket principle have been utilized for probing the upper atmosphere to heights that are beginning to be in excess of 100 kilometers. These rockets are able to transport to these elevations various scientific apparatus capable of measuring temperature, density, sun's spectra, ozone content, cosmic rays, and other phenomena of interest. The information obtained can be telemetered to the ground station, and in some cases the original records recovered after the instrument section has been parachuted to earth. By a system of radio direction finders and a radio beacon in the missile, the height and speed of the missile can be determined. Since these missiles travel at several times the speed of sound and are therefore at any one level for a very short time, the problem of measuring a quantity such as air temperature is difficult and often subject to large error. In spite of these difficulties, however, reliable measurements of the properties of the upper atmosphere are made.

As an illustration of the philosophy of measurement from instrumented rockets, we shall discuss two of the many types of measurements possible; the first will be a measurement of air temperature and the second an air-density measurement. Direct measurements of temperature by conventional thermometers are not possible, because the high thermal lags and the large errors introduced by the compression of the air on the leading edge of the thermometer invalidate the results.

The first and perhaps oldest method of temperature measurement is made through the use of the hydrostatic equation in the differential form of Eq. 11.28. This equation may be expressed as

$$T = -\frac{g}{R_G} p \frac{dz}{dp} = -\gamma_A \frac{dz}{d \ln p}. \qquad (11.32)$$

γ_A, as always, is the autoconvective lapse rate. It is possible to locate the barometric instruments in such a position on a rocket that the change in static pressure can be obtained. The height of the missile can also be determined continuously. Equation 11.32 indicates that the slope of the curve of height versus the logarithm of the pressure is proportional to the temperature.

A second method of measurement uses the Bernoulli equation to relate the ram or dynamic pressure† built up at the nose of a high-speed missile to the air density. If p' is the stagnation pressure built up when air traveling at velocity v relative to the rocket is brought to rest at the nose of the rocket and p is the static (ambient) pressure, the air density at any height where the pressure is p is

$$\rho = 2\frac{p' - p}{v^2}. \tag{11.33}$$

Since p', p, and v are measurable, ρ and p are known and the temperature can be found from Eq. 11.1. The practical difficulties in determining the proper values of p and p'‡ have subjected this measurement to relatively large probable errors.

In the main, the rocket measurements have substantiated the temperature-height profiles as deduced from the more indirect acoustical, meteor, and ozone determinations.

Other temperature measurements

Other inferences of the temperature of the upper atmosphere than the ones described are obtained by even more indirect means. In general, these methods of determining temperature are subsidiary to the main problem under consideration, e.g., the equilibrium amount of ozone, where the temperature enters through temperature-sensitive coefficients. In order for the theoretical solution of a given problem to give results consistent with observation, the proper value of coefficients must be selected, which, if the coefficients are temperature dependent, determines a temperature profile for the atmosphere. A particularly valuable example of this is in the theory of the photochemical equilibrium of ozone.

The photochemical equilibrium of ozone is discussed in Chapter 10,

† The dynamic pressure is the difference between the stagnation and static pressures in horizontal flow; i.e., $p' - p = \frac{1}{2}\rho v^2$.

‡ For example, the precession of a rocket about its line of flight causes the pressure force on a measuring instrument to vary with time in a fashion that may not be completely determinable and therefore may permit only approximate corrections.

Eqs. 10.22 through 10.32. The ratio k_{12}/k_{13} of Eq. 10.28 is sensitive to changes in temperature. Calculations of the equilibrium ozone distribution caused by the photochemical effect of sunlight require an increasing temperature above 35 kilometers. It has also been found experimentally that in the Huggins region of ozone, the spectroscopic determination of the absorption coefficients in the regions of minimum absorption are a linear function of temperature. The amount of absorption decreases with a decreasing mean temperature of the ozone layer traversed by a beam of solar ultraviolet light of the proper frequency.

In the ionosphere, the mean temperature of a layer can be determined from the scale height, Eq. 10.3 of Chapter 10. The mean temperature of a layer can be found from values of the recombination coefficients of the ions. For example, an empirical expression for the recombination coefficient Γ for ions depends on the absolute temperature in the following fashion:

$$\Gamma = 1.02 \times 10^{-11} \frac{Z^2}{T^{1/2}} \text{ cubic centimeters per second.} \quad (11.34)$$

Z is the number of nuclear charges in an atom (7 for the nitrogen and 8 for the oxygen atom).

The green line of atomic oxygen ($\lambda = 0.5577\,\mu$) emitted from the 200–300 kilometer level of the atmosphere gives evidences of temperatures of the order of 1200° K at these levels. This high temperature is deduced from the relationship of the broadening of an atomic line by the Doppler effect, wherein molecules under thermal agitation cause frequency shifts because of a component of motion toward or away from the observer. The fractional change in wavelength $\Delta\lambda/\lambda$ is related to the absolute temperature T by

$$\frac{\Delta\lambda}{\lambda} = 7.2 \times 10^{-7} \sqrt{\frac{T}{M}}. \quad (11.35)$$

M is the atomic mass of the emitting atom in grams.

Concept of temperature above 100 kilometers

Until observations of pressure and temperature were extended well into the stratosphere, there was no need to question the conventional meteorological concepts of the atmosphere being a thermodynamic system obeying the classical equations of motion of a gas whose equation of state is given by Eq. 11.1. However, we have known since the middle of the nineteenth century that when gases are maintained at very low pressures (< 0.01 millibar), our concept of temperature becomes the statistical averaging of the momentum exchange between the gas molecules and a thermometer surface. The statistical theory of the distribution of speeds of non-ionized gases was developed by Maxwell in England and Boltzmann in Germany. The theory is discussed in any textbook on statistical mechanics.† We shall borrow without proof the results of the theory.

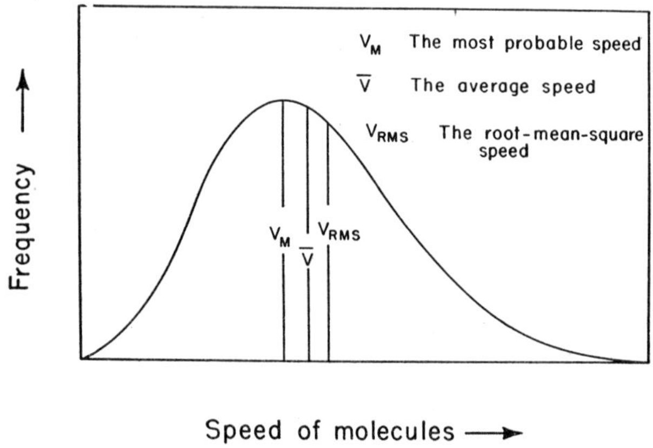

Fig. 11.6 A Maxwell-Boltzmann distribution of molecular speeds in a gas with a uniform ambient temperature. v_M, $\overline{v}$, and v_{RMS} are velocity parameters dependent on the temperature of the gas. The Maxwell-Boltzmann distribution shifts along the velocity axis as the temperature is changed.

A Maxwell-Boltzmann distribution of speeds contains all speeds between zero and infinity. Such a distribution is shown as Fig. 11.6. Certain parameters of this distribution can be expressed, among them the arithmetic mean speed, the most probable speed, and the root-mean-square speed. The theory shows that each distribution of these speeds is a unique function of the absolute temperature of the gas. For example, if $\bar{v}$ is the arithmetic mean speed of a gas whose absolute temperature is T, the two quantities are related by

$$\bar{v} = \sqrt{\frac{8}{\pi}\frac{kT}{M}}. \tag{11.36}$$

k is the Boltzmann constant and M is the mass of a molecule of the particular gas under consideration. Thus, the temperature of a gas may

† See F. W. Sears, *An Introduction to Thermodynamics, the Kinetic Theory of Gases, and Statistical Mechanics,* Addison-Wesley Press (1950); or, for a more advanced discussion, see E. H. Kennard, *Kinetic Theory of Gases,* second edition, McGraw-Hill Book Co., New York (1938).

be thought of as being a parameter that describes the mean speed of the molecules of the gas.

The molecules in their motion collide with other molecules. The number of collisions per a unit time suffered by the n molecules per unit volume of the gas when all molecules are moving with a speed equal to the arithmetic mean speed $\bar{v}$ is called the collision frequency. The collision frequency ν is given by

$$\nu = \pi n \sigma^2 \bar{v}. \tag{11.37}$$

σ is the collision cross section which, for hard spheres, would be equal to the sum of the radii of two spheres. If force fields are present the collision cross section may be other than the sum of the experimentally determined radii.

The average distance between collisions is called the mean free path. It is the total distance covered by a representative molecule traveling at speed $\bar{v}$ for t units of time divided by the number of collisions suffered in time t. Letting $\bar{l}$ be the mean free path,

$$\bar{l} = \frac{\bar{v}t}{\nu t} = \frac{1}{\pi n \sigma^2}, \tag{11.38}$$

with the aid of Eq. 11.37. When the molecules in a gas have a Maxwellian distribution of speeds rather than a uniform speed $\bar{v}$, the mean free path is given by

$$\bar{l} = \frac{1}{\sqrt{2}\,\pi n \sigma^2}. \tag{11.39}$$

Using these equations, one can show that the mean free path of an air molecule is of the order of 10^{-5} centimeter at sea level, 10^{-2} centimeter at 50 kilometers, 1 centimeter at 100 kilometers, and 10^4 centimeters at 400 kilometers.

A material body, such as a conventional mercury in glass thermometer, when subjected to molecular bombardment at sea level, undergoes in unit time so many collisions with the air molecules that a temperature equivalent to the statistical mean speed is always reached. However, when the mean free path of the molecules reaches a dimension comparable to the dimensions of the device used to indicate temperature, the collisions are less frequent, and a correspondingly longer period of time is required for a statistically representative number of molecular collisions to occur. Only after this minimum period of time can a body be said to have the same temperature as the surrounding air. In fact, as the molecular mean free path becomes long and the collision frequency correspondingly less, the temperature of a body is determined by radiative equilibrium with its surroundings rather than by the conduction of heat to or from the surrounding air through molecular bombardment. It is because of these factors that at heights in excess of 100 kilometers temperature measurements by means acceptable at higher air densities fail as indicators of the ambient air temperature in this thin atmosphere. Consequently, above 100 kilometers, atmospheric temperatures are deduced from spectroscopic observations of the aurora and the airglow, from theories of the radiative equilibrium of the atmospheric constituents, and from the temperature dependence of ionospheric parameters.

The temperature distribution of the atmosphere from sea level to 100 kilometers

The various methods of measuring the temperature of the upper atmosphere have indicated that on the average there is a temperature increase in the stratosphere above 30 kilometers. The temperature profile reaches a maximum at 50 to 60 kilometers, a minimum temperature at 80 to 85 kilometers, and then increases in temperature above this level at a rate of about 4° C per kilometer. The temperature increases cited are due to the selective absorption of solar ultraviolet radiation by ozone in the 30- to 70-kilometer region† and by the selective absorption of solar ultraviolet radiation by the *D* layer of the ionosphere in the 60- to 90-kilometer region. Above 100 kilometers, large fluctuations in temperature between day and night are to be expected on the basis of the large variations in ionospheric concentrations of ions between day and night. The highest temperatures would be encountered in the daylight hours.

On the basis of the available information, the National Advisory Committee for Aeronautics has issued an NACA standard atmosphere that lies between the extremes of temperature reported. Figure 11.7 is a plot of the NACA standard atmosphere. Dashed lines are included on either side of the standard lapse of temperature to show the probable variation from the standard that might be encountered on any one thermal sounding of the atmosphere at any single time and geographical position in the atmosphere.

Vertical pressure distribution in the atmosphere

Figure 11.8 shows the pressure distribution with height consistent with the NACA standard atmosphere. The pressure distribution was computed from the proper application of Eq. 11.32.

† Although the greatest mass of ozone is found below 30 kilometers, the production and destruction of ozone occurs in the main above 30 kilometers. The equilibrium rate of ozone production is temperature dependent (see Chapter 10).

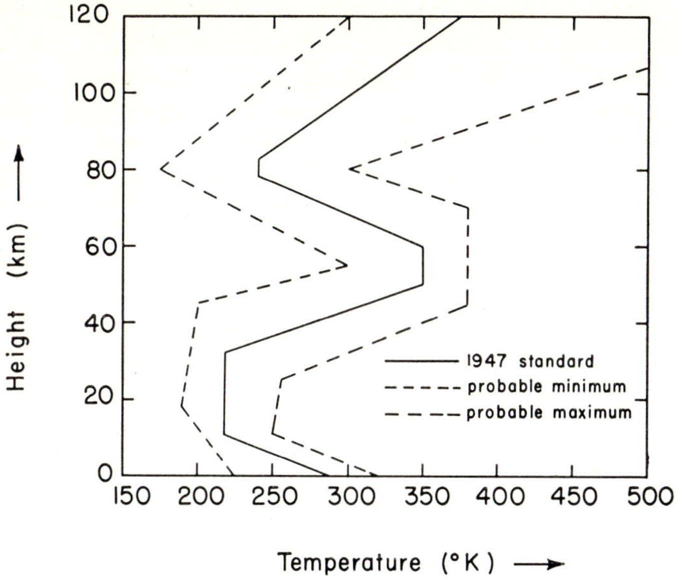

Fig. 11.7 The National Advisory Committee for Aeronautics (NACA) standard atmosphere. The limits of probable deviation from the standard are indicated by the dashed lines.

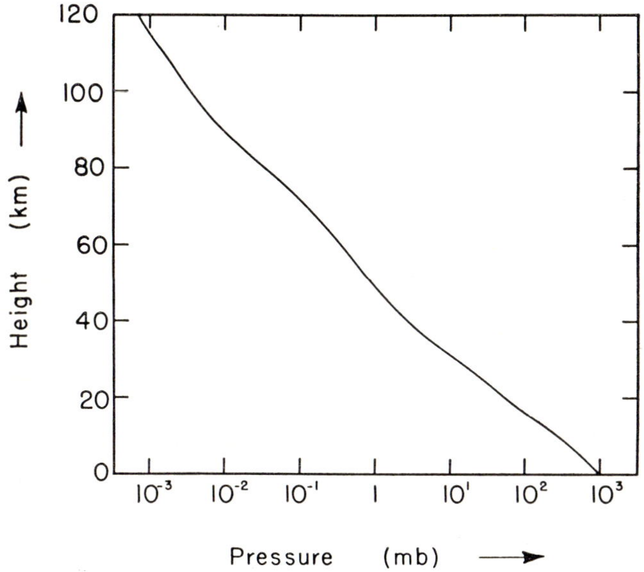

Fig. 11.8 The pressure distribution in an NACA standard atmosphere.

Table 11.1

Upper-Air Temperatures (°K) as a Function of Height Derived from Various Techniques of Measurement

The values at any height are only typical. Wide variations exist within a particular system of measurement as well as between the different measuring techniques. The measurements are so few that a synoptic representation, a diurnal representation, and a seasonal representation of temperature over the world can only be estimated. The temperature variations may be exceedingly large, making it difficult to assume an average condition that is meaningful.

	Balloons		Ozone		Sound		Meteors	
Height, km								
0–20	Highly Variable							
20	217	219	230	250	218	193		
25	225	220	250	265	218	195		
30	230	223	280	290	223	205		
35	238		305	315	233	214		
40	250		365	385	245	235		
45			410	420	268	255		
50			435	445	293	263		
55			450	465		313		271
60							288	271
65							281	277
70							281	279
75							288	281
80							283	276
85							274	274
90							295	273
95							295	
100								
105								
110								

	Rockets		Aurora		*E*-layer		NACA
Height, km							A standard
0–20							exists†,‡
20	210	215					218.0
25	215	220					218.0
30	220	235					218.0
35	245	245					240.0
40	305	250					276.7
45	300	260					313.3
50	290	260					350.0
55	280	245					350.0
60	240	255					350.0
65	210	275					319.4
70		240					288.9
75		205					258.3
80							240.0
85							247.3
90							265.5
95							283.8
100			228				302.0
105							320.3
110				347	350		338.5
115							356.8
120							375.0

† Diehl, W. S., "Standard Atmosphere—Tables and Data," NACA Report No. 218 (1925).

‡ Brombacher, W. G., "Altitude-pressure Tables based on United States Standard Atmosphere," NACA Report No. 538 (1935).

Water vapor and liquid water in the stratosphere

The question of the normal amount of liquid water and water vapor in the stratosphere is still (1953) in a state of indeterminacy because of lack of sufficient observational data. The amounts of water vapor measured to date are exceedingly small. For comparison with experi-

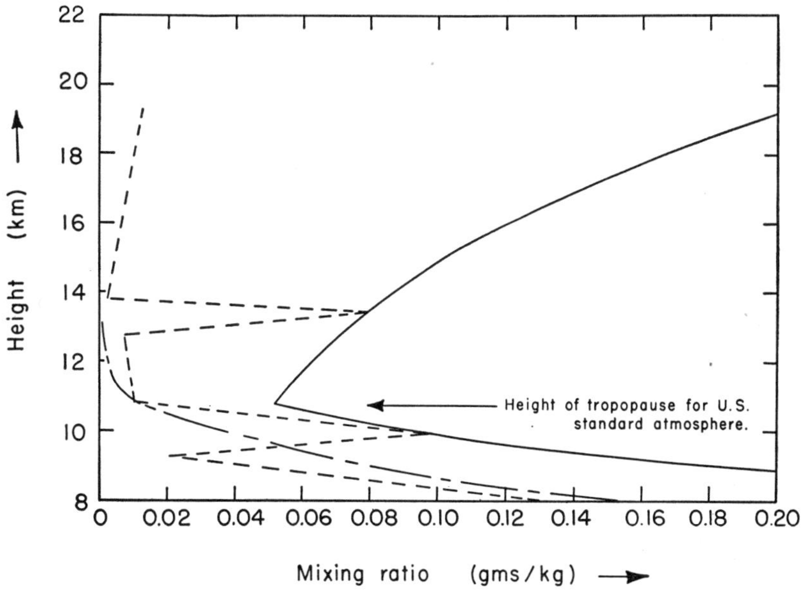

Fig. 11.9 The solid line is a plot of the saturated mixing ratio as a function of height in an NACA standard atmosphere. The curve with center-line notation (—— – ——) represents average moisture conditions in the stratosphere over England as reported by Dobson, and the dashed line (– – –) indicates the results of one balloon ascent over America as reported by Barrett. The data have been reduced to a standard atmosphere representation in order to show trend and approximate values only. Note that although the stratosphere is normally very dry, saturated layers can exist.

mental data, Fig. 11.9 shows the saturation mixing ratio as a function of height in an NACA standard atmosphere. The values were computed from the approximate formula for w_s, the saturation mixing ratio, in units of grams of water vapor per kilogram of air,

$$w_s = 622 \frac{e_s}{p} \cdot \qquad (11.40)$$

The amount of water vapor and cloud at the tropopause is highly variable, ranging from nearly dry (relative humidities as low as 3 per cent have been observed) to water saturation. An average value would be near 50 per cent relative humidity. Measurements of water vapor above the tropopause have been made in both England[5] and America,[1] where the average height of the tropopause is near 11 kilometers, but as yet no measurements have been made in extreme polar or tropical regions. In England, the humidities have been measured to heights of 13 kilometers using a frost-point hygrometer. In America, Barrett[1] and coworkers have reported a successful ascent where the temperature and moisture content of the atmosphere to 30 kilometers were measured. A saturated layer at 16 kilometers was reported, with very dry air in other levels.

The results to date indicate that the lower stratosphere is normally very dry, with the relative humidity falling to 2 or 3 per cent in respect to ice at 1 or 2 kilometers above the tropopause. These results are in strong contrast to the view that is often held that the average moisture condition of the stratosphere is one of near saturation. The evidence usually quoted for a saturated stratosphere is the presence of nacreous (mother-of-pearl) clouds that occasionally appear at heights of about 25 kilometers in high latitudes. It is argued that the spectral colors observed with these clouds indicate water droplets of 1 to 10 micron diameter. The clouds are used as evidence that the stratosphere must be saturated at these levels. This view does not seem unreasonable, especially in the light of Barrett's work. It is premature, however, to consider the origin and even composition of nacreous clouds as definitely determined.

In the very high atmosphere, during the early evening hours when the earth is dark but direct sunlight illuminates the upper atmosphere, a rare cloud is observed on occasion at a mean height of 82 kilometers. The cloud is named a noctilucent cloud because it is observed at night. No satisfactory explanation for its origin is given, although it may be composed of meteoric dust. Eighty-two kilometers coincides with the minimum temperature level in the upper atmosphere and because of the stability of the layer of air above would be an optimum region for maintaining a high concentration of particulate matter.

References

1. Barrett, E. W., Herndon, L. R., Jr., and Carter, H. J., "A Preliminary Note on the Measurement of Water Vapor Content in the Middle Stratosphere," *J. Meteorol.,* *6,* 367–368 (1949).
2. Brasefield, C. J., "Winds and temperatures in the lower stratosphere," *J. Meteorol.,* *7,* 66–69 (1950).
3. Cox, E. F., and others, "Upper-Atmosphere Temperatures from Helgoland Big Bang," *J. Meteorol.,* *6,* 300–311 (1949).

4. Crary, A. P., "Stratosphere Winds and Temperatures in Low Latitudes from Acoustical Propagation Studies," *J. Meteorol., 9,* 93–109 (1952).
5. Dobson, G. M. B., Brewer, A. W., and Cwilong, B., "Meteorology of the Lower Stratosphere," *Proc. Roy. Soc.,* (A) *185,* 144–175 (1946).
6. Elterman, L., "The Measurement of Stratospheric Density Distribution with the Searchlight Technique," *Geophysical Research Papers,* No. 10, Air Force Cambridge Research Center, Cambridge, Massachusetts (1951).
7. Grimminger, G., "Analysis of Temperature, Pressure and Density of the Atmosphere Extending to Extreme Altitudes," Report No. 105, Rand Corp., Santa Monica, California (1948).
8. Gutenberg, B., "Propagation of Sound Waves in the Atmosphere," *J. Acoust. Soc. Amer., 14,* 151–155 (1942).
9. Hulburt, E. O., "The Upper Atmosphere of the Earth," *J. Opt. Soc. Amer., 37,* 405–415 (1947).
10. Jacchia, L. G., "Ballistics of the Upper Atmosphere," Harvard College Observatory and M. I. T. Center of Analysis, Tech. Report No. 2 (1948).
11. Lindemann, F. A., and Dobson, G. M. B., "Theory of Meteors and Density and Temperature of the Outer Atmosphere to which it Leads," *Proc. Roy. Soc.,* (A), *102,* 411–437 (1922–23).
12. McKinley, B. M., and McKinley, D. W. R., "Photoelectric Meteor Observations," *Can. J. Phys., 29,* 111–121 (1951).
13. Nazarek, A., "The Temperature Distribution of the Upper Atmosphere over New Mexico," *Bull. Am. Meteorol. Soc., 31,* 44–50 (1950).
14. Newell, H. E., Jr., "A Review of Upper Atmosphere Research from Rockets," *Trans. Amer. Geophys. Union, 31,* 25–34 (1950).
15. Scultetus, H. R., "Ältere Beobachtungen von Leuchtstreifen," *Z. Meteorol.,* Heft *8/9,* Jahrgang 3 (August—September, 1949).
16. Störmer, C., "Mother-of-Pearl Clouds," *Weather, 3,* 13–18 (1948).
17. The Rocket Panel, "Pressures, Densities, and Temperatures in the Upper Atmosphere," *Phys. Rev., 88,* 1027–1032 (1952).
18. Veldkamp, J., "On the Propagation of Sound Over Great Distances," *J. Atm. and Terrest. Phys., 1,* 147–151 (1951).
19. Vestine, E. H., "Noctilucent Clouds," *J. Royal Astron. Soc. Canada, 28,* No. 6; 249–272, No. 7: 303–317 (1934).
20. Warfield, C. N., "Tentative Tables for the Properties of the Upper Atmosphere," *Nat. Advisory Comm. Aeronaut.,* Tech. Note. No. 1200 (January, 1947).
21. Whipple, F. J. W., "The High Temperature of the Upper Atmosphere as an Explanation of Zones of Audibility," *Nature, 111,* 187 (1923).
22. Whipple, F. J. W., "The Propagation of Sound to Great Distances," *Quart. J. Royal Meteorol. Soc., 61,* 285–308 (1935).
23. Whipple, F. L., "Meteors and the Earth's Upper Atmosphere," *Rev. Mod. Physics, 15,* 246 (1943).
24. Whipple, F. L., "Results of Rocket and Meteor Research," *Bull. Am. Meteorol. Soc., 33,* 13–25 (1952).
25. Williams, G. C., "Nacreous Clouds Observed in Southeastern Alaska, January 24, 1950," *Bull. Am. Meteorol. Soc., 31,* 322–323 (1950).

Source Books

B1. Fleming, J. A., *Terrestrial Magnetism and Electricity,* McGraw-Hill Book Co., New York (1939).

B2. Kuiper, G. P., *The Atmospheres of the Earth and Planets,* University of Chicago Press, Chicago (1949).

B3. Mitra, S. K., *The Upper Atmosphere,* Royal Asiatic Society of Bengal, Calcutta (1948).

From the *Compendium of Meteorology,* T. F. Malone, editor, American Meteorological Society, Boston (1951).

C1. Dobson, G. M. B., and Brewer, A. W., "Water Vapour in the Upper Air," 311–319.

C2. Gutenberg, B., "Sound Propagation in the Atmosphere," 366–375.

C3. Newell, H. E., Jr., "Temperatures and Pressures in the Upper Atmosphere," 303–310.

C4. Whipple, F. L., "Meteors as Probes of the Upper Atmosphere," 356–365.

Problems

11.1 The following data (arbitrary units) are the result of reducing the photographic record of the flight of a single meteor as it burns in the earth's atmosphere.

Time	1	2	3	4	5	6	7	8	9	10
Velocity	1.00	0.98	0.97	0.96	0.95	0.94	0.93	0.92	0.91	0.90
Intensity	1.0	1.9	2.7	3.3	4.3	5.0	5.6	6.2	6.8	7.3

a. What is the mass of the meteor at the 6th instant if the initial mass were determined to be 0.83 g?

b. What is the density of the air at the 6th instant if the density at the zero instant is known to be 3.3×10^{-9} g cm^{-3}?

11.2 From viscosity measurements, the collision cross sections of nitrogen and oxygen are found to be 3.75×10^{-8} cm and 3.61×10^{-8} cm, respectively. What is the mean free path, the average speed, and the collision frequency of the molecules at a level where the pressure and temperature are 5×10^{-4} mb and 2° C, respectively?

11.3 On a day in which the temperature of the upper atmosphere is to be determined by the anomalous propagation of sound, a radiosonde ascent provides the following data. All measurements are referred to sea level.

Height, km	0	2	4	6	8	9.2	10	12
Temperature, °C	15°	−1°	−17°	−30°	−43°	−50°	−50°	−50°

Height, km	14	15	16	18	20	22	24
Temperature, °C	−50°	−50°	−48°	−42°	−36°	−29°	−21°

The wind velocity will be considered negligible on this particular day. The sound detector is located at sea level, 225 km from the source of the explosion.

Find the temperature distribution in the upper atmosphere if the sound of the explosion were detected 740 seconds after the explosion occurred. The angle of descent of the acoustic wave was measured to be 28 degrees at the ground.

11.4 From a suitably instrumented rocket, the following atmospheric data was obtained.

Height, km	48	55	62	70
Ambient pressure, mm of Hg	1.00	0.32	0.16	0.05

Compute and plot the temperature profile from 48 to 70 km.

APPENDIX

A Short Table of Useful Constants

	Mean radius of earth	6.3712×10^3 km
	Mean radius of sun	6.953×10^5 km
	Mean distance earth to sun	1.4968×10^8 km
g	Acceleration of gravity at latitude 45°	
	At sea level	980.616 cm sec^{-2}
	At 50 km	965.4 cm sec^{-2}
	At 100 km	950.5 cm sec^{-2}
c	Velocity of light	2.9978×10^{10} cm sec^{-1}
h	Planck's constant	6.610×10^{-27} erg sec
k	Boltzmann's constant	1.381×10^{-16} erg deg^{-1}
σ	Stefan's constant	5.735×10^{-5} erg cm^{-2} sec^{-1} deg^{-4}
	Avogadro's number	6.023×10^{23} mole^{-1}
e	charge on the electron	4.803×10^{-10} e.s.u.
R_D	Gas constant, dry air	2.870×10^6 erg g^{-1} deg^{-1}
R_W	Gas constant, water vapor	4.615×10^6 erg g^{-1} deg^{-1}
R	Gas constant (air containing w g kg^{-1} water vapor)	approximately $R_D(1 + 0.0006w)$
γ_A	Autoconvective lapse rate	34.17 deg km^{-1}
γ_D	Dry adiabatic lapse rate	$\frac{2}{7}\gamma_A = 9.76$ deg km^{-1}
$\frac{R_D}{c_p}$	Constant in Poisson's equation	$\frac{2}{7} = 0.286$
$\frac{c_p}{c_v}$	Ratio of specific heats of dry air	$\frac{7}{5} = 1.400$

Molecular Weights

	Dry Air	28.966
H_2O	Water, ice or water vapor	18.016
O	Atomic oxygen	16.000
O_2	Oxygen	32.000
O_3	Ozone	48.000
N	Atomic nitrogen	14.008
N_2	Nitrogen	28.016
CO_2	Carbon dioxide	44.00
NaCl	Sodium chloride	58.454
AgI	Silver iodide	234.80

The Saturation Vapor Pressure of Water Vapor

The unit of the pressure of saturated water vapor is the millibar (mb) and the unit of temperature is the degree Celsius, i.e., the degree Centigrade (°C). (From the *Smithsonian Meteorological Tables*, 1951.)

Over Water

(°C)	0	1	2	3	4	5	6	7	8	9
40	73.777	77.802	82.015	86.423	91.034	95.855	100.89	106.16	111.66	117.40
30	42.430	44.927	47.551	50.307	53.200	56.236	59.422	62.762	66.264	69.934
20	23.373	24.861	26.430	28.086	29.831	31.671	33.608	35.649	37.796	40.055
10	12.272	13.119	14.017	14.969	15.977	17.044	18.173	19.367	20.630	21.964
0	6.108	6.566	7.055	7.575	8.129	8.719	9.347	10.013	10.722	11.474

	Over Water					Over Ice				
(°C)	0	−2	−4	−6	−8	0	−2	−4	−6	−8
0	6.108	5.275	4.545	3.906	3.348	6.107	5.173	4.372	3.685	3.097
−10	2.863	2.441	2.076	1.760	1.488	2.597	2.172	1.811	1.506	1.248
−20	1.254	1.054	0.883	0.737	0.613	1.032	0.850	0.699	0.572	0.467
−30	0.509	0.421	0.346	0.284	0.232	0.380	0.308	0.249	0.200	0.161
−40	0.189	0.153	0.124	0.100	0.080	0.128	0.102	0.081	0.064	0.050

The NACA Standard Atmosphere (Lower Levels)

(The NACA standard atmosphere for selected heights as given in the *Smithsonian Meteorological Tables*, 1951.)

Altitude, km	Temperature, °C	Pressure, mb	Altitude, km	Temperature, °C	Pressure, mb
0	15	1013.25	10.769	−55	234.53
2	2	794.90	12	−55	193.38
4	−11	616.29	14	−55	141.35
6	−24	471.65	16	−55	103.30
8	−37	355.82	18	−55	75.53
10	−50	264.19	20	−55	55.21

Index